GENETIC ALGORITHMS FOR VLSI DESIGN, LAYOUT & TEST AUTOMATION

Pinaki Mazumder
Elizabeth M. Rudnick

Prentice Hall PTR
Upper Saddle River, NJ 07458
www.phptr.com

ISBN 0-13-011566-5

Editorial/production supervision: *Vincent J. Janoski*
Acquisitions editor: *Bernard Goodwin*
Marketing manager: *Kaylie Smith*
Manufacturing manager: *Alan Fischer*
Cover design director: *Jerry Votta*
Cover designer: *Scott Weiss*

Cover concept: *Pinaki Mazumder*
Cover art: *Alejandro Gonzalez*
Interior illustrations: *Elizabeth M. Rudnick & Pinaki Mazumder*

©1999 by Prentice Hall PTR

Published by Prentice Hall PTR
Prentice-Hall, Inc.
A Simon & Schuster Company
Upper Saddle River, NJ 07458

Prentice Hall books are widely used by corporations and government agencies for training, marketing, and resale.

The publisher offers discounts on this book when ordered in bulk quantities. For more information, contact: Corporate Sales Department, Phone: 800-382-3419; Fax: 201-236-7141; E-mail: corpsales@prenhall.com; or write: Prentice Hall PTR, Corp. Sales Dept., One Lake Street, Upper Saddle River, NJ 07458.

All products or services mentioned in this book are the trademarks or service marks of their respective companies or organizations.

TRADEMARKS: Xilinx and XC3000 are trademarks of Xilinx, Inc. PAL is a trademark of Advanced Micro Devices, Inc.

All rights reserved. No part of this book may be reproduced, in any form or by any means, without permission in writing from the publisher.

Printed in the United States of America
10 9 8 7 6 5 4 3 2 1

ISBN   0-13-011566-5

Prentice-Hall International (UK) Limited, *London*
Prentice-Hall of Australia Pty. Limited, *Sydney*
Prentice-Hall Canada Inc., *Toronto*
Prentice-Hall Hispanoamericana, S.A., *Mexico*
Prentice-Hall of India Private Limited, *New Delhi*
Prentice-Hall of Japan, Inc., *Tokyo*
Simon & Schuster Asia Pte. Ltd., *Singapore*
Editora Prentice-Hall do Brasil, Ltda., *Rio de Janeiro*

To our parents

Animesh and Chhaya
Stanley and Virginia

CONTENTS

PREFACE   xi

1 INTRODUCTION   1
- 1.1 Introduction to GA Terminology   3
- 1.2 The Simple GA   5
- 1.3 The Steady-State Algorithm   7
- 1.4 Genetic Operators   9
  - 1.4.1 Selection   9
  - 1.4.2 Crossover   10
  - 1.4.3 Mutation   12
  - 1.4.4 Fitness Scaling   13
  - 1.4.5 Inversion   14
- 1.5 GA Example   16
- 1.6 GAs for VLSI Design, Layout, and Test Automation   18
  - 1.6.1 Partitioning   20
  - 1.6.2 Automatic Placement   22
  - 1.6.3 Automatic Routing   26
  - 1.6.4 Technology Mapping for FPGAs   30
  - 1.6.5 Automatic Test Generation   32
  - 1.6.6 Power Estimation   34

2 PARTITIONING   37
- 2.1 Problem Description   38
  - 2.1.1 Partitioning Algorithms   41
  - 2.1.2 Taxonomy of Partitioning Algorithms   42
- 2.2 Circuit Partitioning by Genetic Algorithm   46

		2.2.1	Incorporation and Duplicate Check	52
		2.2.2	Mutation	53
		2.2.3	Genetic Multiway Partitioning	55
		2.2.4	Evaluation	56
		2.2.5	Results	58
	2.3	Hybrid Genetic Algorithm for Ratio-Cut Partitioning		60
		2.3.1	Genetic Encoding	61
		2.3.2	Selection, Crossover, and Mutation	61
		2.3.3	Local Improvement	62
		2.3.4	Preprocessing	64
		2.3.5	Weighted DFS Reordering (WDFR)	64
		2.3.6	Time Complexity	65
		2.3.7	Results	65
		2.3.8	Comparison of GA with Other Methods	66
	2.4	Conclusion		67
3	**STANDARD CELL AND MACRO CELL PLACEMENT**			**69**
	3.1	Standard Cell Placement		71
		3.1.1	GASP Algorithm	72
		3.1.2	Crossover Operators	76
		3.1.3	Optimizing the Genetic Algorithm	78
		3.1.4	Comparison with Simulated Annealing	80
		3.1.5	Experimental Results	81
	3.2	Macro Cell Placement		85
		3.2.1	Unified Algorithm	87
		3.2.2	Application to Macro Cell Placement	90
		3.2.3	Experimental Results	98
		3.2.4	Limitations and Enhancements	104
4	**MACRO CELL ROUTING**			**107**
	4.1	The Steiner Problem in a Graph		109
		4.1.1	Problem Definition	110
		4.1.2	Description of the Algorithm	111
		4.1.3	Experimental Method	118
		4.1.4	Results	121
		4.1.5	Summary and Conclusion	134
	4.2	Macro Cell Global Routing		135
		4.2.1	Routing Phase I : Applying the GA for the SPG	135

	4.2.2		Routing Phase II : Minimizing Layout Area	136
	4.2.3		Experimental Results	137
	4.2.4		Summary and Conclusion	138
5	**FPGA TECHNOLOGY MAPPING**			**140**
	5.1		Problem Description	141
	5.2		Circuit Segmentation and FPGA Mapping	143
		5.2.1	TLU-Based FPGA Architecture	143
		5.2.2	Mapping Using Circuit Segmentation	146
	5.3		Circuit Segmentation for Pseudo-Exhaustive Testing	150
	5.4		Experimental Results	151
		5.4.1	Results for FPGA Technology Mapping	151
		5.4.2	Experimental Results of Segmentation for PET	154
	5.5		Summary	156
6	**AUTOMATIC TEST GENERATION**			**158**
	6.1		Problem Description	160
	6.2		Test Generation in a GA Framework	165
		6.2.1	Encoding Alphabet	167
		6.2.2	GA Parameters	168
		6.2.3	Fitness Function	168
		6.2.4	An Implementation	171
		6.2.5	Evaluation of GA Parameters	174
	6.3		Test Generation for Test Application Time Reduction	178
	6.4		Deterministic/Genetic Test Generator Hybrids	184
		6.4.1	GA-HITEC Hybrid	186
		6.4.2	ALT-TEST Hybrid	190
	6.5		Use of Finite State Machine Sequences	199
		6.5.1	Algorithm Overview	201
		6.5.2	Single Time Frame Mode	206
		6.5.3	Test Generation Procedure	207
		6.5.4	Experimental Results	210
	6.6		Dynamic Test Sequence Compaction	212
		6.6.1	Overview	214
		6.6.2	Experimental Results	218
	6.7		Conclusions	224

7	**PEAK POWER ESTIMATION**	**227**
	7.1 Problem Description	229
	7.1.1 Adding Delay Models to the Problem	232
	7.2 Application of Genetic Algorithms to Peak Power Estimation	234
	7.3 Estimation of Peak Single-Cycle and n-Cycle Powers	235
	7.3.1 Resolving the Problem of State Reachability	236
	7.4 Peak Sustainable Power Estimation	237
	7.5 Experimental Results	239
	7.6 Summary	249
8	**PARALLEL IMPLEMENTATIONS**	**252**
	8.1 Wolverines: Standard Cell Placement on a Network of Workstations	254
	8.1.1 Standard Cell Placement Using the Genetic Algorithm	255
	8.1.2 Analysis of Serial Algorithm	258
	8.1.3 Distributed Placement Algorithm	259
	8.1.4 Experimental Results	264
	8.1.5 Conclusions	280
	8.2 Parallel Genetic Algorithms for Automatic Test Generation	280
	8.2.1 Sequential GA-Based ATG	281
	8.2.2 Parallel GA-Based ATG	284
	8.2.3 Experimental Results	288
	8.2.4 Conclusions	293
9	**CONCLUSION**	**295**
	9.1 Problem Encoding	296
	9.2 Fitness Function	297
	9.3 Type of GA	298
	9.4 GA Parameters	298
	9.5 Genetic Algorithms vs. Conventional Algorithms	300
	9.6 Concluding Remarks	301
GLOSSARY		**305**
BIBLIOGRAPHY		**314**
INDEX		**332**
ABOUT THE AUTHORS		**336**

PREFACE

This book describes how genetic algorithms (GAs) can be utilized for developing efficient computer-aided design (CAD) tools for performing VLSI design optimization, layout generation, and chip testing tasks. It is written primarily for practicing CAD engineers and academic researchers who want to apply GAs and analyze their performance in solving large VLSI/CAD optimization problems.

Although GAs were developed over twenty-five years ago, not much research and experimental work have been done to ascertain their capabilities in solving complex and extremely large constrained combinatorial optimization problems that one generally encounters in designing VLSI/CAD tools. Unlike graph theoretic approaches, integer/linear programming, simulated annealing, and a host of other optimization techniques that have been quite successfully deployed as core problem solving methods in various VLSI/CAD tools, GAs are not yet as widely used. We hope that this book will motivate readers to widely apply GAs in developing VLSI/CAD tools.

For this purpose, we have carefully selected a few important VLSI design automation problems with unique problem solving features, and we have shown how in each case, various aspects of the GA, namely its chromosome, crossover and mutation operators, etc., can be separately formulated to solve these problems. In order to estimate the effectiveness of GAs, we have compared their performance with conventional algorithms. While most of the solution techniques proposed in this book have been developed in an ad hoc and exploratory manner, the basic formulations of the GAs are, nevertheless, applicable to a range of related problems. However, further experimentation is needed to find better settings of GA parameters for each problem. If the empirical study is also combined with insightful mathematical modeling, then we strongly believe that the performance of the genetic-based tools can be further improved and real payoffs of the use of GAs in CAD tools can be demonstrated.

The main objectives of this book are: to aggregate various genetic-based research work performed by the authors and their coresearchers at The University of Michigan, Ann Arbor, and the University of Illinois, Urbana, as well as by colleagues at the University of Iowa, Iowa City; to educate readers in the VLSI/CAD community about the merits of GAs by demonstrating some sample solution techniques;

to motivate readers to develop improved techniques with appropriate mathematical formulations; and finally, to encourage readers working in other fields of science and engineering to explore the GA as a powerful method for solving problems in their areas of work. We have included sufficient introductory material to enable a reader who is not well-versed in GAs to know how to use them effectively. It is our sincere hope that in the future, GAs will prove to be a general-purpose heuristic method for solving a wider class of engineering and scientific problems.

Another purpose of this book is to foster research work on the development of distributed CAD tools that run efficiently on a network of workstations. Originally, Prof. Mazumder's research group was intrigued by the intrinsic parallelism of GAs, and the group embarked upon this research work with a view toward developing a suite of VLSI layout tools that can efficiently utilize the distributed resources of a network of workstations loosely connected through a local area network. With the availability of inexpensive personal computers and workstations that can be linked via an Ethernet type network, the CAD tool development environment has dramatically shifted from a single powerful uniprocessor to a cluster of networked desk-top computers. The main goal for developing this suite of distributed layout tools was to demonstrate that GAs are uniquely suited for running concurrently on a number of workstations without requiring much communication overhead. In the recent past, some existing layout tools have been successfully modified to run efficiently on tightly coupled shared-memory (e.g., Sequent's bus-based Balance) and message-passing (e.g., Intel's hypercube) machines. In order to achieve high speedup, these algorithms require frequent data exchange between two or more processors in a cluster. However, by and large, conventional layout algorithms are not amenable to parallelism on a network of workstations. As VLSI chips are reaching the integration level of one hundred million transistors and more, chip design tasks are becoming extremely complex and computation intensive. New generation CAD tools must be able to run in parallel over a large number of inexpensive computers interconnected together by a local area network. It will therefore be worthwhile to invest an effort in developing genetic-based CAD tools.

Organization of the Book

There are three distinct classes of VLSI problems which the book addresses: (1) the layout class of problems, such as circuit partitioning, placement, and routing; (2) the design class of problems, including power estimation, technology mapping, and netlist partitioning; and, finally, (3) reliable chip testing through efficient test vector generation. All these problems are intractable in the sense that no polynomial time algorithm can guarantee optimal solution of the problems, and they actually belong to the dreadful NP-complete and NP-hard categories. The book is organized as follows.

Chapter 1 provides an introduction to the two basic types of GAs: the simple genetic algorithm and the steady-state algorithm. GA terminology is introduced, and genetic operators are discussed. A simple test generation example is used to illustrate the operation of a GA, and then GAs for problems in VLSI Design, Layout, and Test Automation are introduced.

Chapter 2 addresses the problem of circuit partitioning. It begins with a review of previous approaches used and then describes a steady-state GA for solving the problem. Experimental results are presented, and a hybrid GA that incorporates local optimization is described.

Chapter 3 focuses on automatic placement for standard cells and macro cells. A GA for standard cell placement is described, results are presented, and the genetic approach is compared to simulated annealing. In addition, an algorithm that combines a GA and simulated annealing for macro cell placement in discussed.

Chapter 4 discusses problems encountered in macro cell routing. It begins by addressing the Steiner problem in a graph. Previous approaches are reviewed, a GA to solve the problem is described, experimental results are presented, and comparisons are made to previous work. Finally, the GA for the Steiner problem in a graph is applied to the macro cell routing problem, and results are presented to demonstrate the effectiveness of this approach.

Chapter 5 describes a GA for FPGA technology mapping, which is a key phase of logic synthesis and involves partitioning the circuit into a number of subcircuits that are not necessarily disjoint. Application of circuit partitioning to pseudo-exhaustive testing is also addressed, and experimental results are given for FPGA technology mapping.

Chapter 6 discusses the problem of automatic test generation. A GA framework for test generation is presented, and results of experiments to evaluate various GA parameters are given. Integration of GAs with deterministic algorithms and incorporation of problem-specific knowledge into a GA are discussed. The chapter concludes by describing how the GA framework can be applied to the problem of test sequence compaction.

Chapter 7 deals with power estimation for VLSI circuits. In particular, it describes a GA for estimating the peak power dissipation in a circuit. The peak power estimates provide a tight lower bound on the actual peak power and are significantly more accurate than previous approaches. The actual sequences of vectors that achieve these bounds are also generated by the GA. The effects of the delay model used on the quality of the results are also discussed.

Chapter 8 explores parallel implementations of GAs for standard cell placement and test generation. The migration operator for parallel GAs is introduced in this chapter. GAs that require little communication between processors and are therefore suitable for a network of workstations are described. Experimental results are presented to illustrate the effects of various communication patterns. Very good speedups are achieved, as demonstrated for several benchmark circuits.

Chapter 9 concludes the book by giving guidelines for devising a GA to solve a new problem in the area of VLSI design, layout, and test automation or in another domain of science and engineering. Problem encoding, fitness function, type of GA, and GA parameters are addressed, and the genetic approach is compared to conventional approaches.

Applicability of the Book

This book is intended for design engineers and researchers in the fields of VLSI and CAD. The book introduces the main concepts of GAs in a simple and easily understandable way that may encourage first time users to learn various aspects of GAs quickly from the first chapter and then go on to study the detailed applications in other chapters more carefully. This book also provides college and university students a systematic exposure to a wide spectrum of design, physical layout, and chip testing problems that form integral parts of digital testing and CAD for VLSI system design courses. The breadth and depth of issues presented justifies using this book as a state-of-the-art reference source in the above courses. The book also includes a chapter on parallel implementations of GAs for layout and test generation. The parallel GAs demonstrate how uniprocessor algorithms can be accelerated linearly with the number of loosely connected computer workstations deployed. Different communication structures that have been used in message passing between different processes are compared.

To Readers

This book presents a number of VLSI/CAD applications of GAs that have primarily originated from research work performed at the universities where we are teaching. The book is being published nine years after we first started working in this field and reported work in international conferences and archival journals. Meanwhile, many other researchers have been inspired by some of the early success stories and subsequently applied GAs very successfully to a large number of VLSI/CAD problems. In order to contain the size of the book, we have had to exclude much of the excellent work done by others. Therefore, we would like to ask readers to search the following Internet sites for comprehensive listings of publications on GAs and evolution theory that not only refer to other key papers pertaining to VLSI/CAD problems but also include important papers illustrating the use of GAs in solving problems in various other fields of engineering, science, business, etc.:

- http://www.dai.ed.ac.uk/groups/evalg
- http://www.bioele.nuee.nagoya-u.ac.jp/wsc1/papers
- http://www-illigal.ge.uiuc.edu/

Acknowledgments

Work on GAs for VLSI layout described in this book was started by Prof. Mazumder and his doctoral student Khushro Shahookar in 1989 when they developed the

GASP genetic algorithm for standard cell placement. Subsequently in 1991, Mohan Sundarar, when he was working on his doctoral studies with the author on a different topic in VLSI design, developed Wolverines, the first distributed cell placement tool. Henrik Esbensen from Aarhus University, Denmark was inspired to work on GAs and later to finish his doctoral studies under Prof. Mazumder's supervision after he attended an international CAD conference where the author presented some of the early work of his research group. Prof. Mazumder would like to acknowledge the contributions of these researchers in carrying out the work on GAs. Their unbridled enthusiasm has indeed enabled him to deepen his research interest in the layout automation field in addition to his own research work in semiconductor memory testing and quantum circuit design. He would especially like to thank Prof. Ernest S. Kuh of the University of California, Berkeley for his continued encouragement and good advice that inspired the author to carry on his research work in VLSI layout automation. Above all, he deeply acknowledges his gratitude to his family members for their unabated emotional support and understanding, when in the early years of his teaching career, he could spend very little time at home with his wife and two small kids. Ironically, he was working on GAs that heavily spend time on offspring development and evaluation, for superior performance!

Prof. Rudnick began working on GAs for VLSI test automation in early 1992 as part of her doctoral research and was inspired by several factors: her participation in a project on simulation-based test generation for Prof. Janak Patel's class on Digital System Testing and Testable Design, a course on GAs taught by Prof. David Goldberg at the University of Illinois, and Prof. Daniel Saab's pioneering work in applying GAs to test generation, also at the University of Illinois. She would like to acknowledge the unwavering support of her research advisor, Prof. Janak Patel, throughout the course of this research. She would also like to acknowledge the significant contributions of Dr. Michael Hsiao, who carried out much of the research on test generation and power estimation as part of his doctoral research, and Dr. Dilip Krishnaswamy, who lead the research effort on parallel GAs for test generation as part of his doctoral research under the direction of Prof. Prith Banerjee. She would like to thank Prof. Irith Pomeranz of the University of Iowa and Venkataramana Kommu of Synopsys for providing the chapter on FPGA technology mapping. In addition, she would like to thank her family and friends for their support and encouragement.

Finally, we would like to thank the anonymous reviewers, whose insightful comments enabled us to improve the quality of the book significantly, and the following members of Prentice Hall: Bernard Goodwin for undertaking the publishing of this book and Vincent Janoski, Diane Spina, and Lisa Iarkowski for their assistance in preparing the manuscript.

Pinaki Mazumder
University of Michigan, Ann Arbor
September, 1998

Elizabeth M. Rudnick
University of Illinois, Urbana

CHAPTER 1
at a glance

- Introduction to Genetic Algorithms (GAs)
- GA terminology
- Genetic operators
  - Crossover
  - Mutation
  - Inversion
- EDA problems solved by GAs

Chapter 1

INTRODUCTION

The Genetic Algorithm (GA) was invented by Prof. John Holland [95] at the University of Michigan in 1975, and subsequently it has been made widely popular by Prof. David Goldberg [79] at the University of Illinois. The original GA and its many variants, collectively known as genetic algorithms, are computational procedures that mimic the natural process of evolution. The theories of evolution and *natural selection* were first proposed by Darwin to explain his observations of plants and animals in the natural world [48]. Darwin observed that, as variations are introduced into a population with each new generation, the less-fit individuals tend to die off in the competition for food, and this *survival of the fittest* principle leads to improvements in the species. The concept of natural selection was used to explain how species have been able to adapt to changing environments and how, consequently, species that are very similar in adaptivity may have evolved.

Much has been learned about genetics since the time of Charles Darwin. All information required for the creation of appearance and behavioral features of a living organism is contained in its chromosomes. Reproduction generally involves two parents, and the chromosomes of the offspring are generated from portions of chromosomes taken from the parents. In this way, the offspring inherit a combination of characteristics from their parents. GAs attempt to use a similar method of inheritance to solve various problems, such as those involving adaptive systems [95]. GAs have also been applied to optimization problems [54], and the applications that we will address in this book fall into this category. The objective of the GA is then to find an optimal solution to a problem. Of course, since GAs are heuristic procedures, they are not guaranteed to find the optimum, but experience has shown that they are able to find very good solutions for a wide range of problems.

GAs work by evolving a population of individuals over a number of generations. A fitness value is assigned to each individual in the population, where the fitness computation depends on the application. For each generation, individuals are selected from the population for reproduction, the individuals are *crossed* to generate new individuals, and the new individuals are mutated with some low mutation probability. The new individuals may completely replace the old individuals in the population, with distinct generations evolved [79]. Alternatively, the new individuals may be combined with the old individuals in the population [95]. In this case, we may want to reduce the population in order to maintain a constant size, e.g., by selecting the best individuals from the population. Both of these approaches have been used for applications that will be described in this book, and both approaches have yielded good results. The choice of which approach is used may depend on the application. Since selection is biased toward more highly fit individuals, the average fitness of the population tends to improve from one generation to the next. The fitness of the best individual is also expected to improve over time, and the best individual may be chosen as a solution after several generations.

GAs use two basic processes from evolution: *inheritance*, or the passing of features from one generation to the next, and competition, or *survival of the fittest*, which results in weeding out the bad features from individuals in the population. The main advantages of GAs are:

- they are adaptive, and learn from experience;

- they have intrinsic parallelism;

- they are efficient for complex problems; and

- they are easy to parallelize, even on a loosely coupled Network Of Workstations (popularly known as NOW), without much communication overhead.

Various different representations and operations have been used in genetic algorithms, and many of these variations are described in detail in existing texts [79], [95]. Our goal in this chapter is to provide an overview of the approaches and operations that we have found to be useful for problems in VLSI design and test automation. We begin with a quick introduction to GA terminology. The two basic GA approaches are presented next, first the *simple GA*, also called the *total replacement algorithm*, and then the *steady-state algorithm*, which is characterized by overlapping populations. Detailed descriptions of various genetic operators used are given next, followed by an example illustrating the use of a GA for test generation. We conclude with a brief discussion about the application of GAs to problems in VLSI design and test automation.

1.1 Introduction to GA Terminology

All genetic algorithms work on a *population*, or a collection of several alternative solutions to the given problem. Each individual in the population is called a *string* or *chromosome*, in analogy to chromosomes in natural systems. Often these individuals are coded as binary strings, and the individual characters or symbols in the strings are referred to as *genes*. In each iteration of the GA, a new *generation* is evolved from the existing population in an attempt to obtain better solutions.

The *population size* determines the amount of information stored by the GA. The GA population is evolved over a number of generations.

An *evaluation* function (or fitness function) is used to determine the *fitness* of each candidate solution. The fitness is the opposite of what is generally known as the *cost* in optimization problems. It is customary to describe genetic algorithms in terms of fitness rather than cost. The evaluation function is usually user-defined, and problem-specific.

Individuals are *selected* from the population for reproduction, with the selection biased toward more highly fit individuals. Selection is one of the key operators on GAs that ensures survival of the fittest. The selected individuals form pairs, called *parents*.

Crossover is the main operator used for reproduction. It combines portions of two parents to create two new individuals, called *offspring*, which inherit a combination of the features of the parents. For each pair of parents, crossover is performed with a high probability P_C, which is called the *crossover probability*. With probability $1 - P_C$, crossover is not performed, and the offspring pair is the same as the parent pair.

Mutation is an incremental change made to each member of the population, with a very small probability. Mutation enables new features to be introduced into a population. It is performed probabilistically such that the probability of a change in each gene is defined as the *mutation probability*, P_M.

Inversion is a genetic operator which does not change the solution represented by the chromosome, but rather changes the chromosome itself, or the (binary) representation of the solution. The *inversion probability* is denoted by P_I.

The *generation gap* is the fraction of individuals in the population that are replaced from one generation to the next and is equal to 1 for the simple GA.

A *schema* is a specific set of values assigned to a subset of the genes in a chromosome. It is a partial solution and represents a set of possible fully specified solutions. For example, consider a cell placement problem, in which the solutions are encoded as triples consisting of the cell identifiers and their assigned x- and y-coordinates. A schema in this context represents a subplacement in which a subset of the cells are assigned to specific positions. The remaining cells may be placed in various different positions. A schema with m specified elements and *don't-cares* in the rest of the $n - m$ positions (such as an m-cell subplacement in an n-cell placement problem) can be considered to be an $(n - m)$-dimensional hyperplane in the solution space. All points on that hyperplane (i.e., all individuals contain the given subplacement) are *instances* of the schema. Note here that a subplacement does not have to contain physically adjacent cells, such as a rectangular patch of the chip area.

For a given problem, various genes may be linked, and specific values may be required for groups of genes in order to obtain a good solution. These *schemata* represent the various features of the candidate solutions. GAs implicitly operate upon the various schemata in parallel, which is why they are so successful in solving complex optimization problems. The genetic operators create a new generation of individuals by combining the schemata of parents selected from the current generation. Due to the stochastic selection process, the fitter parents, which are expected to contain some good schemata, are likely to produce more offspring. At the same time, the bad parents, which contain some bad schemata, are likely to produce fewer offspring. Thus, in the next generation, the number of instances of good schemata tends to increase, and the number of instances of bad schemata tends to decrease. The fitness of the entire population is therefore improved.

In a typical, binary-coded GA, where the chromosomes are bit strings, each string in the population is an instance of 2^L schemata, where L is the length of each individual string. The number of different strings or possible solutions to the problem is also 2^L, and the total number of different schemata contained in all these strings is 3^L, since each gene in a schema may be 0, 1, or don't care, x. Thus, the population represents a very large number of schemata, even for relatively small population sizes. By evaluating a new offspring, we get a rough estimate of the fitness of all of its schemata. The numbers of these schemata present in the population is thus adjusted according to their relative fitness values. This effect is known as the intrinsic parallelism of the GA. As more individuals are evaluated, the relative proportions of the various schemata in the population reflect their fitness values more and more accurately. When a better schema is introduced into the population through one offspring, it is inherited by others in the succeeding generation, and thus, its proportion in the population increases. It starts driving out the less fit schemata, and the average fitness of the population improves.

1.2 The Simple GA

The simple GA (also referred to as the total replacement algorithm) is illustrated in Fig. 1.1 and also in Fig. 1.2 in flowchart form. The simple GA is composed of

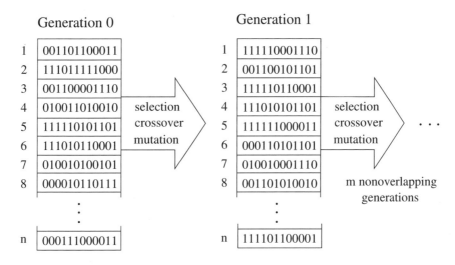

Figure 1.1. The simple genetic algorithm

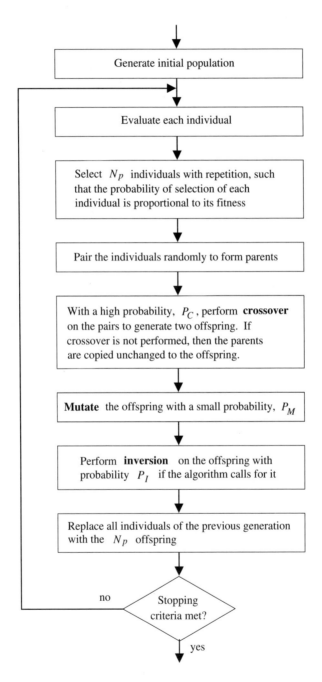

Figure 1.2. Flowchart of the simple genetic algorithm

populations of strings, or *chromosomes*, and three evolutionary operators: *selection, crossover,* and *mutation* [79]. The chromosomes may be binary-coded, or they may contain characters from a larger alphabet [69], [80]. Each chromosome is an encoding of a solution to the problem at hand, and each individual has an associated fitness which depends on the application. The initial population is typically generated randomly, but it may also be supplied by the user. A highly fit population is evolved through several generations by *selecting* two individuals, *crossing* the two individuals to generate two new individuals, and *mutating* characters in the new individuals with a given mutation probability. Selection is done probabilistically but is biased toward more highly fit individuals, and the population is essentially maintained as an unordered set. Distinct generations are evolved, and the processes of selection, crossover, and mutation are repeated until all entries in a new generation are filled. Then the old generation may be discarded. New generations are evolved until some stopping criterion is met. The GA may be limited to a fixed number of generations, or it may be terminated when all individuals in the population converge to the same string or no improvements in fitness values are found after a given number of generations. Since selection is biased toward more highly fit individuals, the fitness of the overall population is expected to increase in successive generations. However, the best individual may appear in any generation.

1.3 The Steady-State Algorithm

In a GA having overlapping generations, only a fraction of the individuals are replaced in each generation [55], [236]. The steady-state algorithm is illustrated in Fig. 1.3. In each generation, two *different* individuals are selected as parents, based on their fitness. Crossover is performed with a high probability, P_C, to form offspring. The offspring are mutated with a low probability, P_M and inverted with probability P_I, if necessary. A duplicate check may follow, in which the offspring are rejected without any evaluation if they are duplicates of some chromosomes already in the population. The offspring that survive the duplicate check are evaluated and are introduced into the population only if they are better than the current worst member of the population, in which case the offspring replaces the worst member. This completes the generation. In the steady-state GA, the generation gap is minimal, since only two offspring are produced in each generation.

Duplicate checking may be beneficial because a finite population can hold more schemata if the population members are not duplicated. Since the offspring of two identical parents are identical to the parents, once a duplicate individual enters the population, it tends to produce more duplicates and individuals varying by only slight mutations. Premature convergence may then result. Duplicate checking is advantageous under the following conditions:

- the population size is small;
- the chromosomes are short; or
- the evaluation time is large.

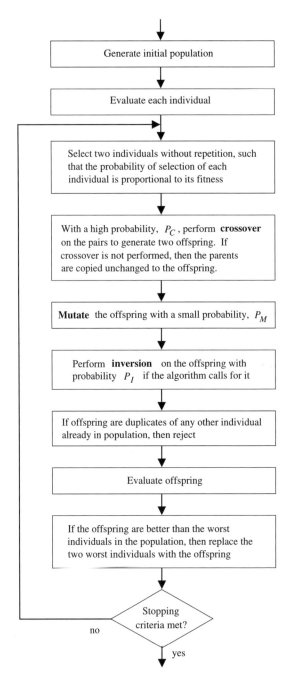

Figure 1.3. Steady-state genetic algorithm

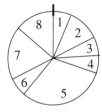 

(a) Roulette Wheel Selection (b) Stochastic Universal Selection

Figure 1.4. Proportionate selection schemes

Each of the above conditions reduces the duplicate checking time in comparison to the evaluation time. If the duplicate checking time is negligible compared to the evaluation time, then duplicate checking improves the efficiency of the GA.

The steady-state GA is susceptible to stagnation. Since a large majority of offspring are inferior, the steady-state algorithm rejects them, and it keeps making more trials on the existing population for very long periods of time without any gain. Because the population size is small compared to the search space, this is equivalent to long periods of localized search.

1.4 Genetic Operators

The genetic operators and their significance can now be explained. The description will be in terms of a traditional GA without any problem-specific modifications. The operators that will be discussed include selection, crossover, mutation, fitness scaling, and inversion.

1.4.1 Selection

Various selection schemes have been used, but we will focus on *roulette wheel selection, stochastic universal selection,* and binary *tournament selection* with and without replacement [13], [81]. As illustrated in Fig. 1.4(a), roulette wheel selection is a proportionate selection scheme in which the slots of a roulette wheel are sized according to the fitness of each individual in the population. An individual is selected by spinning the roulette wheel and noting the position of the marker. The probability of selecting an individual is therefore proportional to its fitness. As illustrated in Fig. 1.4(b), stochastic universal selection is a less noisy version of roulette wheel selection in which N equidistant markers are placed around the roulette wheel, where N is the number of individuals in the population. N individuals are selected in a single spin of the roulette wheel, and the number of copies of each individual selected is equal to the number of markers inside the corresponding slot. In binary tournament selection, two individuals are taken at random, and the better individual is selected from the two. If binary tournament selection is being done without replacement, then the two individuals are set aside for the next selection operation, and they are not replaced into the population. Since two individuals are removed from the population for every individual selected, and the population size remains constant from one generation to the next, the original population is restored after the new population is half-filled. Therefore, the best individual will be selected twice, and the worst individual will not be selected at all. The number of copies selected of any other individual cannot be predicted except that it is either zero, one, or two. In binary tournament selection with replacement, the two individuals are immediately replaced into the population for the next selection operation.

The objective of the GA is to converge to an optimal individual, and *selection pressure* is the driving force which determines the rate of convergence. A high selection pressure will cause the population to converge quickly, possibly at the expense of a suboptimal result. Roulette wheel selection typically provides the highest selection pressure in the initial generations, especially when a few individuals have significantly higher fitness values than other individuals. Tournament selection provides more pressure in later generations when the fitness values of individuals are not significantly different. Thus, roulette wheel selection is more likely to converge to a suboptimal result if individuals have large variations in fitness values.

1.4.2 Crossover

Once two chromosomes are selected, the *crossover* operator is used to generate two offspring. In *one-* and *two-point crossover,* one or two chromosome positions are randomly selected between one and $(L-1)$, where L is the chromosome length, and the two parents are crossed at those points. For example, in one-point crossover, the first child is identical to the first parent up to the crossing point and identical to the second parent after the crossing point. An example of one-point crossover is shown in Fig. 1.5. In *uniform crossover,* each chromosome position is crossed with some probability, typically one-half.

Parent 1:	*1 0 1 1 0 1 1 0 1*	1 1 1 0 0 1 1 0 0
Parent 2:	0 0 1 1 0 1 1 0 0	1 0 0 1 0 1 0 0 0
Offspring 1:	*1 0 1 1 0 1 1 0 1*	1 0 0 1 0 1 0 0 0
Offspring 2:	0 0 1 1 0 1 1 0 0	*1 1 1 0 0 1 1 0 0*

Figure 1.5. One-point crossover

Crossover combines the schemata or building blocks from two different solutions in various combinations. Smaller good building blocks are converted into progres-

1.4. Genetic Operators

sively larger good building blocks over time until we have an entire *good* solution. Crossover is a random process, and the same process results in the combination of bad building blocks to result in poor offspring, but these are eliminated by the selection operator in the next generation. An example of this progressive growth of schemata is given in Fig. 1.6. In the figure, let us assume that the fitness function is such that consecutive strings of 1's result in a higher fitness. Thus, schemata of the form xx111xx are good. The figure shows the offspring from one crossover operation taking part in further crossovers. Although the sequence of events shown may occur over many generations, it illustrates how strings of consecutive 1's can be gradually accumulated.

	Chromosome	Schema
Parent 1:	*01011* |*0110100*	xxx*11*xxxxxxx
Parent 2:	00110 | 1110010	xxxxx111xxxx
Offspring:	*01011* | 1110010	xxx*11*111xxxx
Parent 1:	*01011111* | *0010*	xxx*1111*xxxx
Parent 2:	10011000 | 1101	xxxxxxxx11xx
Offspring:	*01011111* | 1101	xxx*1111*11xx

Figure 1.6. Advantage of crossover: Schema growth

At the same time, schema destruction also results from crossover, as illustrated in Fig. 1.7. Any schemata spanning the cut point of the crossover operation are destroyed; i.e., they are not inherited by either of the offspring. This destruction introduces new random schemata instead. The destruction probability of a schema is proportional to its length. Since long schemata span a large part of the solution string, they are cut by any random cut point with a high probability. Holland's schema theorem analyzes these two processes and concludes that the net result is a gain of good schemata [95].

	Chromosome	Schema
Parent 1:	*11010* | *0010011*	*11*xxxxxxxx*11*
Parent 2:	10110 | 1110000	
Offspring:	*11010* | 1110000	

Figure 1.7. Destruction of long schemata during crossover

The end result of crossover is that schemata from the two parents are combined. If the best schemata from the parents can be combined, the resulting offspring may be even better than the two parents. Since, highly fit individuals are more likely to be selected as parents, the GA examines more candidate solutions in good regions of the search space and fewer candidate solutions in other regions. The optimum solution may therefore be discovered relatively quickly. The performance of the GA depends to a great extent on the performance of the crossover operator used.

The amount of crossover is controlled by the *crossover probability*, which is defined as the ratio of the number of offspring produced in each generation to the population size. A higher crossover probability allows exploration of more of the solution space and reduces the chances of settling for a false optimum. A lower crossover probability enables exploitation of existing individuals in the population that have relatively high fitness.

1.4.3 Mutation

As new individuals are generated, each character is mutated with a given probability. In a binary-coded GA, mutation may be done by flipping a bit, while in a nonbinary-coded GA, mutation involves randomly generating a new character in a specified position. Mutation produces incremental random changes in the offspring generated through crossover, as shown in Fig. 1.8. When used by itself, without any crossover, mutation is equivalent to random search, consisting of incremental random modification of the existing solution, and acceptance if there is improvement. However, when used in the GA, its behavior changes radically. In the GA, mutation serves the crucial role of replacing the gene values lost from the population during the selection process so that they can be tried in a new context, or of providing the gene values that were not present in the initial population.

Before Mutation: 110100010011
After Mutation: 110000010011

Figure 1.8. Mutation operator

For example, say a particular bit position, bit 10, has the same value, say 0, for all individuals in the population. In such a case, crossover alone will not help, because it is only an inheritance mechanism for existing gene values. That is, crossover cannot create an individual with a value of 1 for bit 10, since it is 0 in all parents. If a value of 0 for bit 10 turns out to be suboptimal, then, without the mutation operator, the algorithm will have no chance of finding the best solution. The mutation operator, by producing random changes, provides a small probability that a 1 will be reintroduced in bit 10 of some chromosome. If this results in an improvement in fitness, then the selection algorithm will multiply this chromosome, and the crossover operator will distribute the 1 to other offspring. Thus, mutation makes the entire search space reachable, despite a finite population size. Although the crossover operator is the most efficient search mechanism, by itself, it does not guarantee the reachability of the entire search space with a finite population size. Mutation fills in this gap.

The mutation probability P_M is defined as the probability of mutating each gene. It controls the rate at which new gene values are introduced into the population. If it is too low, many gene values that would have been useful are never tried out. If it is too high, too much random perturbation will occur, and the offspring will lose their resemblance to the parents. The ability of the algorithm to learn from the history of the search will therefore be lost.

1.4.4 Fitness Scaling

If a proportionate selection scheme is used, the GA selects individuals with probability proportional to their fitness. If the raw fitness values are used for this probabilistic selection, without any scaling or normalization, then one of two things can happen. If the fitness values are too far apart, then it will select several copies of the good individuals, and many other *worse* individuals will not be selected at all. This will tend to fill the entire population with very similar chromosomes and will limit the ability of the GA to explore large amounts of the search space. On the other hand, if the fitness values are too close to each other, then the GA will tend to select one copy of each individual, with only random variations in selection. Consequently, it will not be guided by small fitness variations and will be reduced to random search. Fitness scaling is used to scale the raw fitness values so that the GA sees a *reasonable* amount of difference in the scaled fitness values of the best versus the worst individuals. Thus, fitness scaling controls the *selection pressure* or discriminating power of the GA.

The following fitness scaling algorithm applies to evaluation functions that determine the *cost*, rather than the *fitness*, of each individual. From this cost, the fitness of each individual is determined by scaling as follows.

A reference *worst cost* is determined by

$$C_w = \overline{C} + S\sigma$$

where $\overline{C}$ is the average cost of the population, S is the user-defined sigma scaling factor, and σ is the standard deviation of the cost of the population. In case C_w is less than the real worst cost in the population, then only the individuals with cost lower than C_w are allowed to participate in the crossover.

The fitness of each individual is determined by

$$F = \begin{cases} C_w - C & \text{if } C_w > C \\ 0 & \text{otherwise.} \end{cases}$$

This scales the fitness such that, if the cost is $\pm k$ standard deviations from the population average, the fitness is

$$F = (S \pm k)\sigma.$$

This means that any individuals worse than S standard deviations from the population mean ($k = S$) are not selected at all. The usual value of S reported in the literature is between 1 and 5.

The cost vs. fitness is illustrated for three possible cost distributions in Fig. 1.9. If S is small, the ratio of the lowest to the highest fitness in the population increases, and the algorithm becomes more selective in choosing parents (Fig. 1.9(a)).

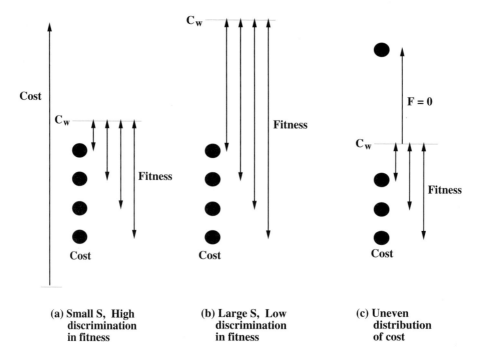

Figure 1.9. Effect of sigma scaling on fitness

On the other hand, if S is large, then C_w is large, and the fitness values of the members of the population are relatively close to each other (Fig. 1.9(b)). This causes the difference in selection probabilities to decrease and the algorithm to be less selective in choosing parents. Fig. 1.9(c) shows a case where one individual is much worse than the rest of the population. The fitness of this individual is set to zero, and the algorithm still has good discrimination between the rest of the individuals.

An alternative way to control the selection process is to use *ranking* [79]. Individuals are sorted according to their fitness values, and the number of copies of each individual selected for the next generation depends on its rank. Ranking is used in the GAs for macro cell routing, as will be described in Chapter 4.

1.4.5 Inversion

The inversion operator takes a random segment in a solution string and inverts it end for end (Fig. 1.10). This operation is performed in a way such that it does

1.4. Genetic Operators

not modify the solution represented by the string. Instead, it only modifies the *representation* of the solution. Thus, the symbols composing the string must have an interpretation independent of their position. This can be achieved by associating an identification number with each symbol in the string and interpreting the string with respect to these identification numbers instead of the array indices. When a symbol is moved in the array, its identification number is moved with it, and therefore, the interpretation of the symbol remains unchanged.

For example, Fig. 1.10 shows a chromosome. Let us assume a very simple evaluation function such that the fitness is the binary number consisting of all bits of the chromosome, with bit 0 being the least significant, and bit-9 being the most significant. Since the bit identification numbers are moved with the bit values during the inversion operation, bit 0, bit 1, etc., still have the same values, although their sequence in the chromosome is different. Hence, the fitness value remains the same. The inversion probability is the probability of performing inversion on each individual during each generation. It controls the amount of group formation.

Before Inversion:	B_9	B_8	B_7	B_6	B_5	B_4	B_3	B_2	B_1	B_0
	1	1	0	0	0	0	1	1	0	0
After Inversion:	B_9	B_8	B_7	B_2	B_3	B_4	B_5	B_6	B_1	B_0
	1	1	0	1	1	0	0	0	0	0

Figure 1.10. Inversion operator

Inversion changes the sequence of genes randomly, in the hope of discovering a sequence of linked genes placed close to each other. That is, for long schemata consisting of many *don't-cares* in the middle, it tries to resequence the genes to bring them to neighboring locations. This operation, illustrated in Fig. 1.11, has the effect of shortening the schema length. Longer schemata have a higher probability of being cut and destroyed by the one-point crossover operator, and hence, long schemata with *don't-cares* embedded will not be explored efficiently. By reducing their lengths, and grouping genes of a schema in close proximity in the chromosome, the inversion operator greatly improves the efficiency of the one-point crossover operator.

Before Inversion: 11xxxx11xx
After Inversion: 11x11xxxxx

Figure 1.11. Shortening of schema by inversion

However, inversion is a random process, and where there is a probability of reducing schema length, there is also a probability of increasing it. Such offspring with increased schema length will prove to be less fit in the long run and will be weeded out by the selection algorithm. Hence, the useful effect of inversion will dominate. If the genes are arranged in the chromosome such that related genes are close together from the beginning, then the good schemata will be short without the need for inversion. Such a genetic coding will be highly beneficial. In problems where such a genetic coding can be derived, the inversion operator is not used.

1.5 GA Example

Many of the optimization problems encountered in VLSI design, layout, and test automation can be solved with high-quality results using genetic algorithms. Let us consider the problem of test generation for a very simple circuit and observe how a GA behaves when applied to this problem. Suppose that we want to generate a test to detect a fault that causes node Y to be stuck at logic 0 (denoted Y s-a-0) in Fig. 1.12. To excite the fault, node Y must be driven to logic 1 in the fault-free

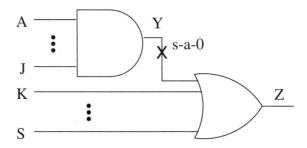

Figure 1.12. GA-based test generation example

circuit, so that there will be a difference between the fault-free circuit and the faulty circuit. Nodes A through J must therefore be set to 1. To propagate the fault effects to output node Z, nodes K through S must be set to 0.

The required test vector is nearly impossible to find using a random approach. Specific values are required on 19 circuit inputs, and it is very unlikely that the correct combination will be generated randomly. Consider how the problem could be solved with a GA. If the GA fitness function simply indicates whether or not the fault is detected, the test is very unlikely to be found. However, if the fitness function favors setting nodes A through J to 1 and setting nodes K through S to 0, the test has a good chance of being found.

A simple GA was implemented with the following fitness function to solve this problem:

$$\text{fitness} = \text{ (1 if fault is excited or 0 otherwise) } + \\ \text{(fraction of inputs A–J set to 1) } + \\ \text{(fraction of inputs K–S set to 0)}$$

A population size of 32 was used, and the GA was run for 16 generations. Tournament selection, uniform crossover, and crossover and mutation probabilities of 1.0

1.5. GA Example

Table 1.1. Best Individual by Generation for GA Example

Generation	Vector	Fitness
0	0011010111000001000	1.5
1	0101011111000000000	1.7
2	1111111111101000000	2.8
3	1111111111100001100	2.7
4	1111111111000000011	2.8
5	1111111110000000000	1.9
6	1101111011000000000	1.8
7	1111111111000000000	3.0

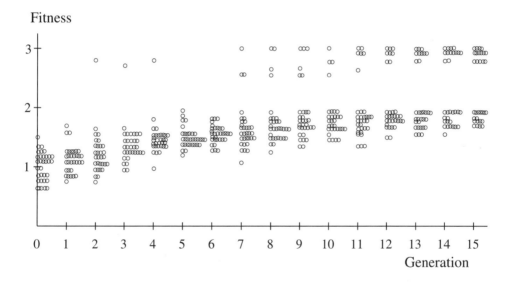

Figure 1.13. Fitness distribution by generation for GA example

and 0.03 were used. The best individuals found in the first eight generations are shown in Table 1.1, and the distribution of individuals in each generation is shown in Fig. 1.13. The GA finds some good individuals as early as generation 2, but the best individuals are lost in generation 5. The solution is finally found in generation 7. For this circuit, 256 individuals must be evaluated in order to find the solution; this is far fewer than the number of random vectors that would be required. The average fitness of the population gradually improves from 1.0 in generation 0 to 2.3 in generation 15. Once the solution is found in generation 7, it appears in every subsequent generation, and by generation 15, it is replicated four times in the population.

The actual fitness distribution by generation for a given GA will depend on a number of factors, including type of GA, selection and crossover schemes, GA parameters, and fitness function used. A properly designed GA is expected to converge to the solution if it is allowed to run for a sufficient number of generations. Whether or not the GA is allowed to run until convergence depends on the application. If solution quality is critical, the GA should be allowed to run for a large number of generations. On the other hand, if execution time is a limiting factor, a suboptimal solution may be selected after only a small number of generations.

1.6 GAs for VLSI Design, Layout, and Test Automation

Due to the complexity of VLSI designs, Electronic Design Automation (EDA) tools have become a necessity. EDA tools exist to automate the design process so that even novice designers can produce good quality designs in a reasonable amount of time. One example is an automatic placement and routing package. Other EDA tools simply facilitate the design process, leaving the important decisions up to the designer. For example, a full custom layout tool enables the designer to specify the exact placement of transistors and metal interconnections for a design, but the mundane tasks of generating the actual mask specifications for wafer fabrication are left to the tool. Even when the automatic tools are used, the manual tools may still be necessary. If the automatic routing tool is unsuccessful in routing some of the interconnections, the routing must be performed manually by the designer.

The key steps in the VLSI design process are illustrated in Fig. 1.14. Design entry is the first step in the design flow. It usually starts with a high-level design specification, e.g., written in English. From the high-level design specification, a behavioral description may be created using a hardware description language such as Verilog or VHDL. If a high-level behavioral description is available, behavioral synthesis tools may be used to generate a Register Transfer Level (RTL) design. Otherwise, the designer enters the circuit specification manually at the register transfer level. In this case, a Verilog or VHDL description may be entered using a text editor, or a block diagram schematic editor may be used. Logic synthesis tools are commonly used to generate gate-level descriptions from RTL descriptions, although designs can also be entered directly at the gate level, either using a text editor and a hardware description language or a graphical schematic capture tool. As the design is entered, it must be verified for correct functionality at every level. Functional simulation has been used for this purpose, combined with electrical-level simulation of low-level components and timing verification. Recently, formal verification tools have become available to perform some of the verification tasks. For designs that require minimal power consumption, power analysis must also be performed.

Following the design entry and verification is the task of layout generation, which provides all the information necessary for generating masks for wafer fabrication. The layout may be generated automatically, using automatic placement and routing tools, or it may be generated manually, as mentioned previously. Layouts must be

1.6. GAs for VLSI Design, Layout, and Test Automation

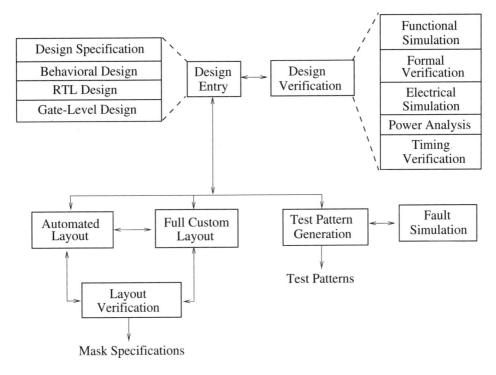

Figure 1.14. VLSI design process

verified to ensure that they conform to design rules before the masks are generated. Test patterns are necessary to test the chips once they are fabricated, and these tests may be generated once the design entry is completed. An Automatic Test Generator (ATG) may be used for this purpose; alternatively, the designer may generate tests manually. In either case, a fault simulator is used to evaluate the quality of the tests generated in terms of fault coverage.

Several of the tasks involved in the VLSI design process involve optimization problems. For example, an automatic placement tool must decide the optimal position in which to place each component. The specific problems are usually NP-complete; therefore, heuristic techniques have been used to obtain solutions. However, even if adequate approaches have been devised in the past, design complexity continues to increase with the continuing improvements in technology. Therefore, new approaches may be warranted over time, and GAs are often a good choice. Much research has been done in applying GAs to various tasks in the VLSI design process. Some of the techniques developed are being integrated into commercial EDA tools. The rest of this book will provide details about some of the EDA applications where GAs have been used. These applications include partitioning, automatic placement and routing, technology mapping for FPGAs, automatic test generation, and power estimation. One chapter is devoted to each of these topics, and an additional chapter is included covering parallel implementations for two of these applications, namely, automatic placement and automatic test generation. Indeed, a key feature of GAs is that they are easily parallelized, and parallelization may be necessary for handling large designs in a reasonable amount of time for many EDA applications. Our objective is to provide examples where GAs have been successfully applied in the past so that the reader will be able to apply similar techniques in solving his or her own problems. We will now proceed to the EDA problems to which GAs have already been applied, concluding the chapter with a brief description of each.

1.6.1 Partitioning

Partitioning is used in various EDA applications, e.g., as a tool for placement algorithms or for assigning circuit elements to blocks that can be packaged separately. In VLSI design applications, partitioning algorithms are used to achieve various objectives.

1. Circuit Layout: A class of placement algorithms, *min-cut placement*, is based on repeated partitioning of a given network, so as to minimize the size of the cut set at each stage, where the cut set is the set of nets that connect two partitions. At each partitioning stage, the chip area is also partitioned, e.g., alternately in the vertical and horizontal directions, and each block of the network is assigned to one region on the chip. This process is repeated several times, until each block consists of only one cell. The resulting assignment of cells on the chip gives the final layout.

2. Circuit Packaging: Semiconductor technology places restrictions on the total number of components that can be placed on a single semiconductor chip. Large circuits are partitioned into smaller subcircuits that can be fabricated on separate chips. Circuit partitioning algorithms are used to obtain such subcircuits, with a goal of minimizing the cut set, which determines the number of pins required on each chip. This technique is gaining renewed importance these days for Field Programmable Gate Arrays (FPGAs), which require automatic software for partitioning and mapping large circuits on several FPGA chips for rapid prototyping.

3. Circuit Simulation: Partitioning has also been used to split a circuit into smaller subcircuits which can be simulated independently. The results are combined to study the performance of the overall circuit. This speeds up the simulation process by several times and is used in relaxation-based circuit simulators. This process is also used for simulating circuits on multiprocessors.

Partitioning algorithms are broadly divided into two classes: constructive and iterative improvement. Constructive algorithms may start from empty initial partitions, and they generally grow clusters of well-connected components around one or more seed nodes or components, selected on the basis of some user-defined criteria, namely, number of fanin and fanout lines associated with a node. The quality of solutions generated by this class of partitioning algorithms is quite poor and degrades significantly as the partitioning problem size increases. The main advantage of constructive algorithms is that they are very fast and they usually scale well with the problem size. Therefore, these algorithms are frequently used to create an initial partitioning that may be further improved by applying other types of algorithms.

Iterative improvement algorithms, in contrast to constructive techniques, start with an initial partitioning, accomplished through some user-defined methods or randomly, and the algorithms tend to incrementally refine the initial partitioning through successive iterations; i.e., at any stage, a complete solution is always available. The algorithms tend to terminate when no more improvements can be found. Thus, these algorithms often terminate at local optima that are closely tied with the initial partitioning. Certain stochastic algorithms may give better results than deterministic algorithms. Iterative algorithms are widely used for high-quality two-way as well as multiway partitioning.

Chapter 2 gives a short overview of popular partitioning algorithms before discussing the genetic-based partitioning algorithms. In general, graph algorithms are highly successful in solving partitioning problems. The original graph approach, proposed by Kernighan and Lin [114] for solving telephone routing problems, was later adapted by Schweikert and Kernighan [205], to solve circuit partitioning by introducing a new representation, known as the net-cut or hypergraph model. This early version of circuit partitioning used a sorting algorithm to identify which pair of nodes (one from each block) should be swapped so that the cut-set size is minimized. Since the average run time for sorting n items is $O(n \log n)$, the overall complexity of the algorithm is worse than quadratic to the problem size, and it is unacceptable for large circuits. Fiduccia and Matthyeses [70] later developed an efficient bucket data structure to reduce the complexity of the partitioning algorithm, and by carefully modifying the original Kernighan-Lin algorithm, they developed a linear time circuit partitioning algorithm. However, this algorithm often gives solutions of inferior quality, especially if the circuit netlist is such that multiple ties occur during selection of the cell to be moved. Krishnamurthy [125] improved the performance of the Fiduccia-Mattheyses-type partitioning algorithm by adding a multilevel lookahead cell gain mechanism that enables the algorithm to more accurately select the cell which minimizes the cut-set size, considering future movement of cells. Dutta and Deng [63] pursued a more powerful method by adding a probability-based heuristic to the Fiduccia-Mattheyses algorithm and further improved the quality of solutions. Sanchis [200] has extended these approaches to perform multiway partitioning. Far superior results are achieved for recursive or repeated applications of the Kernighan-Lin or Fiduccia-Mattheyses algorithms, as was originally suggested by Kernighan and Lin in their seminal paper [114].

Besides the graph algorithms, a number of other combinatorial optimization techniques for circuit partitioning have been described in the literature, e.g., stochastic (simulated annealing [115]), adaptive (evolution methods [198]), neural [243], and spectral [8], [9]. In Chapter 2, we demonstrate how genetic algorithms can be designed to formulate partitioning problems. Two different types of GAs are discussed to show how genetic algorithms can be used to perform high-quality bipartitioning and multiway partitioning. Location-based GA encoding has been used in which the chromosome is a string of binary bits, where a 0 represents that a cell is in the left partition and a 1 represents that the cell is in the right partition. Both breadth-first and depth-first orderings are discussed in these two different implementations. The first version uses the classical cut-set minimization criterion for dividing the cells in two (for bipartitioning) or more (for multiway partitioning) partitions. The second version of the GA uses the ratio-cut metric, which tends to detect the clusters in a circuit netlist (owing to the hierarchical nature of real-world electrical circuits), and divides the cells into two or more blocks based on the cell clusters. This version of the GA employs a Fiduccia-Mattheyses-type graph-based local improvement algorithm in order to obtain fast solutions to large problems. Ratio-cut and Fiduccia-Mattheyses-based graph algorithms are compared with these two versions of genetic algorithms, and it is shown that the GA-based partitioning techniques give superior results for bipartitioning and multiway partitioning.

1.6.2 Automatic Placement

Usually in a custom design of chip layout, two different styles of cell placements are adopted: *standard cell placement* and *macro cell placement*. Standard cells are relatively smaller blocks available in a library to implement various functions: from basic NAND, NOR, NOT, XOR, etc., of several fanin capabilities to complex decoders, multibit magnitude comparators, programmable counters, 4-bit ALU functions, etc. These cells are optimized carefully by the cell library vendors and are provided to VLSI designers in the form of a library pertaining to a specific process technology. Standard cells are so called since they have constant cell height, but variable cell width that depends on the functional complexity of a cell. Macro blocks, on the other hand, are parameterizable, have variable dimensions, and often have arbitrary rectilinear shapes. Since the designer compiles these blocks depending on user requirements, the blocks are usually significantly larger than standard cells. The macro cell layout style is the most flexible layout style. It allows inclusion of compiled blocks, such as RAM or ROM arrays, PLAs, etc., in arbitrary sizes, which are difficult to include in the standard cell layout style. A typical VLSI design may include standard cells, macro cells, and full custom logic.

Standard cells are usually aggregated laterally in rows of cells. They are permuted between themselves in such a way that the overall amount of interconnections required to wire them together to satisfy the given netlist will be minimum and also that the overall chip size will be minimum. Standard cells are usually designed so that the power and ground interconnections run horizontally through the tops and bottoms of the cells. When the cells are placed adjacent to each other, these interconnections form a continuous track in each row. The logic inputs and outputs of a module are available at *pins*, or terminals, along the top or bottom edge (or

1.6. GAs for VLSI Design, Layout, and Test Automation

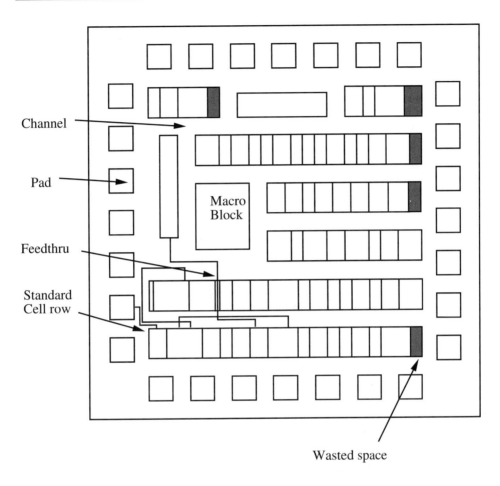

Figure 1.15. Standard cell layout style with macro blocks

both). Earlier generations of standard cell layouts set aside predefined horizontal routing zones, more appropriately called channels, between the rows of standard cells, as illustrated in Figure 1.15. These channels had horizontal tracks which were used for metal interconnection wires corresponding to the nets in a netlist. Connections from one row to another were done either through vertical wiring channels at the edges of the chip or by using feedthrough cells, which are standard height cells with just a few interconnections running through them vertically. At that time, VLSI technologies could support only one or two layers of metal interconnections in addition to polysilicon wires that were mostly used for short local connections. Over-the-cell routing was not possible at that time. Now with the availability of multilayer metal interconnects, over-the-cell routing is possible. The resulting routing styles are not necessarily of the classical channel routing genre where largely horizontal segments of tracks containing metal wires are used to define the interconnects between different cells. Nevertheless, the main objective of a standard cell placement algorithm remains the same: to permute the cells between themselves and assign them to various rows in such a way that the objective function involving the overall interconnect length and the chip size is minimized without violating the non-overlapping cell requirement; i.e., no two adjoining cells should overlap laterally or between the edges of two successive rows.

Macro cells are usually rectilinear in shape and are much larger than standard cells. Not only can these cells be placed anywhere within the perimeter of a chip layout, but they can also be rotated in various orientations, as long as all rectilinear edges of a macro cell remain strictly parallel to one of the chip axes. Fig. 1.16 shows a chip composed entirely of macro cells with no standard cells. Each cell is surrounded by routing channels. The layout design objectives of standard cell

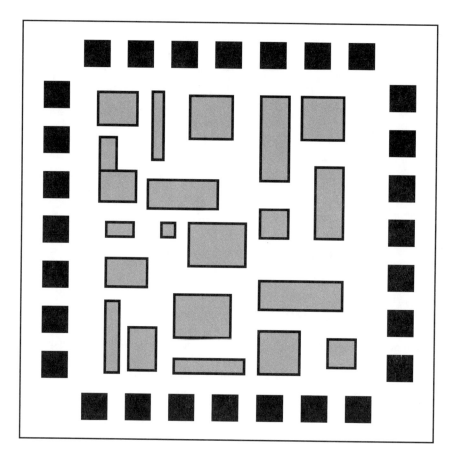

Figure 1.16. Macro cell layout style

placement and the mode of operation of the placement algorithm are significantly different from the layout optimization criteria in a macro cell layout design. The larger macro cells have many terminals that require interconnections. Thus, a macro cell placement tool must account for the additional channel area that is required to place the maze of interconnects around each macro cell. This estimation process is reasonably compute intensive. It involves determination of cell placement and orientation to minimize channel area, overall interconnect length, and chip area. The fact that all macro cells are not strictly rectangular makes the optimal placement of macro cells computationally intractable. These rectilinear shapes may require a major reallocation of positions and orientations of other cells whenever a big non-rectangular cell is rotated.

Cell placement problems can be broadly divided into four different classes based on the optimization techniques applied [212]. Graph-based approaches typically use a partitioning method known as the *min-cut* algorithm, originally proposed by Breuer [21]. By applying an alternating sequence of horizontal and vertical slicings that will partition components such that the cut-set size is minimized (hence, the name min-cut), one can find the correct placement of cells by recursively using a partitioning algorithm. Surais and Kedem [221] have extended this approach to concurrent two-dimensional slicing or sectioning so that the resulting partitioning is divided into four different quadrants. This graph-based approach is called *quadrisectioning* and is found to be very fast. The *second* class of algorithms is based on physical phenomena that are dictated by the laws of physics. Three well-known algorithms belonging to this class are: (i) *simulated annealing*, which mimics the well-known annealing process that blacksmiths apply to manufacture metallic implements [115], (ii) *generalized force-directed placement*, which is based on the well-known Hooke's law in elasticity [86], and iii) power dissipation (i.e., distribution of voltages and currents) in a resistive linear passive network of electrical components [35]. Although these physical phenomena range in different aspects from the Maxwell-Boltzmann distribution of molecules and lattice structures in a solid object to distribution of electrons in an electrical circuit, they all tend to arrange the natural system such that the potential energy of the system is always minimized and the system is in a stable state. In a cell-placement problem, this minimization of potential energy of a physical system is akin to minimizing its cost or objective function and is tantamount to simultaneous minimization of chip area and improvement of chip speed. The *third* class of algorithms is based on numerical optimization, which may involve postulating the cell placement problem as a set of equations to be solved by finding the eigenvalues and eigenvectors of a block diagonal matrix. Eigenvectors corresponding to a nontrivial minimum eigenvalue pertain to the minimized objective function and can denote the cell positions with reference to their (x, y) coordinates. Similar to spectral-based partitioning algorithms [8], [9], placement tools developed using numerical optimization techniques have been found to give very high quality solutions [226]. Other algorithms in this class also utilize integer programming and linear programming based optimization techniques. The *fourth* class of algorithms is based on evolution algorithms.

In Chapter 3, genetic-based cell placement algorithms are discussed for both standard cell and macro cell layouts. For standard cell placement, a permutation class of genetic representation is used that is akin to the one used in solving the traveling salesman problem (TSP). In order to eliminate conflicts and inconsistencies that may occur due to application of a simple crossover operator with random cut and paste, three different crossover operators are introduced which systematically generate offspring that represent valid placement solutions. The three crossover operators are called *order*, *PMX*, and *cycle* crossover. A meta-genetic optimization process is used to select the various tuning parameters, such as crossover rate, inversion rate, and mutation rate. Cycle crossover has been found to be superior to other crossover operators. For macro cell placement, a two-dimensional bin-packing chromosomal representation is used, where the cells are arranged in the form of a priority queue. Since macro cell algorithms are very computation intensive, the genetic algorithm is transformed into simulated annealing by decreasing the population size according to some temperature schedule, and mutation operations are increased in the later phase of the algorithm. The mixed genetic algorithm (GA) and simulated annealing (SA) approach, called *SAGA*, is shown to provide superior results to the conventional optimization techniques used in macro cell placement tools for several benchmark circuits.

1.6.3 Automatic Routing

The objective of an automatic routing tool is to connect the cell pins while meeting any layout constraints. Typical constraints are to minimize the overall area, avoid meandering paths and thereby reduce delays, and minimize the lengths of critical nets. Interconnection wires are assumed to have widths that meet the design rules imposed by the technology used and to be adequately separated from adjacent wires.

Modern high-performance, multimillion transistor VLSI chips with multiple layers of interconnects are extremely dense and often too complex to have interconnects routed in their entirety by a single pass of a grid-based algorithm, such as a maze router, which is commonly used for Integrated Circuit (IC) and Printed Circuit Board (PCB) wire routing. Consequently, due to the high degree of complexity, most VLSI routing algorithms operate in a two-phase mode. First, a *global* or *loose* routing is performed that determines the topologies of nets and their relative positions with respect to objects which act as obstacles to the routing. This is followed by a *detailed* routing phase, where the final interconnection specifications among the module terminals are determined in terms of layers, vias, and track assignments. Examples of detailed routers are channel routers, maze-type routers, river routers, and switchbox routers, and effective heuristics have been developed for such routers. Specialized routers are used for power, ground, and clock routing, which are generally separated from signal layers and are critically routed to meet the constraints of a high-performance chip.

The main task of global routing is to break up a larger, more complex problem into several smaller ones that are more tractable in nature. In this book, we discuss only the problem of global routing, and in particular, global routing for macro cells. In a simultaneous place-and-route algorithm, often global routing is blended with the cell placement algorithm in an intertwined fashion. Therefore, the genetic-based automatic global routing technique described in Chapter 4 is a natural sequel to Chapter 3, which describes the genetic-based macro cell placement algorithm.

Depending upon the problem complexity, global routing can be broken into two subproblems: (i) global routing graph construction and (ii) wiring the nets within the routing regions. Most global routers perform routing in terms of a global routing graph, which is extracted from the given placement and the routing region definitions. The edges of the graph correspond to future routing regions, while the vertices correspond to intersections of routing regions. While the graph is highly regular for gate arrays and most PCB routing, the graph is irregular for macro cell routing. The routing graph of a simple placement of a macro cell design, assuming suitable routing region definitions, is shown in Fig. 1.17.

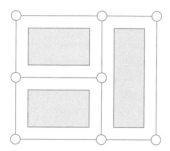

Figure 1.17. Initial global routing graph for a small macro cell problem

Further, the structure of the global graph can also be augmented by information on available feedthroughs. A feedthrough represents a routing area that can pass through a particular cell or macro and is either deliberately introduced as a mechanism for reducing wire lengths of global nets or can be constructed from unused space within a cell. In addition, in order to ease the task of detailed routing, several optimality criteria are used by the global router, such as minimizing total wire length, minimizing total area, minimizing local densities, etc. Every edge, therefore, is assigned one or more cost values typically representing the length of the associated routing region and/or the capacity of the region, i.e., the number of nets which can pass through the region.

To compute a global route for a specific net, vertices representing the terminals are added at appropriate locations, as illustrated in Fig. 1.18. Here, for each terminal, the location of the corresponding terminal vertex is determined by a perpendicular projection of the terminal onto the edge representing the appropriate routing channel. Finding a global route then becomes equivalent to finding a minimum-cost subtree in the routing graph which spans all the terminal vertices. This tree finding can be done using standard shortest-path, spanning-tree, or Steiner-tree algorithms with modifications to account for features peculiar to a certain layout style. Subse-

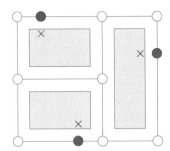

Figure 1.18. Routing graph augmentation due to terminals

quent to global routing, each net that spans more than one region (edge in the global routing graph) is decomposed into subnets, one per region. Between neighboring regions, pseudo-pins are introduced to indicate the intersection of global paths with the boundaries of the local regions.

In macro cell layout, the routing region definition is more involved. Routing region definitions consist of partitioning the routing area into a set of nonintersecting rectangular regions called channels. Channel definition and ordering are key steps in the overall placement, global routing, and detailed routing. After routing regions are defined, the routing graphs are generated according to one of three different models: (i) the rectilinear adjacency graph model, (ii) the polar graph model, or (iii) the hybrid graph model. The rectilinear adjacency graph and horizontal and vertical polar graphs are illustrated in Fig. 1.19. Global routing can be broadly divided into two main classes: sequential approaches and concurrent approaches.

1. **Sequential Approaches:** As the name suggests, these routers construct a complete global routing by considering one net at a time. These can be further categorized based on the number of alternative solutions they consider for each net and the number of passes they perform. The actual algorithms depend on the nature of the global routing graph. For instance, for grid graphs, the maze routing algorithm and its variants are often used. However, for more general routing graphs, particularly for macro cell layout, general shortest-path or Steiner-tree algorithms are required.

2. **Concurrent Approaches:** A main drawback of sequential routers is that the result quality is highly dependent on the order in which nets are routed and in general, it is difficult to devise a good net ordering a priori. Consequently, in concurrent routing, all nets are considered simultaneously and the net-ordering problem is bypassed. Two popular solution techniques that fall into this category include:

    (a) **Integer Programming:** This method formulates the global routing problem as an integer programming problem. Methods based on relaxation or simulated annealing have been used to help solve the integer programs.

1.6. GAs for VLSI Design, Layout, and Test Automation

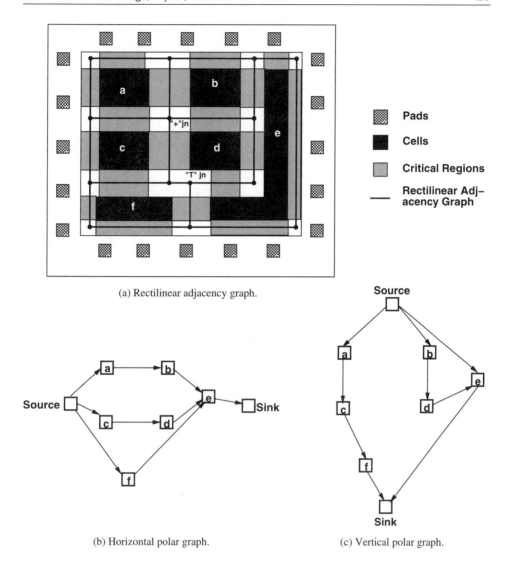

(a) Rectilinear adjacency graph.

(b) Horizontal polar graph.

(c) Vertical polar graph.

Figure 1.19. Routing graphs for 6-cell macro cell layout

(b) **Hierarchical Methods:** These approaches either proceed in a top-down or a bottom-up fashion, obtaining complete solutions at each level of the hierarchy and using the solutions obtained to influence the routing at the next level. Often the problems that arise at any one level of the hierarchy are simple enough to be solved exactly.

Chapter 4 describes a new genetic-based Steiner global routing algorithm. Numerous algorithms of various kinds have been developed for the Steiner Problem in a Graph (SPG) that find a minimum cost subgraph spanning a set of designated vertices on a given graph. Exact algorithms can be found in [11], [14], [40], [60], [90], [140], [216]. However, since the SPG is NP-complete [111], these algorithms have exponential worst case time complexities. Therefore, a significant research effort has been directed toward polynomial time heuristics [11], [121], [172], [184], [222], [239]. Simulated annealing has also been applied to SPG [59]. A first effort to apply genetic algorithms to SPG is due to Kapsalis, et al. [110]. The genetic-based algorithm for SPG described in Chapter 4 differs from [110] in a number of ways: by enforcing constraint satisfaction, the proposed algorithm eliminates the need for imposing penalties for invalid solutions; the new algorithm uses an inversion operator; it also uses a more efficient method for performance evaluation; and unlike in [110], which requires problem-specific parameter settings, a fixed set of parameters are used in the algorithm described in Chapter 4. Consequently, from experimental results, it can be seen that the GA-based global routing algorithm discussed in Chapter 4 outperforms the GA-based algorithm in [110] and an iterated version of the shortest path heuristics proposed by Winter and Smith [239]. It has been demonstrated that the GA is capable of finding optimal solutions in more than 77% of all test cases run and is within 1% of the optimal result in more than 92% of all test runs.

1.6.4 Technology Mapping for FPGAs

Field Programmable Gate Arrays (FPGAs) are devices that can be configured by the user to implement a required logic function. FPGAs are being used extensively for fast system prototyping due to their low cost and the ability to produce the final circuit within a relatively short time. An FPGA can be viewed as a modification of a mask-programmable gate array, where the reprogramming time and cost are drastically reduced, or as a Programmable Array Logic (PAL) device whose size is increased by an order of magnitude. A PAL device contains an array that implements AND/OR logic and is restricted to a small number of equivalent logic gates.

Figure 1.20 shows the structure of a typical FPGA, which consists of logic blocks surrounded by routing resources. Each logic block consists of a combinational logic part and a sequential part (a latch). There are two main classes of FPGA architectures: one is Lookup Table (LUT) based, and the other is multiplexer based. The LUT-based FPGA is widely available commercially through several FPGA manufacturers. In a LUT-based FPGA device, the programmable logic block consists of a K-input lookup table, also called a configurable logic block (CLB), which can implement any Boolean function of up to K variables.

A K-input lookup table is a digital memory in which K inputs are used to address a 2^K by 1-bit memory that stores the truth table of the Boolean function. The XC3000 device from Xilinx is an FPGA device in which the basic programmable

1.6. GAs for VLSI Design, Layout, and Test Automation

logic block is a K-input lookup table which can implement any Boolean function of up to K variables. The CLB for the XC3000 device is shown in Figure 1.21 [24]. Here, a CLB can represent either a 5-variable function or two 4-variable functions which together depend on at most 5 variables.

The technology mapping problem for a LUT-based FPGA device is to transform a Boolean network into a functionally equivalent network of K-input LUTs (in general, technology mapping is a logic synthesis step in which equations and latches are implemented by choosing gates from a fixed library of primitive components). The objective of the technology mapping tool for FPGAs is to find the best mapping of logic onto FPGA logic blocks. Here, the quality of the mapping is measured in terms of the size of the required FPGA and the routability of the design, as well as other factors, such as delays. The technology mapping problem for the XC3000 device is to map the design onto CLBs in the device.

There are several approaches for solving the FPGA mapping problem. Some approaches [72], [112], [152], [202], [240] map the logic based on algebraic decomposition. Other approaches [130], [168] are based on function decomposition, which is a Boolean optimization method. All of the above approaches to solving the mapping problem map the logic onto LUTs and then pack these LUTs into CLBs. As the number of inputs to a LUT increases, the search space increases exponentially, and these methods, particularly those based on Boolean optimization, do not scale well.

Chapter 5 describes a genetic algorithm for technology mapping to FPGAs. The algorithm can target either the mapping onto LUTs or can directly map onto CLBs.

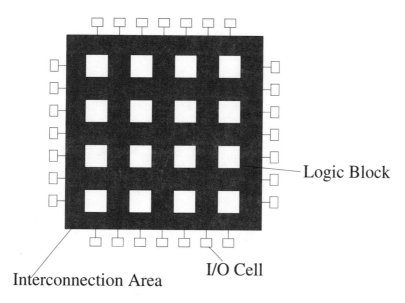

Figure 1.20. FPGA structure

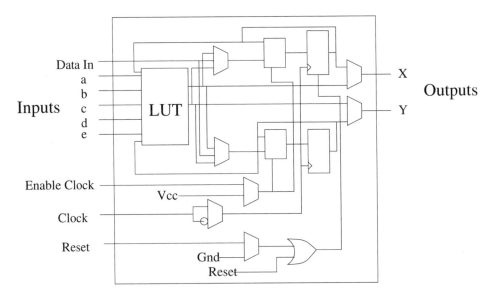

Figure 1.21. Xilinx XC3000 CLB

Designing the GA such that it maps directly onto CLBs leads to far better CLB mappings. Furthermore, as the number of inputs to a LUT increases, the GA-based approach exhibits superior results compared to other approaches.

1.6.5 Automatic Test Generation

Once a chip is fabricated, it must be tested to ensure that it functions as designed. The assumption is made that the design is correct. The set of tests applied should be able to uncover all possible defects that could occur in the manufacturing process. That is, the output response of a defective chip should be different from the outputs of a good chip. Then the defective chips can be weeded out so that they are not shipped to customers.

Consider the defect between the output of the NOR gate and V_{ss} as shown in Fig. 1.22(a). A transistor-level representation is shown in Fig. 1.22(b), and two possible logical manifestations of the defect are shown in Fig. 1.22(c). If the resistance is low, the output of the NOR gate will behave as if it is stuck at logic 0. If the resistance is high, a timing error may result. We would like to ensure that the test set contains a test to detect this defect in case it occurs. Obtaining a set of tests that will detect most possible defects is a difficult problem, and this is the objective of an automatic test generator. The approach commonly used is to try to generate a test set that detects all single stuck-at faults in the circuit. After a high fault coverage for single stuck-at faults is achieved, additional test vectors may be generated that target other fault models, such as the delay fault model.

1.6. GAs for VLSI Design, Layout, and Test Automation

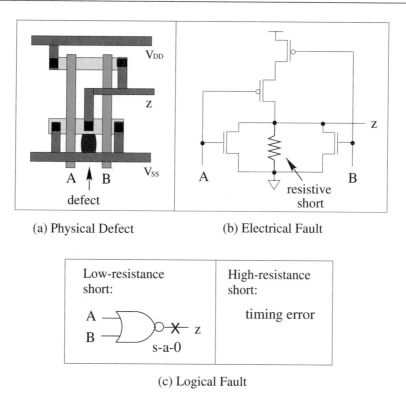

Figure 1.22. Effects of a physical defect

Deterministic test generation algorithms for combinational circuits [85] and sequential circuits [37], [76], [134], [141], [145], [160], [204] have been used in the past, but execution times are often long, due to the large number of backtracks that often occur. Simulation-based approaches have also been used [6], [20], [139], [203], [207], [218], [242] and are effective in reducing execution time. However, fault coverages are often lower. The use of GAs for simulation-based test generation has been shown to provide better fault coverages than deterministic algorithms in some cases and lower execution times [46], [97], [98], [99], [102], [179], [181], [189], [194], [195], [196], [197], [219].

GA-based test generators developed at the University of Illinois in the past few years are described in detail in Chapter 6. A simple GA with nonoverlapping populations has been used, due to the irregular search space. Exploration of a large number of candidate solutions overrides exploitation of known good solutions. Furthermore, in contrast to other applications, the GA is invoked repeatedly for test generation, and the objective with each invocation is to generate a vector or sequence of vectors to aid in detecting one or more faults. Because of the large number of invocations of the GA and the high cost of accurate fitness evaluation, the GA parameters must be set to reduce the computation time to a reasonable level, at the expense of solution quality. In addition, approximate fitness functions can be used to further reduce the computation time. Of course, tradeoffs in computation time vs. solution quality can always be made, depending on which factor is more important in a particular situation. In fact, experimental results published in the literature clearly illustrate the tradeoffs that have been made in various implementations.

GA encoding for the test generation problem has been straightforward. A simple string representation is typically used in which each gene represents the logic value to be applied to a particular input of the circuit at a specific time. Standard crossover and mutation operators can therefore be used. While the first implementations targeted a large number of faults in the circuit concurrently, the trend has been to separate the tasks of fault excitation and fault effect propagation and to target faults individually. This approach was first used in GATTO [181], [46], and it has resulted in the highest fault coverages ever reported for many of the benchmark circuits in STRATEGATE [99]. Although the simple strategy used initially for test generation in GATEST [189], [194] does not achieve such high fault coverages, it is very effective for test sequence compaction, as discussed in Chapter 6.

1.6.6 Power Estimation

Even though VLSI designs are fabricated in CMOS technology, which is a low-power technology, we still have to worry about power for large chips. Power dissipation in CMOS has two components: static, due to leakage current, and dynamic, due to switching activity. The static power is relatively low and is often neglected in power estimation. Dynamic power dissipation occurs for two reasons. First, the short circuit component is due to the current pulse that flows through the p and n transistors from V_{dd} to ground while they are both conducting simultaneously. This component of the power dissipation is small in a well-designed circuit. Secondly, the capacitive component is due to charging and discharging the various capacitances associated with each gate. This capacitive component depends on the node capacitance, the switching frequency, and the supply voltage as follows:

$$Power = \frac{1}{2} \, t \, C \, V_{dd}^2 \, f$$

where V_{dd} is the power supply voltage, C is the node capacitance, f is the clock frequency, and t is the number of toggles between logic 1 and logic 0 that occur within a clock cycle. Estimation of power dissipation for a circuit involves summation of the power dissipated at individual nodes.

The objective of a power estimation tool is to determine the maximum power dissipation that will occur. If the power dissipation is found to be too high, modifications can be made to the design, or appropriate packaging technology can be used. At the gate level, static techniques that use probabilistic information about circuit activity are often employed for estimating power dissipation, e.g., [42]. However, the average power dissipation estimated using this approach does not provide accurate boundary conditions for determining the limits of the design. Dynamic techniques that simulate the circuit using typical input sequences are more effective at estimating the peak power dissipation, which provides a limit for circuit operation. However, a method is needed to provide candidate sequences so that a sequence that leads to maximal power dissipation can be found. Deterministic techniques proposed previously for this problem may be unsuitable for large circuits. Chapter 7 describes a GA that solves this problem and is able to handle large circuits while requiring very little computation time. The peak power estimates provided by the GA achieve much tighter lower bounds on the actual peak power than a random approach that requires considerably more execution time.

The GA begins with random sequences and attempts to evolve a sequence with maximum power dissipation. A simple binary encoding is used in which each individual in the GA population represents the initial circuit state and a sequence of vectors to be applied. The fitness function estimates power dissipation by simulating the sequence and keeping track of the toggle count at each node. This toggle count is weighted by the node capacitance to get power dissipation at each node, and the summation of power dissipation over all nodes is computed.

Power dissipation in a circuit is dependent upon delays of internal circuit nodes. Glitches and hazards result in power dissipation. Therefore, the accuracy of the power estimate is dependent upon the delay model used when simulating each sequence. Peak power estimation under four different delay models is explored in Chapter 7, and comparisons are made among the various models. In general, sequences optimized under one nonzero delay model provide good measures for other nonzero delay models. In addition, various durations are studied, and computation of peak single-cycle, n-cycle, and sustainable power dissipation is addressed.

SUMMARY

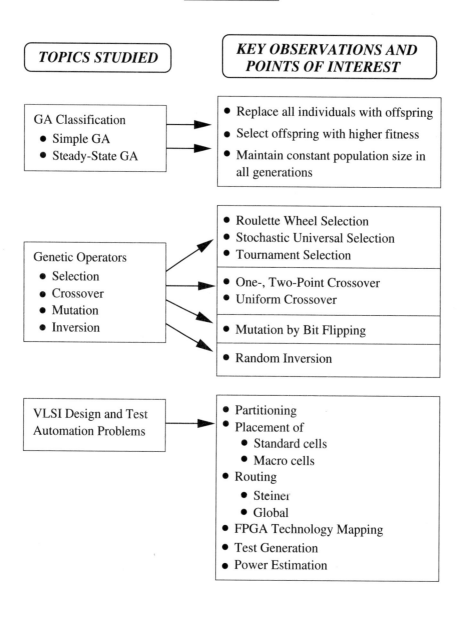

CHAPTER 2
at a glance

- Partitioning taxonomy
- Genetic algorithm for partitioning
    - Bipartitioning
    - Multiway partitioning
- Fast hybrid genetic algorithm using ratio-cut metric for partitioning

Chapter 2

PARTITIONING

P. Mazumder and K. Shahookar

The problem of circuit partitioning is encountered frequently in VLSI design, e.g., in circuit layout, circuit packaging, and circuit simulation. The objects to be partitioned in VLSI design are typically logic gates or instances of standard cells. The objective is to separate the cells into two or more blocks such that the number of interconnections between blocks is minimized. A balance constraint is often imposed which ensures that the blocks contain about the same number of components.

Min-cut placement algorithms perform partitioning repeatedly in an attempt to find a placement that minimizes overall routing. After the circuit is partitioned into two blocks, each block is partitioned into two subblocks, and so on, until each block contains only one or a small number of cells. At this point, each block can be assigned a location on the layout surface. By minimizing the number of interconnections between blocks at each step, the min-cut algorithm is able to minimize the number of nets that must be routed over long distances. Balance is an important criterion in placement algorithms, since cells must be evenly distributed across the layout surface.

This chapter addresses the problem of partitioning, and in particular, the use of genetic algorithms for circuit partitioning. We begin with a formal description of the problem and a brief review of partitioning algorithms. The use of a steady-state GA for circuit partitioning is described next, and experimental results are presented. The chapter concludes with a description of a hybrid GA that performs local optimization in addition to genetic search in order to reduce execution time.

2.1 Problem Description

The circuit partitioning problem can be formally represented in graph theoretic notation as a weighted graph, with the components represented as nodes, and the wires connecting them as edges. The weights of the nodes represent the sizes of the corresponding components, and the weights of the edges represent the number of wires connecting the components.

2.1. Problem Description

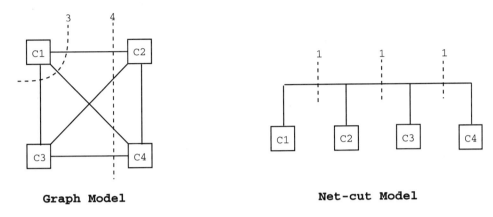

Figure 2.1. Comparison of the net-cut and the graph models for partitioning

In its general form, the partitioning problem consists of dividing the nodes of the graph into two or more disjoint subsets such that the sum of weights of the nodes in each subset does not exceed a given capacity, and the sum of weights of edges connecting nodes in different subsets is minimized.

The problem input consists of a graph $G = (V, E)$, where $v_i \in V$ is a vertex in the vertex-set V, and $e_{ij} = (v_i, v_j) \in E$ is an edge in the edge-set E. Let a_i be the weight (area) of a vertex i, and c_{ij} be the weight or *cost* of an edge e_{ij}. Also given is the number of partitions k, and the capacity of each subset or partition, $A_1, A_2, ..., A_k$. The output consists of disjoint subsets $V_1, V_2, ..., V_k$ such that
$\cup_{n=1}^{k} V_n = V$,
$\forall\, V_n,\ \sum_{\forall v_i \in V_n} a_i < A_n$,
and $\forall\, e_{ij}$ such that $[v_i] \neq [v_j]$, $C = \sum c_{ij}$ is minimized.

$V_n = [v_i]$ represents the subset containing v_i, and C is the *cost* of the cut. The set of edges cut by the partition, e_{ij}, $[v_i] \neq [v_j]$ is called the *cut set*.

The edge weights in the problem input can also be given in the form of a connectivity matrix. The element c_{ij} of the matrix represents the cost of connectivity between v_i and v_j.

Schweikert and Kernighan [205] illustrated the deficiencies of using simple graph models for partitioning circuits. They proposed the *net-cut* circuit model in which a single net represents the connections between all components joined by the same signal line. Consider a net of four components shown in Fig. 2.1. If modeled as a graph with an edge created for every connected pair of components, it would become a fully connected graph of four nodes. If this graph were partitioned, the cut cost would be three or four edges. However, if this circuit was represented by the net-cut model, then the cut cost would be only one net.

For this reason, VLSI circuits are generally represented as bipartite graphs consisting of two sets of nodes, the cells and the nets. Edges connect each cell to several nets, and each net to several cells, as shown in Fig. 2.2. In this case, let

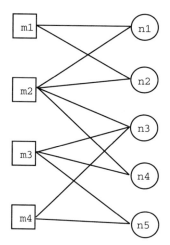

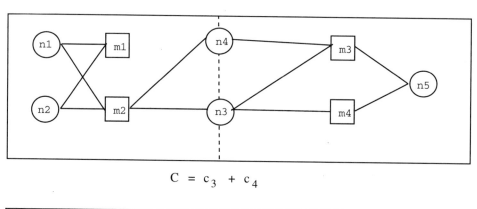

$$C = c_3 + c_4$$

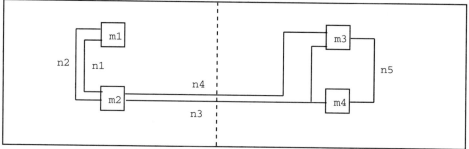

Figure 2.2. Bipartite graph model for partitioning

2.1. Problem Description

$G = (M, N, E)$, $m_i \in M$ is a *cell*, $n_i \in N$ is a *net*, and $e_{ij} = (m_i, n_j) \in E$ is an edge which represents that m_i and n_j are *connected* electrically. For any n_j, for all i for which e_{ij} exists, we say that the cells m_i are connected by net n_j. Conversely, for any m_i, for all j for which e_{ij} exists, we say that the nets n_j are connected to cell m_i. Each cell m_i has an area a_i, and each net n_j has a cost c_j. The edges of the bipartite graph are unweighted.

In this case, the partitioning problem divides the set of cells into disjoint subsets, $M_1, M_2, ..., M_k$, such that the sum of cell areas in each subset M_i is less than a given capacity A_i, and the sum of the costs of nets connected to cells in different subsets is minimized. That is,

$\cup_{n=1}^{k} M_n = M$,

$\forall\, M_n$, $\sum_{\forall m_i \in M_n} a_i < A_i$,

and $\forall\, n_j$, if n_j is connected to cells in p different partitions,

then, $C = \sum (p-1) c_j$ is minimized.

A net n_i is called an *internal* net if all cells connected to it are in one partition only, i.e., $\forall e_{ji}$ and $\forall e_{ki} \in E$, $[m_j] = [m_k]$. A net n_i is called an *external* net if it is connected to cells in different partitions, i.e., if $\exists e_{ji}$ and $e_{ki} \in E$, such that $[m_j] \neq [m_k]$.

2.1.1 Partitioning Algorithms

Partitioning a set of circuit elements into two or more subsets such that connectivities between the weighted edges across the blocks can be minimized is a computationally hard problem. Kernighan and Lin have shown that in the worst case, it will take exponential time to divide a set of circuit elements into k blocks by enumerating all possible permutations in which n circuit components can be divided into k equal blocks of size $p = n/k$. Clearly, there are $\binom{n}{p}$ ways of choosing the first subset, where $\binom{n}{p}$ is the number of combinations of n elements, taken p at a time. Once the first subset is chosen, there are $\binom{n-p}{p}$ ways of choosing the second subset, $\binom{n-2p}{p}$ ways of choosing the third subset, and so on. But this includes all $k!$ permutations of the k partitions. Counting all these permutations as one partition, the total number of unique ways of partitioning the graph is

$$N_k = (1/k!) \binom{n}{p} \binom{n-p}{p} \binom{n-2p}{p} \cdots \binom{2p}{p} \binom{p}{p} = \frac{n!}{k!(p!)^k}.$$

Using Sterling's approximation [120], we get $N_k \approx O\left((n/p)^{(n-n/p)}\right)$. Thus, an algorithm resorting to exhaustive enumeration will take exponential time. The partitioning problem has been expressed in various forms by different researchers and has been shown to be NP-complete. The NP-completeness proof of the following problem was first given by Karp [111] and also appears in Garey and Johnson's treatise [75].

Problem 1: Given a graph G and a positive real size $A(v)$ associated with each vertex $v \in V$, can V be partitioned into two disjoint subsets P and Q such that $V = P \cup Q$ and $\sum_{v \in P} A(v) = \sum_{v \in Q} A(v)$?

This is a decision problem, and just answers whether a given set with elements of arbitrary size can be partitioned into two subsets of equal size or not. Obviously the problem of computing an exact bipartition of a graph with positive real weights attached to the nodes is harder than the above problem.

Hyafia and Rivest [105] have shown the following problem to be NP-complete.

Problem 2: Assume that a given graph G has a positive real size $A(v)$ associated with each vertex $v \in V$, and weight $w(e)$ with each edge $e \in E$. Also assume that two integers J and K are given. The problem is to determine whether there exists a partition of V consisting of m disjoint sets $V_1, V_2, ..., V_m$ such that the sum of the sizes of the vertices belonging to each subset is less than or equal to K and the total cut cost is less than or equal to J.

The above problem remains NP-complete for all constant values of $K \geq 3$ even if all the weights of the vertices and edges are equal to 1. If $K = 2$, the problem can be solved in polynomial time.

Garey et al. [75] have given an alternative proof for the above problem, when the number of subsets in the resulting partition is only two.

2.1.2 Taxonomy of Partitioning Algorithms

Since 1970, when Kernighan and Lin proposed a simple graph-based heuristic that iteratively improves an initial partitioning by using a semi-greedy graph search technique, numerous combinatorial optimization techniques have been applied to solve the graph and circuit partitioning problems. Broadly, these partitioning algorithms can be classified into groups according to the optimization strategy they applied. Some algorithms start with empty sets and build up the partition by adding one element at a time. These are *constructive* algorithms. Others start with an arbitrary initial partition and repeatedly modify it. These are *iterative improvement* algorithms. The following is a broad classification of the partitioning algorithms.

Clustering Algorithms: These constructive algorithms [118] cluster the given set of nodes into small groups. Each cluster should satisfy various criteria, such as total size and the number of external connections. To form a cluster, a seed node is chosen to initialize a cluster, and nodes are repeatedly added to it. If the size or external connections criteria are violated by addition of more nodes, the growth of the cluster is stopped. This is known as a sequential constructive algorithm, since the cluster is grown around a single seed node. This greedy approach is simple and runs fast, but the quality of the result it generates is very poor, since the final partition often is constructed from loose nodes around different clusters. Somewhat better results can be obtained by simultaneously building different clusters around a number of seed nodes [118]. This concurrent approach tends to build partitions in a more uniform manner and has been widely used in the past for constructively building initial partitions, which are then further improved by means of an iterative algorithm. The quality of the partition, however, in constructive techniques, strongly depends on the initial choice of the seeds, and a different choice can give a very different partition.

Graph Algorithms: These algorithms are formulated in terms of operations on the vertices and edges of the graph representing a circuit and generally involve a partition with two subsets. Kernighan and Lin [114] of Bell Telephone Laboratories presented the first iterative improvement algorithm for partitioning graphs.

The Kernighan-Lin (K-L) algorithm starts with a random partition and tries to minimize the cut cost by making small local changes through interchanging pairs of nodes. The algorithm makes several passes. Each pass consists of a series of interchanges of pairs of nodes. The nodes are interchanged in the sequence of the maximum gain in the cut cost. At the end of a pass, all nodes get interchanged, and the cut cost becomes the same as that at the beginning of the pass. The intermediate partitions are examined, and the one with the sequence of pairwise exchanges that yields the smallest cut cost is returned as the outcome of the pass. From empirical studies, it was found that only a constant number of passes (typically between two and four) are required, independent of the graph size. The quality of the final partition often heavily depends on the initial partition. It is, therefore, a good idea to run a heuristic-based algorithm, such as the K-L algorithm, several times with randomly generated initial partitions and to observe the best solution, as well as the standard deviations of the distribution.

Since the K-L algorithm uses the graph model and employs a sorting procedure, to evaluate quickly which pairs may reduce the cut size if they are exchanged, it suffers from an $O(n^2 log\ n)$ time complexity, where n is the number of nodes in a graph. Fiduccia and Mattheyses [70] proposed a more efficient method for implementing Kernighan and Lin's algorithm leading to a fast linear time algorithm for partitioning (the F-M algorithm). This algorithm is an iterative heuristic similar to Kernighan and Lin's algorithm. It improves upon the time complexity of the K-L algorithm by moving only one cell at a time and using efficient data structures in order to search for the best element to move and also to minimize the effort required to update the cells after each move. These data structures eliminate the need for repeated sorting that takes $O(n \log n)$ time in the K-L algorithm. The partitioning is done such that the sizes of the two blocks are in a given ratio to the size of the original circuit, up to a tolerance of ±1 cell. The areas of the cells are considered to be a measure of the size of the partition. In addition, the user can specify some cells as being fixed in either block. Each iteration is linear in the size of the input, $O(P)$, where P is the total number of pins in n circuit elements, and in practice, only a few iterations are required for convergence.

Although the F-M algorithm is fast, it tends to converge to local minima, when ties occur between many cells in the gain lists, and the ties are blindly broken in a random way without giving further consideration to the future effects on other nets when a cell is selected to move. Krishnamurthy refined the technique of choosing the best cell for movement to the other block by adding a Look-Ahead (LA) mechanism [125]. Krishnamurthy's algorithm maintains a multidimensional version of the F-M data structure, which lists expected gains of cells in future moves. Thus, in the event of ties in the current level gain in cut size due to a move, Krishnamurthy's LA algorithm chooses a cell which gives better gain in future moves. Since multilevel gain calculations and updating take a significant amount of time, typically 2- (called LA-2) or 3- (called LA-3) level gain lists are used in practice. Although the timing overhead of 20 runs of LA-3 is comparable with 100 runs of the F-M algorithm, statistically it has been observed that LA-3 gives about 7% improvement in cut size in comparison with the F-M algorithm.

Recently, Dutt and Deng proposed a new probability-based augmentation of the graph partitioning algorithm, called the probabilistic gain computation approach, (PROP) [63], which is capable of capturing the global and future implications of moving a node at the current time. The technique associates with each node j a probability $p(M(j))$, where $M(j)$ indicates the event that in the current pass, node j will actually be moved to the other block. From the value of this probability, $g(u)$, the probabilistic gain of the nodes (for all u's in the graph) are calculated. While calculating $g(u)$ of a node, all nets connected to u contribute to $g(u)$, depending on their individual probabilities pertaining to whether a particular net is on the cut set or not. The probability calculations tend to have a reasonable overhead, and the time complexity of PROP is $O(rqP)$, where P is the total number of pins, q is the average number of pins with which a net is connected, and r (typically less than 5) is the total number of passes required before the algorithm halts.

It has been found that 100 runs of the F-M algorithm take about the same amount of time as 40 runs of LA-2 and 20 runs of PROP. In terms of quality of results, all three algorithms were tested on a large number of benchmark circuits. The totals of the best cut sets obtained by these three algorithms, run as above, are 1776 (F-M), 1898 (LA-2), and 1380 (PROP). Thus, an improvement of 27.3% over LA-2 and 22.3% over F-M can be achieved by PROP. It was seen that although LA-3 requires twice as much time as PROP, it performs inferior to PROP, which has an improvement of 16.6% over LA-3.

A major limitation of all the previously mentioned graph algorithms is that they are suitable for bipartitioning a graph or a circuit network, and not for multiway partitioning. In order to perform a k-way partitioning by the above bipartitioning algorithms, one has to employ recursive bipartitioning if k is exactly equal to or very close to a power of 2. Otherwise, one has to perform about $k(k-1)/2$ separate runs of a bipartitioning algorithm. Nevertheless, neither of these methods yields good results, since they tend to sequentially improve the partitioning between two blocks at a time. In order to obtain a globally optimal multiway partitioning, one has to apply partitioning over the entire graph or net list simultaneously.

Sanchis [200] modified Krishnamurthy's algorithm to perform multiway partitioning. She developed appropriate data structures for representing multiway partitioning of various blocks, and she experimentally proved that the optimal number of gain levels necessary depends on the number of blocks, the net size and degree distributions of the circuit network, but not on the size of the network. Higher levels of gains are increasingly useful as the number of blocks in a partition increases. If the total number of connections is given by P, the maximum number of pins on a cell is given by p, and the number of gain levels by l, then in order to perform a k-way concurrent partitioning, Sanchis' approach takes about $O(lkP(log\ k + p + l))$ time. Clearly, for $k = 2$ and $l = 1$, the algorithm reduces to a linear algorithm similar to the F-M algorithm.

Ratio Cut: Although graph algorithms are highly successful in bipartitioning and multiway partitioning, one limitation they have is that they do not capture the fact that real digital circuits are hierarchical in nature. The presence of hierarchy imposes a certain amount of clustering. Graph algorithms are oblivious of these clusterings and tend to divide the graph in a way of strict balanced (or within the bounds of acceptable imbalance) partitioning, and frequently, the resulting cut sizes are far from minimal. The objective of the ratio-cut algorithm, originally introduced by Wei and Cheng [235], is to identify the natural clusters in the circuit and prevent them from being truncated by the cut set, and thereby the algorithm focuses on finding the best ratio cut, as opposed to the minimal cut size. The ratio cut due to dividing the graph into two blocks is given by the ratio of cut set between two blocks and the product of cardinality (size) of each block. Although the denominator is maximum when both blocks are of equal size, due to the clustering nature, the minimum ratio cut may be obtained for a specific cut set that may divide them into two unequal size blocks. The goal of the ratio-cut algorithm is to heuristically find the best ratio cut, instead of finding the minimum cut size. The ratio-cut approach has been applied to a set of benchmark netlists, and it has been seen that the cut size can be improved up to 70% over the ones obtained by the F-M algorithm.

Stochastic Algorithms: Among the various well-known stochastic optimization methods, the simulated annealing algorithm has been widely used for solving numerous VLSI layout optimization problems [115]. The algorithm starts with a random solution and makes incremental refinements by moving cells from their current location to a new location in order to generate new solutions. Generally, all moves that decrease the cost are accepted, and moves that increase the cost are also accepted according to a probabilistic decision function. This decision is made by comparing a rational number drawn randomly between 0 and 1 with a continuous probability function, whose value reduces according to a preassigned cooling schedule and the magnitude of change in cost, similar to the Maxwell-Boltzmann's molecular distribution. Thus, when the algorithm starts, an initial high temperature setting allows many moves that actually increase the cost from its current value significantly. Subsequently, when the temperature reaches its final value, only very few moves that increase the cost are accepted. Thus, simulated annealing behaves as a stable algorithm that is capable of finding globally optimal solutions. The main problem of simulated annealing is that it takes significantly longer time than a linear time algorithm, such as the F-M algorithm.

In order to reduce the time complexity of a simulated annealing based circuit partitioning algorithm, Greene and Supowit [87] devised a new *rejectionless* method that doesn't reject any moves. They selected moves by attaching weight factors to them, depending on their effect on the cost function. This technique works extremely well for partitioning a circuit into a small number of clusters, especially if the average degree of the nodes is small. The run time of such a *rejectionless* simulated annealing algorithm is linear.

Neural Algorithms: Yih and Mazumder [243] have presented a neural network model for circuit partitioning, using iterative improvement techniques. In their representation, a set of component neurons describes which cells are in which blocks in the partition, depending on whether a component neuron is in a firing state (1) or in a nonfiring state (0). In order to represent the critical nets with k component neurons (i.e., nets which are entirely in one block, or nets which have exactly one cell in one block and the rest in the other), a set of of k noncomponent neurons, connected to component neurons with fixed negative synaptic weights, and 2 noncomponent neurons, with fixed positive synaptic weights, were carefully added such that the neural net will try to handle the critical nets appropriately. The neural net was designed to meet the balancing criteria, and the performance of the neural net was tested for several benchmark netlists and compared with the results of the F-M algorithm. For all benchmark circuits, the distribution of cut sizes obtained for 300 random initial partitions for the neural-net tallied very close to the F-M algorithm. The key advantage of the neural net technique is that it can be implemented in hardware to achieve a tremendous speedup over software algorithms.

2.2 Circuit Partitioning by Genetic Algorithm

After a brief overview of the genetic algorithm that has been designed to solve the circuit partitioning problem, details of all genetic operators will be discussed. Basically, this version of the GA can be described as a steady-state GA [50], with simulated-annealing-like mutation of the population. The algorithm is illustrated in Fig. 2.3.

The GA starts with several alternative solutions to the optimization problem, which are considered as individuals in a *population*. These solutions are coded as binary strings, called *chromosomes*. The initial population is constructed randomly. These individuals are *evaluated*, using a problem-specific fitness function. The fitness values are scaled using a suitable function, *sigma scaling*.

The GA then uses these individuals to produce a new *generation* of hopefully better solutions. In each generation, two of the individuals are selected probabilistically as *parents*, with the selection probability proportional to their fitness. Therefore, a fitter individual, hopefully containing some useful features, has a higher probability of propagating itself. *Crossover* is performed on these two individuals to generate two new individuals, called *offspring*, by exchanging parts of their structure. Thus, each offspring inherits a combination of features from both parents. This enables

2.2. Circuit Partitioning by Genetic Algorithm

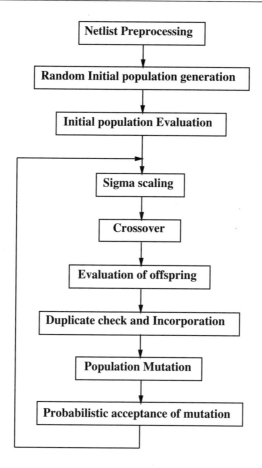

Figure 2.3. Flowchart of genetic algorithm for circuit partitioning

the GA to try out various features in different combinations and see whether the individuals still retain their fitness.

The new individuals are evaluated and then compared with the rest of the population. If they are better than the worst individual, and if they are not identical to any other individual, then they replace the worst individuals in the population.

The next step is *mutation*. An incremental change is made to each member of the population, with a small probability. After mutation is performed on an individual, it no longer has just the combination of features inherited from its two parents, but also incorporates this additional change caused by mutation. This ensures that the GA can explore new features that may not be in the population yet. It makes the entire search space reachable, despite the finite population size. The change in fitness due to this mutation is evaluated. If the fitness is increased, the mutation is always accepted, and if the fitness is reduced, then the mutation is accepted probabilistically, based on the amount of reduction in fitness. This prevents the population from stagnating for long periods of time, when most of the offspring produced by crossover are worse than the population and get rejected.

Netlist Preprocessing, and I/O Pad Processing: At first, the input netlist is modified in order to speed up the algorithm and improve its performance. The netlist is then mapped to a new netlist, on which the genetic algorithm is run. The netlist preprocessing step is identical for both bipartitioning and multiway partitioning algorithms. Single-pin nets are removed from the netlist. Single-pin nets may result from unconnected pins on standard cells, which are assigned dummy net names.

Nets larger than a certain user-specified size, N_{max}, are also removed from the netlist. Sometimes, very large nets, such as clock or power/ground nets, are included in the netlist, and such nets connect to most of the cells in the circuit. It is pointless to consider the cut cost of such nets, since no matter how the circuit is partitioned, the net will cut all or most of the partitions. In fact, in a bipartitioning problem, any net connecting more than half of the cells will always suffer a cut in a balanced partition. Besides being useless, large nets also take up much more evaluation time in multiway partitioning (the evaluation time does not depend on net size in bipartitioning). Therefore, the algorithm allows the user to specify the maximum net size of interest in any particular problem. The experimental results given below are obtained by reevaluating the final partition based on the original netlist, without any nets removed.

This algorithm was tested on MCNC standard cell benchmarks, which include I/O pads. Wherever I/O pads are given with fixed coordinates, they are assigned permanently to the partitions along the chip boundaries, based on their coordinates, both in bipartitioning as well as multiway partitioning. The I/O pads do not need to be assigned to the chromosome, since their partitions are already known. However, they are considered in the net cut evaluation function.

Chromosome Representation: Each cell is represented by 1 bit in the chromosome, the value of which determines the partition to which the cell is assigned. The chromosome is stored as an array of 32-bit packed binary words, with some bits in the last word unused. The netlist is traversed in a breadth-first search order, and the cells are assigned to the chromosome in this order. Thus, if two cells are directly connected to each other, there is a high probability that their partition bits will be close to each other in the chromosome. An example of the breadth-first search sequence and the corresponding chromosome is shown in Fig. 2.4.

The advantage of this method is that clusters of cells in the circuit have a higher probability of forming similar clusters of bits in the chromosome. During crossover, such closely placed clusters of bits have a higher probability of obtaining their partition assignments from one parent or another as a group, and thus remain intact. This improves the search efficiency of the genetic algorithm and makes the inversion operator unnecessary for this application.

2.2. Circuit Partitioning by Genetic Algorithm

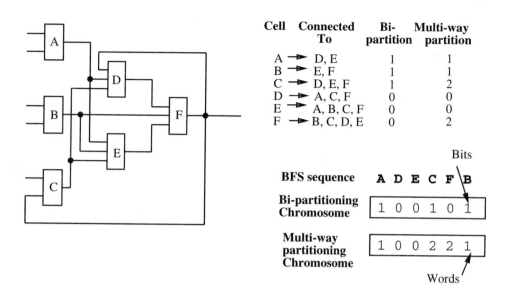

Figure 2.4. Example of breadth-first search ordering of chromosome

Sigma Scaling: In each generation, two parents are selected probabilistically, with the probability of selection proportional to their *fitness*. The evaluation function determines and stores the *cost* or *badness* of each chromosome. From this cost, the fitness of each individual is determined by sigma scaling as follows. A reference *worst cost* is determined by

$$C_w = \overline{C} + S \times \sigma$$

where $\overline{C}$ is the average cost, S is the user-defined sigma scaling factor, and σ is the standard deviation of the population cost. In case C_w is less than the real worst cost in the population, then only the individuals with cost lower than C_w are allowed to participate in the crossover.

The fitness of each individual is determined by

$$F = C_w - C \quad \text{if } C_w > C.$$

Two different individuals are then randomly selected as parents, with selection probability directly proportional to the fitness.

If S is large, C_w is large, and the fitnesses of the members of the population are relatively closer to each other. This causes the difference in selection probabilities to decrease, and the algorithm to be less selective in choosing parents. This may affect the algorithm in many complex ways, such as reduction of premature convergence of the population.

On the other hand, if S is small, the ratio of the least to the highest fitness in the population increases, and the algorithm becomes more selective in choosing parents. In fact, in this case, some high-cost members of the population may not be able to participate in the algorithm at all. This may, in certain conditions, speed up the algorithm, while compromising the final result quality.

Crossover: Crossover is a very efficient bit-mask operation in the case of bipartitioning. Each bit-mask operation processes 32 bits in parallel, without unpacking. The traditional binary one-point crossover operator is used. A random cut point is selected. Offspring 1 inherits the bit string to the left of the cut point from parent 1, and the bit string to the right of the cut point from parent 2. Offspring 2 inherits the remaining parts of both parents.

If the bipartitioning algorithm is being used for an application where the two partitions are interchangeable, then the bitwise complement of the chromosome represents the same partition, so an additional *complementing function* ensures that a chromosome and its complement are treated as the same chromosome. Complementing is not necessary in applications like min-cut placement, where the left and right partitions are not interchangeable (for example, certain cells are forced to be on the left or right, depending on the external signal connections).

If the two parents in a crossover are nearly the complements of each other, then they represent very similar partitions, and the offspring should be very similar to both. However, without the complementing function, the offspring generated are very different from the parents, and in effect, inherit nothing from the parents. An example is illustrated in Fig. 2.5. In order to correct this, one of the parents is complemented if more than half of the bits of the two parents are different. The complementing results in more than half the bits of the two parents becoming similar and thus allows more features to be inherited by the offspring. Since each chromosome consists of only a few 32-bit words, the storage requirements are small, and the complement of each chromosome in the population is permanently stored. The complement is computed after evaluation and acceptance of the new offspring into the population.

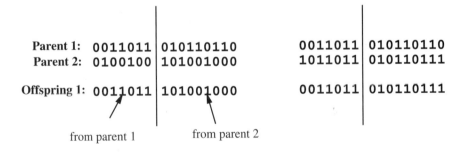

(a) **Before complementing Parent 2:** offspring is similar to neither parent

(b) **After complementing Parent 2:** offspring is similar to both parents

Figure 2.5. Effect of complementing on crossover

2.2. Circuit Partitioning by Genetic Algorithm

The crossover operation is performed as follows. The complete words to the left and right of the cut points are copied from the appropriate parents to the offspring. This leaves the *middle* word (if any) containing the cut point, which is split between two parents. An array of 31 crossover masks and their complements are precomputed and stored, one for each bit position of the cut point. These masks are defined as

$$M_{ci} = M_{ci31} M_{ci30} \cdots M_{ci0}$$

$$M_{cij} = \begin{cases} 0 & \text{if } j \leq i \\ 1 & \text{otherwise} \end{cases}$$

where, M_{cij} is the jth bit in the mask M_{ci}, for the cut point between the ith and $(i+1)$-th bit position. The middle words of the offspring are determined by

$$W_{o1} = W_{p1}.M_{ci} + W_{p2}.\overline{M_{ci}}$$

$$W_{o2} = W_{p2}.M_{ci} + W_{p1}.\overline{M_{ci}}$$

where W_{p1}, W_{p2}, W_{o1}, and W_{o2} are the middle words of the two parents and the two offspring, respectively. This operation is illustrated in Fig. 2.6. The resulting offspring is always legal, but it may be an unbalanced partition, with too many cells in one partition and too few in the other. The partition balance is corrected by a penalty in the fitness function.

Evaluation: Net-cut evaluation is performed using bit-mask operations. For each net, a multiword mask of the size of the chromosome is precomputed, such that if a cell is connected to the net, the corresponding bit-position is set.

$$M_{eij} = \begin{cases} 1 & \text{if } C_j \in N_i \\ 0 & \text{otherwise} \end{cases}$$

where, C_j is the jth cell in the breadth-first search order, M_{ei} is the mask for net N_i, and M_{eij} is the jth bit position of M_{ei}. The following two operations are performed for each net, and in each case, a zero result indicates the absence of the net from the left or right partitions:

$$\mathcal{C}.M_{ei}, \quad \overline{\mathcal{C}}.M_{ei}$$

where $\mathcal{C}$ is the chromosome. If the net is present in both partitions (nonzero result of both operations), or if the net is present in the partition opposite to its I/O pad, then it is said to have a cut.

Partition balance evaluation is done by counting the number of ones in the chromosome. This can be done 8 bits at a time in parallel, using bytewise table lookup. A quadratic penalty has been used for imbalance, so that a large imbalance is penalized more than a small imbalance. The relative weights for cut and imbalance, W_c and W_b, are specified by the user.

Thus, the total cost is

$$C = W_c(\text{cut}) + W_b(\text{imbalance})^2.$$

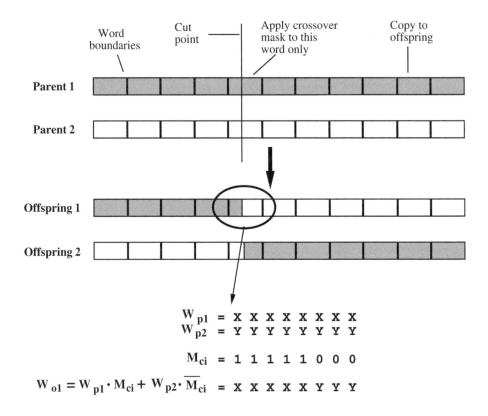

Figure 2.6. Bipartitioning crossover operator

Since mutation is done directly on the population, and not on the offspring, the crossover probability P_C is set to 1, and crossover is always performed on all parent pairs. If this were not so, and the offspring were identical to the parents, they would be rejected in the duplicate check, and the generation would be reduced to a null operation.

2.2.1 Incorporation and Duplicate Check

The two new offspring formed in each generation are incorporated into the population only if they are better than the worst individuals of the existing population.

Before entering a new offspring into the population, it is checked against all other members of the population having the same cost, in order to see whether it is a duplicate. Duplicates can result due to the same crossover operation (same parents and same cut point) occurring twice in the random process. Duplicates have two disadvantages. First, they occupy storage space that could otherwise be used to store a population with more diverse features. Second, whenever crossover

2.2. Circuit Partitioning by Genetic Algorithm

occurs between two duplicates, the offspring is identical to the parents, regardless of the cut point, and this tends to fill the population with even more duplicates.

Duplicate checking both before and after evaluation was tried. In the former case, it saved evaluation time for duplicates, and in the latter case, it saved duplicate checking time for unfit offspring. Duplicate checking after evaluation was a little faster, since the probability of rejection due to duplication is lower than the probability of rejection due to poor fitness. Hence, most of the offspring are rejected just after evaluation, without any duplicate check.

Duplicate checking is fast and efficient in the bipartitioning algorithm, since entire 32-bit words of the chromosome can be compared at a time.

If the two partitions are interchangeable, and the complementing function is in use, then the new offspring must be compared with the members of the population and their complements, in order to eliminate both duplicates and complements of existing members.

2.2.2 Mutation

After crossover and incorporation, mutation is performed on each bit of the population with a very small probability P_M. The traditional binary genetic mutation operator is used. The entire population is treated as one continuous bit-string. We go through the entire population once. For each mutation, the location in bits is determined from the previous location and a random number as follows

$$\text{Next} = \text{Prev} + \lceil \frac{\log(rand[0,1])}{\log(1.0 - P_M)} \rceil$$

where P_M is the mutation probability. From the next location in bits, the chromosome, word, and bit position of the mutation are easily determined. This method is faster than drawing a random number for each bit location.

Mutation is accomplished using a set of 32 precomputed bit masks, defined as follows:

$$M_{mij} = \begin{cases} 1 & \text{if } j = i \\ 0 & \text{otherwise} \end{cases}$$

where M_{mij} is the jth bit of the ith mask M_{mi}. Mutation at bit position i is done by applying the operation W XOR M_{mi}, where W is the word containing the bit to be mutated. This complements bit i. Physically, this causes a random cell to move to the opposite partition.

Each mutation is evaluated and accepted separately, and this process is continued until the end of the population is reached. The acceptance of the mutation operation cannot be completely random, or else the population that has already been improved by the genetic process will go for a random walk. Therefore, the mutation operator acting on the population has to have some probabilistic acceptance characteristics similar to simulated annealing.

The change in cut and balance as a result of mutation is evaluated incrementally; i.e., for only those nets that were connected to the moved cell, the cut is recalculated, and the balance is changed by one. If the change in cost ΔC is negative, signifying that the fitness has increased, the mutation is always accepted, as in simulated annealing. The mutated version replaces the unmutated version of the same individual in the population. Note that it does not replace another less-fit individual; otherwise, the population would contain both the mutated and unmutated versions of the same individual. These would be nearly identical to each other and would cause premature convergence.

If the change in cost is positive, then the acceptance probability is determined by one of the two alternatives in Fig. 2.7. The acceptance probability for the linear case is given by

$$P = 1 - \frac{(1 - P_{av})}{\Delta_{av}} \Delta C$$

where Δ_{av} is the average of cost-increasing moves observed during the run so far, and P_{av} is the mutation acceptance probability assigned to the move that increases the cost by the average amount. P_{av} is user defined and controls the amount of perturbation in the cost of the population. A large acceptance probability for the average cost-increasing move will result in more mutation, and also larger cost increases in the population, which will counteract the cost reductions due to the genetic process. A small P_{av} will result in fewer cost increases in the population, but also more stagnation, as fewer mutations will be accepted.

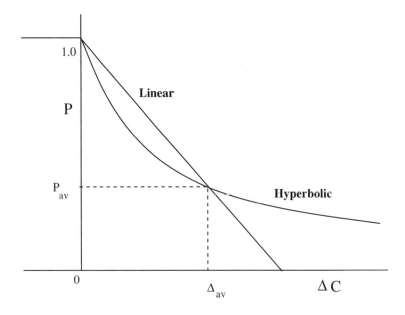

Figure 2.7. Probabilistic acceptance function for population mutation

2.2. Circuit Partitioning by Genetic Algorithm

For the reciprocal case, the acceptance probability is given by

$$P = \frac{1}{1 + \alpha \Delta C / \Delta_{av}}, \qquad \alpha = 1/P_{av} - 1.$$

The linear and reciprocal functions are used as approximations to the exponential function in simulated annealing. This is mainly for speed. The reciprocal function more closely matches the characteristics of the exponential function for large cost increases; i.e., for very large cost increases, there is still a small but finite probability of acceptance. However, this is of little practical value, since in small mutations, the cost is unlikely to change by large amounts. The linear function, which is the fastest approximation, provides adequate performance, and nothing is gained by using the reciprocal approximation.

The advantage of specifying the acceptance probability in terms of the measured Δ_{av} is that it is scaled to be problem-independent and can be set to a fixed value for all problems, which is determined experimentally. Δ_{av} is initially set to zero, thus rejecting all of the first few cost-increasing mutations. As a few samples of ΔC are averaged, Δ_{av} quickly rises to its true value.

This method of mutation is similar to a small amount of continuous constant-temperature simulated annealing on each member of the population. The intention is not to optimize the individuals using this simulated annealing; hence, the temperature is not varied. If the individuals were optimized (or significantly changed) using simulated annealing, then the information inherited by the offspring from the parents would be lost, and the genetic algorithm would fail to perform its function. This is why a crude approximation of the exponential probabilistic acceptance function has been used, since its shape is not critical to the operation of the algorithm.

The objective here is to provide a small constant perturbation in the population, that is so small in amount that its effect is felt only during long periods of stagnation in the genetic algorithm. When the population is already so good that the GA repeatedly generates offspring that are rejected, then a way to proceed further would be to perturb the population slightly until some offspring do get accepted, and the GA gets out of the local optimum.

By setting $P_{av} = 1.0$, all mutations can be accepted, regardless of cost increase, thus making it a truly random mutation operator. Experiments show that this setting significantly degrades the result quality.

2.2.3 Genetic Multiway Partitioning

Chromosome Representation: The partition assigned to each cell is represented by a multibit code, stored as a single word of convenient size. Since parallelism cannot be exploited in this case, packing the codes into a bit string was not performed, in order to save packing/unpacking time at every step. Breadth-first search ordering was used to assign cells to the chromosome, as described for the bipartitioning algorithm above.

Genetic Operators: Crossover and mutation are performed the traditional way. The chromosomes are cut at word boundaries, and the left part of one parent is combined with the right part of the other to make an offspring. Another complementary offspring is made by combining the remaining parts of the parents. This causes some cells to inherit their partition assignments from one parent, and some from the other. The resulting offspring is always legal, but it may be an unbalanced partition, with too many cells in some partitions and too few in others. The partition balance is corrected by a penalty in the fitness function.

The sigma scaling, incorporation, and duplicate checking are the same as in the bipartitioning algorithm.

Mutation is done by choosing a random word and changing it to a random value different from the current value, i.e., moving a randomly chosen cell to a new random partition. The method for choosing random words with probability P_M is the same as that described above for bipartitioning. The entire population is considered as one continuous *word* string (instead of bit string), and the next location for mutation is determined by the equation given above. The probabilistic acceptance method for mutation is also the same as in the bipartitioning algorithm.

2.2.4 Evaluation

There are several ways to evaluate a multiway partition. In this research, a method compatible with min-cut placement was chosen. The partitions numbered 0 to n are assumed to be in spatial order from left to right. If P_{min} is the leftmost partition and P_{max} is the rightmost partition containing cells and I/O pads connected to a net, then the net is cut by all the partition boundaries between P_{min} and P_{max}. This is illustrated in Fig. 2.8 This approximates the one-dimensional wire length when the chip is partitioned and the cells are placed in these partitions. This algorithm can be easily extended to two-dimensional multiway partitioning and cut evaluation to simulate two-dimensional rectilinear wire length evaluation. The main advantage of the genetic algorithm is that, unlike any other partitioning algorithm, it is adaptable to any such problem-specific definition of the cut size.

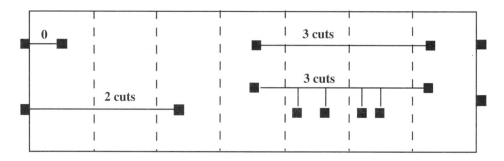

Figure 2.8. Net-cut cost for multiway partitioning

2.2. Circuit Partitioning by Genetic Algorithm

In order to make the recursive F-M algorithm compatible with this cost function so that the results can be compared, recursive F-M partitioning was done, as shown in Fig. 2.9. The partitions were numbered consecutively from left to right, and in the end, the evaluation function of the GA was used to determine the cost of the partition produced by the F-M algorithm.

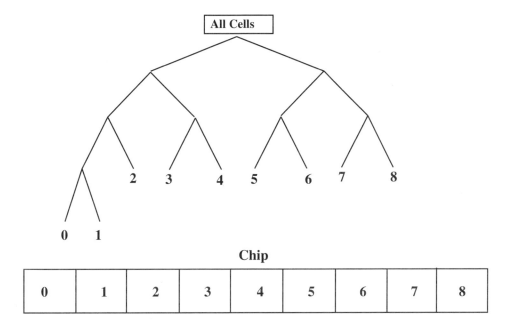

Figure 2.9. Recursive F-M partitioning

The ideal number of cells in each partition is the total number of cells divided by the number of partitions. Partition imbalance is calculated as the sum of squares of the deviation from this ideal. Thus, both overfilled and underfilled partitions are penalized, but the penalty is small for small deviations, and much larger for large deviations. The cut cost and the sum-square balance cost are weighted by user-defined factors, as described for the bipartitioning algorithm above. The objective once again is to be compatible with min-cut placement, where partition balance is important in terms of the number of cells in each partition, not in terms of the area. Each partition can occupy any area that is necessary.

The recursive F-M algorithm maintains a balance of ±1 cell at each stage of bipartitioning, but when the number of partitions is not a power of two, it may have different levels of partitioning in different branches of the tree shown in Fig. 2.9. In such cases, it has no control over the overall multiway balance. The final balance for recursive F-M is evaluated by the evaluation function of the GA. Thus, balance-wise, the F-M algorithm is at a great disadvantage.

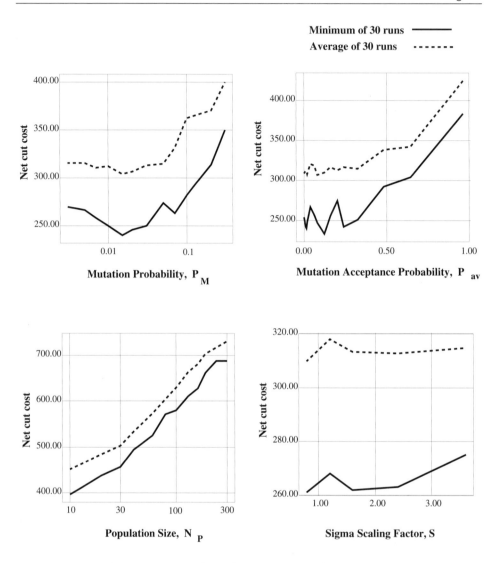

Figure 2.10. Effect of varying the GA parameters

2.2.5 Results

The bipartitioning and multiway partitioning programs were tested on the MCNC standard cell placement benchmark netlists. Fig. 2.10 shows the performance of the algorithm for various values of the genetic algorithm parameters. Table 2.1 shows the genetic algorithm parameters used for the rest of the experiments. In Fig. 2.10, one parameter is varied at a time, while the others remain fixed, as shown in Table 2.1.

2.2. Circuit Partitioning by Genetic Algorithm

Table 2.1. Genetic Algorithm Parameters

Population size	N_P	10
Mutation probability	P_M	0.02
Mutation acceptance probability	P_{av}	0.1
Sigma scaling	S	1.5
Maximum generations per cell	G	300
Max net size	N_{max}	10

Table 2.2 shows the results, including cut and partition imbalance, for the bipartitioning algorithm, compared with the recursive F-M algorithm. Tables 2.3–2.5 show results for the multiway partitioning algorithm. The genetic algorithm gives a better cut size for most of the cases. Note that the balance value can be adjusted for the genetic algorithm by adjusting the penalty weights, but not for the F-M algorithm. Also, the F-M algorithm gives a perfect balance for two-way partitioning, but not for multiway partitioning.

Table 2.2. Results for the Bipartitioning Algorithm

Netlist	Cells	Genetic Algorithm				F-M Algorithm			
		Min Cut	Ave Cut	Max Cut	Ave Imbalance	Min Cut	Ave Cut	Max Cut	Ave Imbalance
Fract	125	8	15.1	26	1.4	9	20.2	32	1
Primary1	752	60	95.0	120	2.9	77	103.9	121	1
Struct	1888	131	166.9	192	1.0	158	187.2	221	1

Table 2.3. Multiway Partitioning Results for **Fract** Benchmark

Fract Benchmark: 125 cells, 147 nets									
Partitions	Cut/balance weight	Genetic Algorithm				F-M Algorithm			
		Min Cut	Ave Cut	Max Cut	Ave Imbalance	Min Cut	Ave Cut	Max Cut	Ave Imbalance
4	100/15	24	41.0	60	6.1	36	49.9	67	91.7
8	100/20	69	97.7	136	18.6	80	101.3	138	158.3
16	100/30	160	197.8	286	77.5	172	206.4	246	100.2
32	100/75	328	383.8	487	141.4	305	409.5	558	100.4

Table 2.4. Multiway Partitioning Results for **Primary1** Benchmark

Primary1 Benchmark: 752 cells, 904 nets								
		Genetic Algorithm				F-M Algorithm		
Parti-tions	Cut/balance weight	Min Cut	Ave Cut	Max Cut	Ave Im-balance	Min Cut	Ave Cut	Max Cut	Ave Im-balance
10	300/10	482	640.7	734	599.6	585	659.3	800	3417.9
20	100/10	1093	1353.5	1623	647.1	1060	1284.0	1475	1577.9
50	70/25	2785	3272.9	3771	1083.1	2609	3161.3	3868	962.3
100	50/50	5102	6114.9	7099	1160.3	5443	6409.4	7356	644.7
4	400/10	221	259.7	332	49.7	208	254.7	288	674.2
8	300/10	443	525.0	662	313.2	445	505.7	586	1039.1
16	200/10	771	1017.9	1152	1351.9	880	1038.8	1229	928.5
32	100/10	1547	1819.6	2055	3507.4	1625	2094.0	2398	1055.1
64	50/10	2538	2994.5	3223	4557.4	3439	4040.0	4939	841.8
128	25/10	4664	5003.3	5682	3921.6	6752	8000.4	9116	659.1

Table 2.5. Multiway Partitioning Results for **Struct** Benchmark

Struct Benchmark: 1888 cells, 1920 nets								
		Genetic Algorithm				F-M Algorithm		
Parti-tions	Cut/balance weight	Min Cut	Ave Cut	Max Cut	Ave Im-balance	Min Cut	Ave Cut	Max Cut	Ave Im-balance
4	200/10	393	439.6	486	7.5	417	480.9	550	307.6
8	500/10	764	913.4	1021	308.2	875	986.5	1053	452.5
16	200/10	1792	1964.9	2098	479.0	1605	1854.8	2097	466.3

2.3 Hybrid Genetic Algorithm for Ratio-Cut Partitioning

The above implementation of the GA is slower than conventional partitioning algorithms and becomes highly time consuming for examples having tens of thousands of cells and nets, since the GA will take a large number of generations to converge to a good set of solutions. Hybrid GAs, which perform local optimization in every generation, can be used to achieve faster convergence, and hence the number of generations can be considerably reduced [25]. Although the running time can be shortened in this way, it is still usually significantly longer than for other algorithms, since each invocation of local optimization can take a considerable amount of time. This section introduces an efficient hybrid genetic algorithm, called the Genetic Ratio-Cut Algorithm (GRCA), for hypergraph bipartitioning using ratio cut as the metric for evaluation of solutions. The technique was developed by Bui and Moon in 1994 [25], and the pseudo-code of the GRCA procedure is shown in Fig. 2.11.

The algorithm employs ratio cut as a metric for partitioning. The *ratio cut* of a partition (A, B) is defined as the ratio $r_{AB} = c(A, B)/(|A| \times |B|)$, where $c(A, B)$ is the cut size of the partition and $|A|$ and $|B|$ denote the sizes of the blocks A and

2.3. Hybrid Genetic Algorithm for Ratio-Cut Partitioning

```
preprocess;        /* optional */
create initial population randomly;

do {
    choose parent1 and parent2 from population;  /* REPRODUCTION */
    offspring = CROSSOVER (parent1, parent2);
    mutation (offspring);
    local_improvement (offspring);          /* modified F-M algorithm */
    if suited (offspring) then
        replace (offspring);
} while (stopping criterion not satisfied);
```

Figure 2.11. Pseudo-code for GRCA

B, respectively. It has been demonstrated by Wei and Cheng [235] that, in linear time, the ratio-cut algorithm can locate clusters buried in hierarchical circuits and can divide the circuits into two or more partitions in a very natural way. The ratio-cut algorithm is broadly divided into three distinct phases corresponding to *initialization, iterative shifting*, and *group swapping*.

2.3.1 Genetic Encoding

In the above GRCA procedure, a chromosome pertaining to a partitioning solution is represented as a string of 0's and 1's such that the number of *genes* in a chromosome is equal to the number of cells in the circuit, and a *gene* has an *allele* 0 if it is in Block A (say) and 1, otherwise. This type of encoding, where each gene location has an explicit meaning, is called *location-based encoding*. Fig. 2.12 shows an example chromosome in which blocks 1, 4, 6, 95, 98, ... are in block A, and 0, 2, 3, 5, 96, 97, ... are in Block B.

Figure 2.12. Location-based encoding

2.3.2 Selection, Crossover, and Mutation

The GRCA procedure employs the genetic selection operator in which individual strings are chosen according to their objective function values, f. To select two parents required for reproduction, a proportional selection scheme as suggested by Goldberg [79] is used. The selection operator may be implemented in algorithmic form in a number of ways. The easiest is to create a biased roulette wheel where each current chromosome in the population has a roulette wheel slot sized in proportion to its fitness. For reproduction, this weighted roulette wheel is spun twice.

Crossover operators are used to create a new offspring by combining parts of the two parent chromosomes. Single-point crossover is quite common, but Bui and Moon [25] have shown that five-point crossover yields good results. In such a crossover operation, five crossover points are chosen randomly on the chromosome, dividing it into six parts, as shown in Fig. 2.13. The contents of the two parents are copied alternately to the offspring. Let this be Offspring 1. In another crossover operation, Offspring 2 is created by copying the *complement values* of Parent 2 (values of Parent 1 unchanged). GRCA selects the better of the two offspring. This can be explained as follows. If the two parents are nearly the complements of each other, they represent the same solution to a great extent (only the blocks are changed). In such a case, Offspring 1 would have a very low fitness value and hence little chance of surviving.

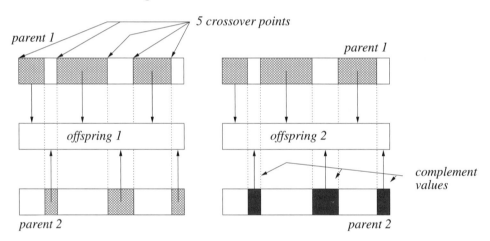

Figure 2.13. Five-point crossover operator

After crossover, GRCA applies the *mutation* operator to the offspring; that is, each bit in the offspring chromosome is complemented with a probability p_m (probability of mutation = 0.015).

2.3.3 Local Improvement

After crossover and reproduction operations are performed, a local improvement algorithm is applied to the selected offspring. The main purpose of this step is to get some visible improvement in the offspring, rather than obtaining a local optimum value, using the offspring as the initial bipartition. The *local improvement* procedure applied is a modified version of the F-M algorithm, which has a complexity of $O(P)$, where P is the total number of pins connected to the cells. The standard F-M algorithm usually takes around 10 passes to converge to a local minimum solution, and there can be potentially up to $N-1$ cell transfers occurring in each pass, where N is the total number of cells. To reduce the complexity of the local improvement procedure, only one pass of the F-M algorithm is usually carried out, and only a fixed number of cell transfers is permitted. This parameter is estimated to be approximately $|N|/6$. This modified version of F-M takes only 5% of the time taken by the standard F-M algorithm. By making the local improvement procedure efficient, the time complexity of the GRCA procedure is greatly reduced.

However, the gain lists are usually maintained by cell gains rather than ratio-cut gains for efficiency. Maintaining gain lists by the latter criterion would require recalculation of gains with every cell transfer, thereby increasing the complexity of the local improvement procedure to $O(|N| \times |P|)$. This is justified by observing that, if the size of module i is not large, it will not contribute much to the denominator in the ratio-cut gain, given by

$$\frac{gain(i)}{(A - size(i)) \times (B + size(i))}. \quad (2.3.1)$$

In the event of ties occurring between the cells that are candidates for transfer, these ties are resolved by using a token strategy. Initially, the token is given to any block at random. In a situation where there are consecutive ties, unless the loser in the previous tie break wins this round, the winner retains the token. This enables the hybrid genetic algorithm to try out more unbalanced partitions. However, it was noticed that using this strategy, maintaining balanced solutions was difficult. This was remedied by allowing only alternate wins by tie breaks. However, unlimited wins were still permitted when consecutive transfers from the same block gave the best gain. This enables the ratio-cut-based genetic algorithm to try both balanced and unbalanced partitions.

Replacement Scheme and Stopping Criterion: After performing local improvement on the offspring, the algorithm has to decide which population member is to be replaced by the offspring. The quality of the solutions produced by a GA is greatly dependent on the replacement scheme adopted. It can be observed that replacing the worst member of the population by the offspring leads the algorithm to lose population diversity, thereby causing performance degradation.

The hybrid genetic algorithm compares the fitness of the offspring to the more similar (in terms of bitwise difference) parent. If the offspring is better than the parent, then it replaces the parent; otherwise, the offspring is discarded. This replacement strategy is called NEAR. However, after some time, GRCA adaptively changes its replacement strategy to a scheme called COMBI. In this scheme, if GRCA fails to replace the first parent, it compares the offspring again with the second parent, and only when the offspring is not better than the second parent does GRCA discard it. This is because using solely the replacement scheme NEAR causes a large number of failed replacements in the latter generations of GRCA, causing it to converge slowly. A *swing* occurs if an offspring is discarded at the end of a generation. Seven consecutive swings constitute an *out* (Baseball batters will like this rule!). Initially, the scheme adopted is NEAR. After two outs, COMBI is used. The algorithm stops either when the number of generations exceeds a prespecified number or two more outs occur.

2.3.4 Preprocessing

In GA terminology, a *schema* represents a pattern of chromosomes. When single-point crossover is applied to two parents, some schemata in the parents survive and some do not. If the crossover point divides the specific symbols of a schema into two parts, only one part of it can be copied to the offspring. Therefore, the survival probability of a schema is anti-proportional to the defining length when single-point crossover is applied (Fundamental Theorem of Genetic Algorithms [95]). However, in the case of multipoint crossover, this is not necessarily true. As observed by De Jong [53], a schema is not disrupted when an even number (including 0) of crossover points fall between the two specific symbols of every pair of adjacent symbols in the schema.

Also, the survival probability of a schema is highly dependent on its inner structure. Let us consider the eighth-order schemata H_1 and H_2 shown below. Specific symbols (indicated by s) are highly clustered in H_2 and evenly distributed in H_1.

```
***s***s***s***s***s***s***s***s***          (H_1)
***sss*s******************s*sss***           (H_2)
```

If we use single-point crossover, the survival probability is the same for the two schemata ($= \frac{6}{34}$). However, for the case of two-point crossover, the survival probability of H_1 is $\frac{57}{361}$, whereas it is $\frac{207}{561}$ for H_2. Thus, we observe that clustered schemata have a higher probability of survival. The aim of preprocessing is thus to obtain highly clustered schemata like H_2.

2.3.5 Weighted DFS Reordering (WDFR)

In location-based encoding, each location on a chromosome has its own meaning. In this case, the ith value on a chromosome specifies the block assigned to cell i. The most natural indexing scheme is to use the indices as given in the netlist input (used by all group-migration methods). However, by grouping the closely linked cells, we can expect highly clustered schemata and therefore better solutions.

Given a hypergraph G, the preprocessor first converts G into a graph by a set of standard clique transformations, where each k-pin net is transformed into a complete subgraph of k vertices, with edge weight $\frac{1}{k-1}$. Then a Depth-First Search (DFS) is carried out on this transformed graph. The only difference from the traditional DFS is in tie breaking. In this case, when the DFS encounters unvisited vertices adjacent to the current vertex, it visits the vertex with the maximum weight. The vertices are reindexed by their DFS numbers. It should be noted that this transformation is transparent to the main genetic algorithm. It is only used for location assignment of modules on the chromosome. This is where GAs differ

2.3. Hybrid Genetic Algorithm for Ratio-Cut Partitioning

from other techniques — preprocessing in the latter is used to provide good initial solutions. However, in GAs, the initial population is randomly generated.

2.3.6 Time Complexity

Each of the processes, *selection*, *crossover*, and *mutation*, used in the Genetic Ratio-Cut Algorithm (GRCA) takes $O|V|$ time. Although the standard F-M algorithm is a linear-time, $O|E|$ algorithm, the constant term within the big O is quite large, and it will take considerable time if it is directly adopted for local improvement. A weak variation of the F-M algorithm, which is used in GRCA for local improvement, also takes $O|E|$ time, but the constant term in this case is about 20 times smaller than in the original F-M algorithm. Therefore, the overall time complexity of GRCA is $O(K \cdot |E|)$, where K is the number of generations. Usually, K ranges from 700 to 1600.

The worst case complexity of the WDFR preprocessing algorithm is $O(|V|^2)$. However, this reduces to $O(|E|)$ when the maximum degree of each cell is bounded by a constant, and the number of pins on a net is also bounded by a constant. The first condition is easily met in VLSI circuits because of the limited fan-in and fan-out capabilities of each cell. The latter condition is due to the fact that a k-pin net induces a complete subgraph of k vertices. This is done by ignoring nets with more than 20 pins in the WDFR. In practice, the preprocessing phase takes no more than 2% of GCRA's total running time.

2.3.7 Results

RCut2.0 is an advanced version of the ratio-cut algorithm that was developed by Wei and Cheng in 1991 [235]. In this section, the performance of GCRA is compared with that of RCut2.0, and the results of comparison for benchmark circuits are tabulated in Table 2.6. It was found that for the various benchmark netlists used for comparison, GCRA takes about 9 times longer than RCut2.0. Therefore, in order to make a fair comparison between them, results of 100 runs of RCut2.0 are compared with the results of 10 runs of GRCA. The 100 runs of RCut2.0 are divided into 10 groups, each consisting of 10 runs. The average of the best of each group is compared with the average of GRCA in Table 2.6 and also shown are the best of 100 runs of RCut2.0 and the best of 10 runs of GRCA. It can be seen that in average results, GRCA performed best for 5 cases (Ckts. 1, 2, 6, 8, and 11) out of 11 shown in the table. Overall, GRCA's average results are 12.3% better than the averages of the best results of RCut2.0 (divided into 10 groups). In best results, GRCA performed the best for 10 out of 11 benchmark circuits that were used to make the comparison. Overall, the best result of GRCA is about 15.6% better than that of RCut2.0. GRCA's run time is comparable with RCut2.0 (run over 10 cases) for small circuits. For larger circuits, GRCA is faster (it takes about 59% of the time) than RCut2.0 run over 10 cases.

Table 2.6. Comparison between GRCA and RCut2.0

Circuit	#modules	#nets	#io pads	RCut2.0 RCut	RCut2.0 CPU	GRCA RCut	GRCA CPU
ckt1	833	902	81	1521	26	694	24
ckt2	3,014	3,029	107	527	166	610	163
ckt3	833	902	81	2670	25	1092	26
ckt4	3,014	3,029	107	621	162	710	167
ckt5	1,663	1,721	70	5.68	67	5.68	65
ckt6	1,607	1,618	65	47.1	77	39.8	74
ckt7	1,515	1,658	36	9.96	55	10.1	64
ckt8	2,595	2,751	63	5.49	133	4.66	102
ckt9	1,752	1,674	69	84.5	105	106	57
ckt10	2,844	3,282	160	323	165	367	174
ckt11	12,142	12,949	0	126	1320	49.1	785

2.3.8 Comparison of GA with Other Methods

The GA differs from the conventional graph optimization techniques like K-L and F-M in the following ways:

- In the GA, search is done from a population, not a single point. In K-L and F-M, a single point in the solution space is iteratively refined to obtain higher fitness values. This often leads to the locations of false peaks in multimodal (many-peaked) search spaces, which is the case in the bipartitioning problem. The GA, on the other hand, is more global, as it handles a large number of points in parallel. By maintaining a population of well-adapted points, the probability of reaching a false peak is reduced. By using local improvement, convergence of the problem-specific F-M technique is also exploited.

- Search in the GA is done using stochastic operators, not deterministic rules as in K-L and F-M. Thus, if the same algorithm is run again, we are likely to obtain different results. The GA uses random choice to guide a highly exploitative search.

- Genetic algorithms usually take more time to converge than deterministic algorithms like K-L and F-M. Thus, their running times are usually greater.

- GRCA uses *ratio cut* as its criterion for evaluation of solutions, unlike F-M and K-L, which employ *cut set cardinality* as the metric. Therefore, the solution generated by GRCA might not satisfy the balance criterion, i.e., that the sizes of the partitions are within some acceptable range, although the ratio cut might be low. However, the iteration procedures in K-L and F-M do guarantee a balanced partition, although minimum cut size may not be achieved.

- One basic assumption of the K-L algorithm is that all cells have the same size. However, this is not the case for practical VLSI circuits, where the size of a 4-input NAND gate is approximately four times that of an inverter. Therefore, although the partitions generated by K-L satisfy a *balance criterion*, they may not fit into blocks of equal size. The GA takes the sizes of cells into account while calculating the fitness function associated with each chromosome.

2.4 Conclusion

This chapter describes the use of a genetic algorithm for circuit partitioning. Partitioning may be required for placing and routing a very large circuit, or for many other applications. The genetic algorithm produces a significant improvement in result quality. It can also optimize a cost function with multiple objectives and constraints. Bipartitioning and multiway partitioning algorithms have been developed using a variant of the steady-state genetic algorithm. The main problem of a pure genetic-based partitioning algorithm is that its run time increases quickly as the problem size increases. In order to tackle the run-time complexity, a fast hybrid genetic algorithm that employs local optimization in every generation was developed, and it was shown to achieve a faster convergence without compromising the quality of the solutions. In order to exploit the hierarchical nature of VLSI circuits, the hybrid genetic algorithm uses the metric of ratio cut, as opposed to cut set size, which is more commonly used in graph algorithms.

The bipartitioning problem is represented as a binary chromosome. Efficient bit-mask operations perform crossover, mutation, and net-cut evaluation 32 bits at a time, without unpacking. The multiway partitioning algorithm has a global view of the problem and generates and optimizes all the necessary partitions simultaneously.

The algorithms were tested on the MCNC benchmark circuits, and the cut size obtained was lower than that for the conventional Fiduccia-Mattheyses algorithm. Unlike the F-M algorithm, the GA allows adjustment of the partition size balance by adjusting the penalty weights. The F-M algorithm gives a perfect balance for two-way partitioning, but not for multiway partitioning. These results show that the genetic algorithm is very well suited for the partitioning problem.

The next version of the hybrid genetic algorithm for ratio-cut partitioning employs a local optimization procedure that is based on a fast variant of the F-M algorithm, and it is combined at each generation with genetic space exploration. The modified F-M algorithm that is used for local optimization is about 20 times faster than the conventional F-M algorithm. Still, it consumes about 71% of the total run time of the hybrid genetic algorithm, and it will be left to readers to develop more efficient local optimization algorithms.

SUMMARY

TOPICS STUDIED

KEY OBSERVATIONS AND POINTS OF INTEREST

Genetic Algorithm for Partitioning with Net-cut Metric
- Representation
- Crossover
- Mutation
- Results

- Each cell is 1 bit
- Breadth-first search ordering

- Binary one-point crossover
- Random cut point
- Complementing function

- Binary genetic mutation
- Probabilistic acceptance as in simulated annealing

- Bipartitioning
- Multiway partitioning

Genetic Algorithm for Partitioning with Ratio-Cut Metric
- Representation
- Crossover
- Local Improvement
- Preprocessing
- Results

- Location-based encoding
- Five-point crossover operation
- FM-based local optimization
- Weighted depth-first search reordering
- Superior performance to Ratio Cut 2.0

CHAPTER 3
at a glance

- Standard cell placement
  - Chromosome representation
  - Permutations of cells by cycle crossover
  - Metagenetic optimization for GA parameters

- Macro cell placement
  - Unified GA and SA algorithm
  - Priority tree crossover
  - Comparison with other approaches

Chapter 3

STANDARD CELL AND MACRO CELL PLACEMENT

K. Shahookar, H. Esbensen, P. Mazumder

Layout synthesis is a key step in the VLSI design process that follows behavioral and logic synthesis. Behavioral synthesis involves a transformation from a behavioral description (e.g., in VHDL or Verilog) into a Register Transfer Level (RTL) representation. During logic synthesis, a circuit, described in RTL language, is optimized and realized at the gate level, by using circuit elements from a given component library. Layout synthesis involves finding a placement of circuit components, which are instances of the library primitives, and an associated routing of the component interconnections that meet the area and delay constraints of the design. The end result of layout synthesis is a placed and routed design, from which the photolithography masks can be derived for chip fabrication. Placement and routing are performed using abstractions of the actual components and metal interconnection lines. Thus, only a component's size, orientation, and terminal (pin) locations are of concern during placement, and the internal cell structure is not considered.

This chapter will discuss the use of genetic algorithms for the placement of standard cells and also for macro cells. In standard cell design, libraries of predefined and precharacterized cells are used, and all cells typically have the same height so that they can easily be placed in rows with power lines in adjacent cells abutting. Rows of cells may be separated by routing channels, and routing channels are sized to accommodate all required interconnections. Even in state-of-the-art high-performance chips, standard cell design is often used for portions of the chip that are not on the critical path. Macro cells, on the other hand, are large irregularly sized circuit modules that are usually generated by a silicon compiler. The irregular sizes of macro cells introduce added complexity to the placement problem.

We begin by addressing standard cell placement and present a genetic algorithm for solving the standard cell placement problem. Results of this approach are compared to those of *simulated annealing*, which has previously been used for the placement problem. Then macro cell placement is discussed, and an algorithm that combines genetic algorithms and simulated annealing is given.

3.1 Standard Cell Placement

The placement problem can be defined as follows. Given an electrical circuit consisting of modules, with predefined input and output terminals, interconnected in a predefined way, construct a layout indicating the positions of the modules, so that the estimated wire length and the layout area are minimized, and any other given constraints are satisfied. The inputs to the problem are the module description, consisting of the sizes and terminal locations, and the *netlist*, describing the interconnections between the terminals of the modules. The output is a list of x- and y-coordinates and orientations for all modules.

The main objectives of a placement algorithm are to minimize the total chip area and the total estimated wire length for all the nets. We need to optimize chip area usage in order to be able to fit more functionality into a given chip size. We need to minimize wire length in order to reduce the capacitive delays associated with longer nets and speed up the operation of the chip. These goals are closely related to each other for the standard cell design style, since the total chip area is approximately equal to the area of the cells plus the area occupied by the interconnect. Hence, minimizing the wire length is approximately equivalent to minimizing the chip area.

In some cases, secondary performance measures may also be needed, such as the preferential minimization of wire length of a few critical nets, at the cost of an increase in the total wire length, as in timing optimized placement, or the minimization of crosstalk, matching of differential nets, and the consideration of the thermal sensitivity of the cells in mixed analog/digital placement.

Another criterion for an acceptable placement is that it should be physically feasible, which means that (1) the cells should not overlap, (2) they should lie within the boundaries of the chip, and (3) they should be confined to rows in predetermined positions.

The cost function in the standard cell placement problem thus consists of the sum of the total estimated wire length, measures of the other quantities of interest, such as the number of crossings of sensitive nets and the horizontal coupling capacitance of adjacent nets for crosstalk, and various penalties for cell overlap, total amount of unused space at the end of each row, total chip area, etc. The goal of the placement algorithm is to determine a placement with the minimum possible cost. This represents the best possible trade-off between the various objectives of placement.

Cell placement is an NP-complete problem, and therefore, it cannot be solved exactly in polynomial time [58], [138], [199]. Trying to get an exact solution by evaluating every possible placement in order to determine the best one would take time proportional to the factorial of the number of cells. This method is, therefore, impossible to use for circuits with any reasonable number of cells. In order to efficiently search through a large number of candidate placement configurations, a heuristic algorithm must be employed. The quality of the placement obtained depends on the heuristic used. At best, we can hope to find a good placement with wire length quite close to the minimum, but with no guarantee of achieving the absolute minimum.

The placement process is followed by *routing*, which is the process of determining the physical layout of the interconnects through the available space. Finding an optimal routing, given a placement, is also an NP-complete problem. Many algorithms work by iteratively improving the placement, and at each step, estimating the wire length of an intermediate configuration. It is not feasible to route each intermediate configuration in order to determine how good it is. Instead we estimate the wire length.

In order to make a good estimate of the wire length, we should consider the way in which routing is actually done by routing tools. Almost all automatic routing tools use Manhattan geometry; i.e., only horizontal and vertical lines are used to connect any two points. Furthermore, two layers are used; only horizontal lines are allowed in one layer, and only vertical lines in the other. An efficient and commonly used method to estimate the wire length is the *semiperimeter method*. The wire length is approximated by half the perimeter of the smallest bounding rectangle enclosing all the pins. For Manhattan wiring, this is a lower bound on the actual wire length.

3.1.1 GASP Algorithm

The GASP Genetic Algorithm for Standard-cell Placement is shown in Fig. 3.1. First, an initial population is constructed *randomly*. Each cell is represented by a set of four integers containing the cell number, the x– and y– coordinates, and a slot identification number (ID), as illustrated in Fig. 3.2. The slot ID number is used to keep track of the approximate slot in the physical layout area to which each cell is assigned. For each configuration in the initial population, the coordinates of the cells are determined by placing them end-to-end in rows, in random sequence. The population size is provided by the user, and it determines the trade-off between processing time and result quality. From a large amount of experimental data, it was observed that a constant population size of 24 gives the best possible solution quality for most of the circuits, when the number of standard cells is within 1000 gates. Each individual is evaluated to determine its fitness. The fitness is the reciprocal of the sum of the bounding rectangle wire lengths. Since cell overlaps are removed before evaluation, as described below, the traditional terms for row length control and overlap are unnecessary. The fitness, f, is given by

$$f = \frac{1}{\sum_{nets}[x(i)W_H(i) + y(i)W_V(i)]}$$

where $x(i)$ and $y(i)$ are the horizontal and vertical spans of net i, and $W_H(i)$ and $W_V(i)$ are the horizontal and vertical weights.

3.1. Standard Cell Placement

Read netlist and cell library files;
Read parameter values, CrossoverRate, MutationRate,
InversionRate, PopulationSize, NumberOfGenerations;
NumberOfOffspring = PopulationSize × CrossoverRate;
NumberOfGenerations = NumberOfGenerations/CrossoverRate;

Generate initial Population randomly;
For j = 1 **To** PopulationSize **Do**
    evaluate (Population, j);

For i = 1 **To** NumberOfGenerations **Do**
    For j = 1 **To** PopulationSize **Do**
        invert (j, InversionRate);

    NewPopulation = ∅
    For j = 1 **To** NumberOfOffspring **Do**
        Select two parents from Population;
        Align slot ID numbers of parents 1 and 2;
        Perform crossover to create one new offspring;
        mutate (offspring, MutationRate);
        Add offspring to NewPopulation;
        evaluate (NewPopulation, j);

    Population = reduce (Population, NewPopulation);
Solution = individual with highest fitness in final Population;

Figure 3.1. GASP algorithm

At the beginning of each generation, *inversion* is performed on each individual, with a probability equal to the *inversion rate*. For this purpose, two cut points are determined randomly, and the segment between them in the cell array is flipped, along with the coordinates and the slot ID numbers, as shown in Fig. 3.2. This leaves the physical placement unchanged.

Crossover then occurs. Two individuals are selected from the population at random, with a probability proportional to their fitness. Before crossover, the slot ID numbers of the second parent are aligned in the same sequence as those of the first parent, so that cells in the same array locations correspond to approximately the same locations on the chip. Then segments are exchanged between parents, so that for each location on the chip, the child inherits a cell from one parent or another. The coordinates and slot ID numbers are inherited entirely from one parent, which means that all the cells will not fit their slots, and there will be some overlap. This process is repeated until the desired number of offspring have been generated. The number of offspring per generation, N_o, is determined by the *crossover rate*:

$$N_o = N_p P_C$$

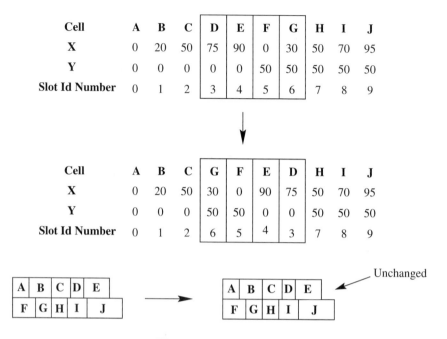

Figure 3.2. Inversion

where N_p is the population size, and P_C is the crossover rate. Since the number of configurations examined is kept constant in all runs for a particular circuit, the actual number of generations is increased as the crossover rate is reduced:

$$N_g = N_{g0} N_p / N_o$$

where N_p is the population size, and N_{g0} is the number of generations specified by the user.

After crossover, the offspring is mutated with low probability. Each mutation consists of a pairwise interchange of two randomly picked cells, leaving the coordinate arrays unchanged (Fig. 3.3). This causes a minor incremental change in the placement. Each mutation therefore affects two genes. This is beneficial for the standard cell placement problem, since it causes less cell overlap (and hence less disruption when the overlap is removed as described below). The *mutation rate* is defined as the percentage of genes in the population which are mutated in each generation. Thus, for an n-cell placement problem, with a population size N_p, the total number of genes is nN_p, and $nN_p P_M/2$ pairwise interchanges are performed for a mutation rate P_M. The mutation rate controls the rate at which new genes are introduced into the population for trial. If it is too low, many genes that would have been useful are never tried out. If it is too high, there will be too much random perturbation, the offspring will start losing their resemblance to the parents, and the algorithm will lose the ability to learn from the history of the search.

3.1. Standard Cell Placement

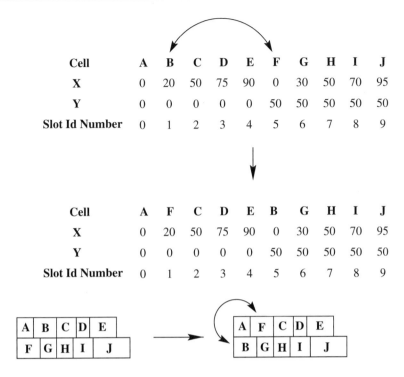

Figure 3.3. Mutation

After crossover and mutation, the fitness of each offspring is evaluated. Simulated annealing-based algorithms traditionally allow, but penalize cell overlaps during the optimization process. After considerable experimentation, it was found beneficial to remove cell overlap in the offspring before evaluation. The GA is unlike simulated annealing, in which only two cells are moved at a time and it is possible to calculate the wire length incrementally. In the genetic algorithm, as many as half the cells in the circuit are moved simultaneously, so the wire length must be computed exhaustively. Thus, it is not necessary to tolerate overlap and preserve the coordinates of most of the cells as required by simulated annealing. It was observed that determining the overlap takes about as much computation time as removing it. It was also observed that using overlap as a penalty and removing overlap at the end of the run resulted in a significant increase in wire length, as much as 10% in some cases, with the genetic algorithm receiving no further chance to work on the circuit. Removing the overlaps after every generation not only gives the algorithm a more accurate picture of the wire length but also gives the algorithm repeated chances to optimize the circuit after it has been perturbed (from the state inherited from the parents) by overlap removal.

After evaluation, the population for the next generation is chosen from the combined set of parents and offspring. Three selection methods have been tried for this reduction step: deterministic reduction, in which the fittest individuals are chosen, random reduction, and random reduction with the retention of the best individual. This completes the processing of one generation.

3.1.2 Crossover Operators

The traditional genetic crossover operator was discussed in Chapter 1. This method works well with the bit string representation. In some applications, where the symbols in the solution string cannot be repeated, this method is not applicable without modification. Placement is a typical problem domain where such conflicts can occur. Thus, we need either a new crossover operator that works well for these problem domains or a method to resolve such conflicts without causing significant degradation in the efficiency of the search process. The performance of the genetic algorithm depends, to a great extent, on the performance of the crossover operator used. Three new versions of the crossover operator that overcome these problems are described here.

As mentioned earlier, crossover is the primary method of optimization in the genetic algorithm, and, in the case of placement, works by combining subplacements from two different parent configurations to generate a new placement. Since traditional crossover can produce conflicts, one must either find a way to combine two different configurations without conflicts or use some method to resolve the conflicts that arise. The performance of three crossover operators was compared. Two of them, order and PMX, differ in their conflict resolution methods, while cycle crossover is a conflictless operator.

Order Crossover: The algorithm is as follows. Choose a cut point at random. Copy the array segment to the left of the cut point from one parent to the offspring. Fill the remaining (right) portion of the offspring array by going through the second parent, from the beginning to the end and taking those elements which were left out, in order. An example is shown in Fig. 3.4(a). This operator conveys a subplacement from the first parent without any changes, and then, to resolve conflicts, compresses the second parent by eliminating the cells conveyed by the first parent and shifting the rest of the cells to the left without changing their order [49]. It then copies this compressed second parent into the remaining part of the offspring array.

PMX: PMX [78] stands for "Partially Mapped Crossover." It is implemented as follows. Choose a random cut point and consider the segments following the cut point in both parents as a partial mapping of the cells to be exchanged in the first parent to generate the offspring. Take corresponding cells from the segments of both parents, locate both these cells in the first parent, and exchange them. Repeat this process for all cells in the segment. Thus, a cell in the segment of the first parent and a cell in the same location in the second parent will define which cells in the first parent have to be exchanged to generate the offspring. An example is shown in Fig. 3.4(b).

3.1. Standard Cell Placement

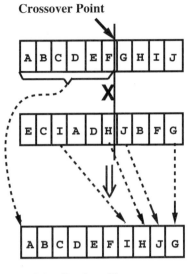

(a) Order Crossover

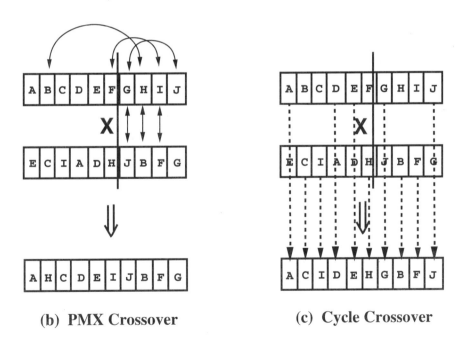

(b) PMX Crossover

(c) Cycle Crossover

Figure 3.4. Crossover operators

Cycle Crossover: Cycle crossover [165] is an attempt to eliminate the cell conflicts that normally arise in crossover operators. In the offspring generated by Cycle crossover, every cell is in the same location as in one parent or the other. The basic idea behind Cycle crossover is to avoid cell conflicts by finding nonoverlapping sets of cells to pass from the two parents. However, such nonoverlapping sets of cells are not contiguous. The algorithm is as follows.

Start with the cell in location 1 of parent 1 (or any other reference point), and copy it to location 1 of the offspring. Now consider what will happen to the cell in location 1 of parent 2. The offspring cannot inherit this cell from parent 2, since location 1 in the offspring has been filled. Therefore, this cell must be obtained from parent 1. Suppose this cell is located in parent 1 at location x. Then it is passed to the offspring at location x. But then the cell at location x in parent 2 cannot be passed to the offspring, so that cell is also obtained from parent 1. This process continues until we complete a cycle and reach a cell that has already been passed. Then we choose a cell from parent 2 to pass to the offspring and go through another cycle, passing cells from parent 2 to the offspring. Thus, in alternate cycles, the offspring inherits cells from alternate parents, and the cells are placed in the same locations as they were in the parents from which they were inherited.

An example is given in Fig. 3.4(c). In this example, we start by passing cell A from parent 1 to the offspring. Since E is located in the same position in parent 2, it cannot be passed from parent 2. Hence, E is also passed from parent 1. D is located in parent 2 in the same position as E in parent 1. Hence D must also be passed from parent 1, and so on until we reach cell A again. This completes the cycle. The next cycle is taken from parent 2 and consists of cells C, B, H, F, and I. The third cycle is again from parent 1 and consists of cells G and J.

3.1.3 Optimizing the Genetic Algorithm

A meta-genetic algorithm [88] was used to optimize the genetic algorithm for cell placement. The three parameters optimized are the crossover rate, inversion rate, and mutation rate. The meta-genetic algorithm, as shown in Fig. 3.5, is itself a genetic optimization process which runs the genetic algorithm to solve a placement problem and manipulates the genetic parameters to optimize the fitness of the genetic algorithm. The individuals in the population of the meta-genetic algorithm consist of three integers in the range [0, 20], representing the mutation rate, inversion rate, and crossover rate for the genetic algorithm. The mutation rate as given by this parameter can vary from 0 to 5% in steps of 0.25%, the inversion rate can vary from 0 to 100% in steps of 5%, and the crossover rate can vary from 20% to 100%, in steps of 4%. The fitness of an individual (a genetic algorithm with a certain parameter combination) is taken to be the fitness of the best placement that the genetic algorithm found in the entire run, using these parameters.

The meta-genetic algorithm population size was 20, and the algorithm was run for 100 generations. The crossover probability was 100%. Crossover consists of selecting each of the three parameters randomly from one parent or the other, with

3.1. Standard Cell Placement

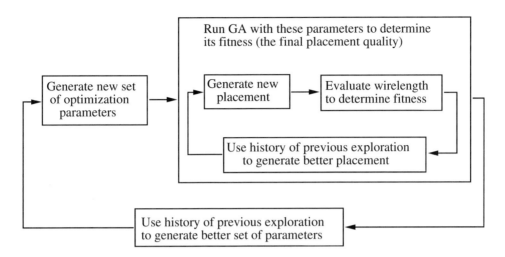

Figure 3.5. The meta-genetic optimization process

equal probability of a parameter value being selected from either parent. Inversion is not used. This is because the length of each individual (three elements) is so small that there is little scope for group formation. After crossover, the offspring are mutated. Mutation consists of adding a random number in the range $[-2, 2]$ to any one parameter in the offspring. The probability of mutation is 1%. The high mutation rate and large number of generations are used so that the performance of the meta-genetic algorithm does not become a bottleneck in the search for the optimum parameters for the genetic algorithm.

The different crossover mechanisms were not included in the optimization. Instead, the optimization was performed separately for each crossover mechanism, as illustrated in Fig. 3.6, and the best results for each were compared to determine an overall winner. This was done because each crossover mechanism was expected to perform best with a slightly different combination of parameters. If the crossover type was included in the meta-genetic search, then the meta-genetic population would consist of individuals with different crossover types, each with its own optimized set of parameters. Unless the meta-genetic crossover operator was constrained to perform crossover between individuals of the same crossover type, it would tend to couple the wrong set of parameters with the wrong crossover type to generate offspring of poor quality. Similarly, optimization was performed separately for each reduction criterion. Thus, only the three rate parameters were included in the meta-genetic optimization process and were optimized for a particular crossover type in each run.

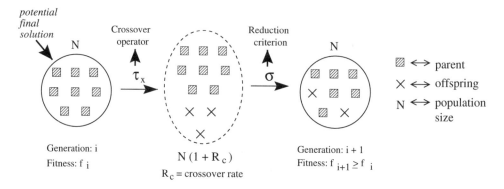

Figure 3.6. Genetic optimization process

3.1.4 Comparison with Simulated Annealing

Simulated annealing is a well-known, high-performance optimization technique for combinatorial problems. The simulated annealing algorithm is presented in Fig. 3.7. The temperature is initialized to a relatively high value, and it is slowly decreased until a "freezing point" is reached. At each temperature, components are selected for possible movement until equilibrium is reached. If movement of the selected components results in an improved placement, the movement is performed. Otherwise, the movement is performed with a probability that decreases exponentially with temperature. Components are typically selected randomly for pairwise exchanges.

```
Temperature = InitialTemperature;
currentPlacement = randomInitialPlacement;
currentScore = score (currentPlacement);
While equilibrium at Temperature not reached Do
    selectedComponents = select (atRandom);
    trialPlacement = move (selectedComponents, atRandom);
    trialScore = score (trialPlacement);
    If trialScore < currentScore then
        currentScore = trialScore;
        currentPlacement = trialPlacement;
    Else
        If uniformRandom (0, 1) <
           e^{-(trialScore - currentScore)/Temperature} then
            currentScore = trialScore;
            currentPlacement = trialPlacement;
    Temperature = Temperature × Alpha;    // Alpha ≈ 0.95
```

Figure 3.7. Simulated annealing algorithm

Both simulated annealing and the genetic algorithm are computation intensive. However, the genetic algorithm has some intrinsic features which, if exploited properly, can result in significant savings. One difference is that simulated annealing operates on only one solution at a time, while the genetic algorithm maintains a large population of solutions which are optimized simultaneously. Thus the genetic algorithm takes advantage of the experience gained in past exploration of the solution space, and it can direct a more extensive search to areas of lower average cost. Since simulated annealing operates on only one solution at a time, it has very little history to use in learning from past trials.

Both simulated annealing and the genetic algorithm have mechanisms for avoiding entrapment at local optima. In simulated annealing, this is accomplished by occasionally discarding a superior solution and accepting an inferior one. The genetic algorithm also relies on inferior individuals as a means of avoiding false optima, but, since it has a whole population of individuals, the genetic algorithm can keep and process inferior individuals without losing the best ones. Furthermore, in the genetic algorithm, each new individual is constructed from two previous individuals, which means that in a few iterations, all the individuals in the population have a chance of contributing their good features to form one super-individual. In simulated annealing, each new solution is formed from only one old solution, which means that the good features of radically different solutions never mix. A solution is either accepted or thrown away as a whole, depending on its total cost.

Simulated annealing is an inherently serial algorithm. We are looking for an algorithm that can be efficiently parallelized on distributed computers, such as a workstation network connected by Ethernet. Such workstation networks, unlike specialized parallel hardware, are very popular today. The genetic algorithm can be parallelized on such loosely coupled distributed computer networks with 100% processor utilization [148]. Substantial research has been done to parallelize simulated annealing [27], [62]. Although some moderately efficient parallel implementations of simulated annealing exist for shared memory machines, it is known to be a serial algorithm that cannot be efficiently parallelized in the distributed workstation environment.

3.1.5 Experimental Results

Meta-Genetic Parameter Optimization

Table 3.1 shows the statistics of the circuits used for the experiment. These circuits were obtained from IBM and have been used by other researchers to evaluate their algorithms [108], [116]. A comprehensive experiment was performed to optimize the parameters of GASP and determine which crossover and reduction methods performed the best. The experimental method was as follows. Initially, the meta-genetic optimization algorithm was run on four test circuits consisting of 72, 100, and 183 cells, respectively. The results of this experiment are reported in Table 3.2(a). This table shows the final optimized values of crossover rate (P_C), inversion rate (P_I), and mutation rate (P_M), the initial and final fitness of the corresponding

Table 3.1. Test Circuits

CELLS	NETS	PINS
100	213	500
183	254	736
469	495	2189
750	1156	3062
800	843	2935

genetic algorithm, and the number of generations required for convergence. Table 3.2(b) shows the ranking of the various crossover and reduction strategies according to the final fitness values. The overall winner in this comparison was cycle crossover with deterministic reduction.

In most cases, either PMX or cycle crossover performed the best, and order crossover performed the worst. Cycle crossover was found to be slightly better than PMX. As mentioned previously, cycle crossover does not suffer from any conflicts, and all the cells in the child are in the same locations as in one parent or another. This might be a factor in its better performance. PMX crossover also results in less disruption of the placement conveyed from the parents as compared to order crossover.

In all cases, deterministic reduction such that the best of the parents and offspring is included in the next generation proved to be better than all other strategies used. The reason for the poor performance of the random reduction methods can be easily explained. Just as it is possible to combine the good features of two parents to form a better offspring, it is also possible to combine the bad features to form a far worse offspring. If these offspring are accepted on a random basis, the best and average cost in the population will oscillate. The losses involved in the random process far outweigh any advantage gained, and the algorithm takes a much longer time to converge. When we allow for the retention of the best solution along with random reduction, the cost of the best solution is seen to decrease monotonically.

The optimum values of the genetic parameters given in Table 3.2 show a large variation. However, these results provide a valuable insight into the acceptable range of parameter values. In all cases, the crossover rate was in the range 20% to 48%. The final population of the meta-genetic algorithm consisted of individuals with several different crossover rates, all with equal fitness. This indicated that the genetic algorithm is not very sensitive to slight variations in crossover rate, as long as a rate in the above range is used. For the mutation rate, in most cases, a very low value of 0.5% to 1.5% proved optimal, 1.5% being preferred only for the smallest circuit. The inversion rate fluctuated the most, but a value of 0 to 30% performed the best. A compromise set of values was adopted that perform reasonably well with most of the circuits tested. These values are: crossover rate = 33%, inversion rate = 15%, and mutation rate = 0.5%.

3.1. Standard Cell Placement

Table 3.2. (a)   Results of the Meta-Genetic Optimization Process

Cells	Reduction Procedure	Crossover Operator	P_C %	P_I %	P_M %	Initial Fitness	Final Fitness	Convergence (Gen.)
	Det.	Order	20-44	10	1.0	1843	1877	18
	Det.	PMX	20-44	70	1.5	1961	2048	24
72	Det.	Cycle	24-44	15	1.5	1954	2109	30
	R + B	Order	20-44	35	0.5	1652	1784	70
	R + B	PMX	24-44	15	0.0	1755	1779	22
	R + B	Cycle	20-44	65	0.5	1751	1880	57
	Det.	Order	48-56	85	0.5	4089	4089	49
	Det.	PMX	20-28	30	1.0	4628	4770	42
	Det.	Cycle	24-44	20	0.5	4632	4789	50
	Random	Order	32-44	20	1.0	1981	2086	29
100	Random	PMX	32-48	30	1.0	1996	2100	32
	Random	Cycle	32-48	50	1.0	1995	2109	46
	R + B	Order	24-44	30	1.0	3490	3495	24
	R + B	PMX	20-28	25	0.5	3570	3711	25
	R + B	Cycle	32-44	55	0.5	3750	3882	28
	Det.	Order	32-40	0	0.5	917	957	23
	Det.	PMX	24-44	0	0.5	958	1101	36
	Det.	Cycle	44-48	20	0.5	1042	1137	35
	Random	Order	72-80	10	0.0	477	518	17
183	Random	PMX	24-36	80	0.0	463	607	86
	Random	Cycle	68-80	5	0.5	695	761	16
	R + B	Order	20-40	40	0.0	827	839	19
	R + B	PMX	24-40	85	0.0	888	922	14
	R + B	Cycle	28-44	20	0.5	883	944	83

(b)   Ranking Based on Final Fitness:

Cells	1	2	3	4	5	6
72	Cycle, Det	PMX, Det	Cycle, R+B	Order, Det	Order, R+B	PMX, R+B
100	Cycle, Det	PMX, Det	Order, Det	Cycle, R+B	PMX, R+B	Order, R+B
183	Cycle, Det	PMX, Det	Order, Det	Cycle, R+B	PMX, R+B	Order, R+B

Det   = Deterministic reduction
Random = Random reduction
R+B   = Random reduction with retention of best string

Table 3.2 also shows that the meta-genetic optimization runs converged in less than 50 generations in most cases. This is because of the small search space. There were only 8000 possible combinations of the three parameters to be searched.

Performance of GASP

The performance of the algorithm was compared with TimberWolf 3.3 (which uses the simulated annealing algorithm) for five circuits ranging from 100 cells to 800 cells. The algorithm used for this purpose was cycle crossover with deterministic reduction (which includes the best configurations from the parent and offspring populations in the next generation), and crossover, mutation and inversion rates of 33%, 0.5%, and 15%, respectively, were used. TimberWolf was run with the attempts per cell parameter set to zero, which allows the algorithm to look up a table for the optimum number of iterations to perform. The genetic algorithm was run until no significant improvement was obtained. The results of this comparison are shown in Table 3.3. The performance measure of the algorithm is taken as the percentage improvement in the wire length starting from a random initial configuration. For determining the percentage improvement in the genetic algorithm, the best individuals in the initial (random) and the final population are considered.

Table 3.3. Comparison of GASP and TimberWolf 3.3

Cells	TimberWolf 3.3				GASP					
	Configs. examined	CPU sec	Δ	Total Memory	Generations	Configs. examined	CPU sec	Δ	M_p	Total Memory
100	1.17M	2102	61.3	384K	7900	62.2K	928	63.1	50K	212K
183	2.14M	2986	58.6	446K	14700	117K	2927	58.8	92K	427K
469	10.97M	40700	69.4	659K	24000	192K	39664	69.9	235K	623K
750	26.33M	44650	70.4	871K	55500	444K	40806	62.0	375K	890K
800	28.08M	49405	64.2	833K	64400	515K	44930	59.0	400K	911K

Δ = Percentage improvement in wire length,
M_p = Estimated memory required to store the population and offspring.

It can be observed from this table that GASP examines 19 to 50 times fewer configurations compared to TimberWolf for achieving the same or better percentage improvement of wire length. This illustrates the search efficiency of the genetic algorithm. The run time, however, is only marginally less than for TimberWolf in all cases. This is due to the increased overhead of wire length evaluation.

Result quality is better in GASP for three circuits and in TimberWolf for two circuits. Thus, we can conclude that the performance of this algorithm is comparable to TimberWolf 3.3 in run time as well as in quality of placement. The memory requirements are also comparable, but this comparison is unfair to TimberWolf, which is a better developed package and stores much more data on the physical structure of the layout. The column M_p of Table 3.3 gives the amount of memory required in GASP to store the population and the offspring of one generation, or in other words, the overhead that the genetic algorithm has over simulated annealing.

Table 3.4 shows the distribution of CPU time between the various functions of GASP. From this distribution, it is seen that wire length evaluation is the bottleneck in the execution time of the algorithm and takes 62% to 67% of the total time. In comparison, crossover takes only 17% to 24% of the total time. Thus, by reducing the evaluation time, we will obtain a greater speedup. Overlap removal takes a very small amount of time, as predicted.

Table 3.4. Distribution of CPU Time (in seconds) for GASP

Cells	Total	Crossover	Inversion	Mutation	Wire Length Evaluation	Overlap Removal	Total Eval	Selection
100	928	161	16	7	625	95	720	17
183	2927	649	37	16	1839	351	2190	32
469	39,664	8349	372	128	26,532	4050	30,582	145
800	44,930	10,674	388	124	28,570	5023	33,593	80

The 183-cell circuit, for which almost identical performance and run times were obtained for both algorithms, was used for the comparison. It was noted that in the first few high-temperature iterations of TimberWolf, the wire length (and the penalties) *increased* significantly over the initial random placement. This increase causes a greatly increased run time but is necessary for avoiding local optima. Additionally, in almost the entire first half of the run time, there is no significant improvement in wire length. In GASP, the wire length decreases monotonically, and the greatest reduction is at the beginning. Thus, a moderate quality placement can be obtained very quickly. The same run can then be continued for a higher quality placement.

The conclusion from this research is that the genetic algorithm converges to a near-global minimum, similar to stochastic approaches such as simulated annealing, but because it examines a large number of configurations concurrently, it searches far more efficiently than algorithms that use pairwise exchange.

3.2 Macro Cell Placement

Macro cells are large irregularly sized circuit modules, which are usually generated by a silicon compiler. The macro cell placement problem can be defined as follows: Given a set of unequally sized Manhattan blocks, connected together as specified in a netlist, determine the X- and Y- coordinates and orientation of each block such that the total wire length and the chip area are minimized, and none of the blocks overlap. The inputs to the problem are the module descriptions, consisting of the shapes, sizes, and terminal locations, and the *netlist*, describing the interconnections between the terminals of the modules. The outputs are a list of x- and y-coordinates and orientations for all modules.

The wire length can be measured as the sum of the half perimeters of the bounding rectangles of the pins on each net. The chip area is the bounding rectangle enclosing all the blocks. The Manhattan blocks are usually rectangles but could also be L- or T-shaped blocks or any arbitrary rectilinear polygons. They may be of fixed shape, or each block may have a number of possible shapes and pin locations.

The macro cell layout style is the most flexible layout style. It allows inclusion of compiled blocks, such as RAM/ROM arrays, PLAs, etc., in arbitrary sizes, which are difficult to include in the standard cell layout style, and impossible in gate arrays. Correspondingly, the macro cell placement and routing problem is the most difficult problem in VLSI layout. Unlike standard cell placement, where the chip area is determined to a large extent by the wire length, in macro cell placement, the chip area is significantly influenced by the wasted area due to the unequal sizes of the modules. Hence, there is a trade-off between minimization of chip area and minimization of wire length. The best placement with respect to the minimum wire length may not have all the modules tightly packed and well fitting, and vice versa.

Rather than simply using a GA for macro cell placement, we have developed a new stochastic optimization algorithm, called SAGA, which is a combination of the genetic algorithm and the simulated annealing algorithm applied to macro cell placement [67]. The aim of this work is to improve the typical convergence rate of the pure GA by combining it with simulated annealing.

The typical GA convergence curve is illustrated in Fig. 3.8. Initially, the solution cost improves very rapidly. However, obtaining further improvements soon becomes difficult, and the majority of the run time is spent in the later phase of the process in which small improvements are obtained very slowly. The work presented here is motivated by the need to overcome this shortcoming of the GA. Our approach is to unify the GA with the Simulated Annealing (SA) algorithm. As can be seen in [209], the typical convergence curve of SA is very different from that of the GA. Initially, SA converges much slower, but in the late phase of the process, SA may be able to obtain improvements faster than the GA. The unified algorithm, called SAGA (an acronym for Simulated Annealing & Genetic Algorithm), is designed in such a way that the initial fast convergence of the GA is combined with the faster convergence of SA in the late phase.

Earlier attempts to combine the GA and SA have been made by some researchers in the fields of applied physics and operations research. Notable work was done by Boseniuk and Ebeling [19], who developed and compared various mixed strategies for the Traveling Salesman Problem (TSP). One of the strategies explored by them is called life cycle. A population of individuals coexists, and mutations are accepted with a certain probability, as in SA. Each individual goes through a life cycle, and as it gets older, its probability of being mutated decreases, while the probability of mating increases. Boseniuk and Ebeling reported that the life-cycle strategy was superior to pure SA on the TSP.

The approach presented here is inspired by Boseniuk and Ebeling's ideas, although many significant refinements have been made to improve the performance of the algorithm. In our approach, the way the GA and SA are mixed is dynami-

3.2. Macro Cell Placement

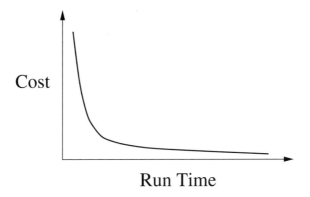

Figure 3.8. Typical convergence of a GA

cally changed during the optimization process, while it is static in [19]. The SAGA algorithm is application independent and highly adaptive. When applied to the macro cell placement problem, SAGA performs better than a pure GA, and results for macro cell placement benchmarks are comparable to or better than previously published results.

3.2.1 Unified Algorithm

SAGA can be viewed as a GA which has been modified in two major ways, to be discussed in detail in the following sections:

1. The mutations performed on an individual are accepted with a certain probability, as in SA. Each individual has its own temperature, and during its lifetime, the temperature is decreased according to its own cooling schedule.

2. Initially, SAGA executes as a pure GA. However, as the GA stagnates, SAGA gradually switches over to SA. The speed of this switch is adaptive, since it is determined by the progress of the optimization itself.

SAGA has two important properties:

- It is application independent in the sense that it can potentially be applied to any optimization problem for which the GA and SA are well suited.

- It unifies the GA and SA in such a way that it can be executed exclusively in GA or SA mode by selecting appropriate values of its control parameters.

Fig. 3.9 shows an outline of SAGA. Lines 3, 6, 8–12, and 17 have to do with the SA part of SAGA. The notation (line x) refers to line x of Fig. 3.9. Initially, the current population is constructed from randomly generated individuals (line 1). Routine *evaluate* (line 4) computes the fitness of each individual, while *bestOf* (line 5) finds the individual with the highest fitness. One execution of the outer **While**

```
01  Generate initial Population of size M = M₀ randomly;
02  For j = 1 to M Do
03      Temperature[Population, j] = InitialTemperature;
04  evaluate (Population);
05  BestIndividual = bestOf (Population);
06  PopsizeReductions = 0;
07  While not stopCriterion() Do
08      If no improvement for R generations then
09          PopsizeReductions = PopsizeReductions + 1;
10          M = max (round(β × M), 1);   // reduce population size M
11          Population = reduce (Population, M);
12          P_M = min(γ × P_M, 1.0);   // increase P_M
13      NewPopulation = ∅;
14      For j = 1 to M Do
15          Select two parents s and t from Population;
16          u = crossover (s, t);
17          Temperature[NewPopulation, j] = InitialTemperature;
18          Add u to NewPopulation;
19      evaluate (Population ∪ NewPopulation);
20      Population = reduce (Population ∪ NewPopulation, M);
21      For j = 1 to M Do
22          SAmutate (j);
23          invert (j, P_I);
24      evaluate (Population);
25      BestIndividual = bestOf (Population ∪ BestIndividual);
26  For j = 1 to M Do
27      optimize (j);
28      optimize (BestIndividual);
29  solution = bestOf (Population ∪ BestIndividual);
```

Figure 3.9. Outline of SAGA

loop corresponds to the processing of one generation (lines 7–25). The population size, M, is kept constant from one generation to the next while improvements are being made (line 20), and the best individual from any generation is saved (line 25). Routine *stopCriterion* (line 7) terminates the simulation when no improvement has been observed for S consecutive generations. Each generation is initiated by the formation of a set of offspring of size M which forms the NewPopulation (lines 13–18). The two mates s and t are selected independently of each other, and each mate is selected with a probability proportional to its fitness (line 15). Routine *reduce* (*population, k*) (line 20) returns the k fittest individuals from the given population. Here we take the M best individuals from the combination of the new and old populations, thereby keeping the population size constant. Following this reduction, all

3.2. Macro Cell Placement

individuals are subjected to possible mutation by routine *SAmutate* (line 22). The routine *invert*(s, P_I) inverts the genotype of s with probability P_I (line 23). Finally, local hill climbing is performed on all existing individuals by routine *optimize* (lines 27–28). The solution is the output of the algorithm (line 29).

The switch toward SA is handled by lines 6, and 8–12 of Fig. 3.9. A step towards SA is taken whenever no improvement has been observed for R consecutive generations, $0 \leq R \leq S$ (line 8). A step consists of reducing the population size M (line 10) and increasing the mutation rate P_M (line 12). In other words, more SA-controlled mutations will be performed on a smaller number of individuals. The population size M after n reductions is

$$M = \max(\text{round}(\beta^n \times M_0), 1),$$

where M_0 is the initial population size, $0 \leq \beta \leq 1$ is a real-valued parameter, and round(x) performs rounding to the nearest integer value of x. Ultimately, we may have $M = 1$, corresponding to a pure SA process. When M is decreased, the M fittest individuals are kept, while the rest are discarded (line 11). Furthermore, the mutation rate P_M is increased so that after n increases, it is given by

$$P_M = \min(\gamma^n \times P_{M,0}, 1.0),$$

where $P_{M,0}$ is the initial mutation rate, and $\gamma \geq 1$ is a real-valued parameter.

Mutation of individual s is performed by routine *SAmutate* as illustrated in Fig. 3.10. Initially, a copy r of individual s is subjected to pointwise mutation; i.e., each component of r is subjected to a random change with probability P_M. If this is the first attempted mutation of individual s, its temperature T_s will be the initial value, cf. (lines 3, 17) of Fig. 3.9, and the variables P_s, T_s, and c_s controlling its cooling schedule are then defined. The mutation is then performed with a temperature-dependent probability as in SA. Specifically, if the cost is decreased, the mutation is always accepted, while if the cost is increased, it is accepted with probability $\exp(\frac{\mathcal{C}(s)-\mathcal{C}(r)}{T_s})$, where $\mathcal{C}(x)$ denotes the cost of solution x.

The absolute values of a suitable temperature schedule are problem dependent. To circumvent this problem, we define a schedule for reducing the probability of accepting cost-increasing mutations. The temperature decrease is then computed so that the specified probability is obtained. More specifically, let P_s be the probability of accepting a mutation on individual s which increases the cost of s by σ, the standard deviation of the cost of all solutions in the search space. From an initial value $\mathcal{P}$, $0 < \mathcal{P} < 1$, P_s is then reduced by a factor α, $0 < \alpha \leq 1$, whenever quasi equilibrium has been obtained. In this implementation, P_s is reduced when λ mutations on s have been accepted at the current temperature, as counted by c_s.

For a given value of P_s, the corresponding temperature T_s is computed as

$$T_s = \frac{-\hat{\sigma}}{\ln(P_s)}$$

where $\hat{\sigma}$ is an estimate of σ computed during generation of the initial population.

```
r = s;
For each component r_i of r Do
    mutate r_i with probability P_M;
If T_s = InitialTemperature then
    P_s = P;
    T_s = -σ̂/ln(P_s);   // temperature
    c_s = 0;   // counter
With probability min (exp (C(s)−C(r)/T_s), 1.0) Do
    s = r;
    c_s = c_s + 1;
    if c_s = λ do :
        P_s = αP_s;
        T_s = -σ̂/ln(P_s);
        c_s = 0;
```

Figure 3.10. Structure of routine SAmutate

SAGA reduces to a pure GA when $M_0 > 1$, $R = S$, $\alpha = 1.0$, and $\mathcal{P}$ is close to 1.0. Pure SA is obtained whenever $M_0 = 1$. In this case, the reproduction step is in general equivalent to a mutation which is accepted if and only if it improves cost. Standard crossover operators as found in [79], [95] have the property that crossover(x, x) always yields the offspring x, in which case the reproduction step becomes equivalent to the empty statement.[1]

3.2.2 Application to Macro Cell Placement

This subsection presents our definition of the macro cell placement problem and describes the specific genetic encoding and corresponding genetic operators developed for this application. The macro cell placement problem is defined as follows.

Given:

- A set of rectangular *cells*, each with a number of *terminals* at fixed positions along the edges of the cell,

- A netlist specifying the interconnections of all terminals, and

- An approximate horizontal length W of the chip under construction,

---
[1] Alternatively, the generation of the new population can be conditioned by $M > 1$, as can the invocation of the inversion operator, if desired.

3.2. Macro Cell Placement

Compute:

- The absolute position of each cell,
- The orientation and possible reflection(s) of each cell, and
- A rectangle B defining the shape of the chip.

The objective is to minimize the area of B subject to the following constraints:

- No pair of cells overlap each other,
- The rectangle B encloses all cells and has approximate horizontal length W, and
- The area within B which is not occupied by cells is sufficiently large to contain all routing needed to implement the required interconnections.

To meet the last constraint, the necessary routing area is estimated during the placement. The estimate is based on the assumptions that two metal layers are used for routing, the area occupied by cells, and the area used for routing are disjoint, and all nets are treated as signal nets.

Genetic Encoding

The genetic encoding is inspired by the two-dimensional bin packing problem, which is the problem of compactly packing a number of rectangular blocks into a bin having fixed width and infinite height in such a way that the distance from the top edge of the highest placed block to the bottom edge of the bin is minimized. The standard algorithm for this problem places the blocks one at a time at the lowest and then at the leftmost position. The placement algorithm is based on a generalization of this scheme. For a given instance of the macro cell placement problem, let a *BL-placement* (Bottom-Left) denote a solution in which no cell can be moved further down or to the left without causing a violation of the routing area estimate. The solution space considered by the algorithm is restricted to the set of all possible BL-placements.

Genotype and Decoder: The genetic representation described here is based on an earlier version presented in [127]. Assume that the given problem has n cells, $b_1, \ldots, b_n$. An example genotype with $n = 7$ cells is shown in Fig. 3.11 together with the corresponding phenotype. A binary tree (V, E), $V = \{b_1, \ldots, b_n\}$, in which the ith node corresponds to cell i, represents the absolute positions of all cells. Two kinds of edges exist, top edges and right edges, so that $E = E_t \cup E_r$, $E_t \cap E_r = \emptyset$. Each node has at most one outgoing top edge and at most one outgoing right edge, and all edges are oriented away from the root of the tree. Let $e_{ij} \in E$ denote an edge from b_i to b_j, and let (b_i^{xl}, b_i^{yl}) and (b_i^{xu}, b_i^{yu}) denote the coordinates of the lower-left and upper-right corners of b_i, respectively. Then $e_{ij} \in E_t$ (E_r) means that cell b_j is placed above (to the right of) b_i in the phenotype. That is,

$$\forall e_{ij} \in E: \quad e_{ij} \in E_t \Rightarrow b_j^{yl} \geq b_i^{yu}, \quad e_{ij} \in E_r \Rightarrow b_j^{xl} \geq b_i^{xu}.$$

The tree is decoded as follows. The cells are placed one at a time in a rectangular area having horizontal length W and infinite vertical length. Each cell is moved as far down and then as far left as possible without violating the routing area estimate described below. The cells are placed in ascending order according to their *priorities* defined by the one-to-one function $\pi : V \to \{1, \ldots, n\}$. Any node has higher priority than its predecessor in the tree. In Fig. 3.11, the priorities are indicated at the top right-hand side of each node. For each cell, one of the eight possible transformations (reflection in one or both dimensions and/or 90 degree rotation) is defined by the function $o : V \to \{0,1,2,\ldots,7\}$, which is also part of the genotype.

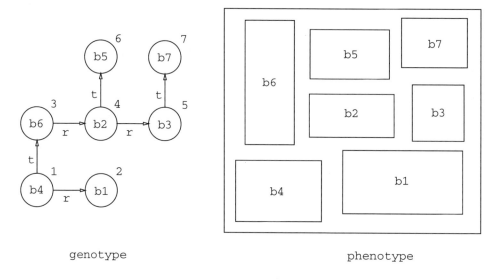

Figure 3.11. An example genotype and the corresponding phenotype

When all cells are placed, the decoder computes the rectangle B. This is done by extending the smallest rectangle enclosing all cells until the routing area estimate is satisfied along all edges of B.

Routing Area Estimation: When the binary tree is decoded, the routing area needed is estimated as each cell is placed. When placing the ith cell, the distance needed in each direction $v \in$ {north, east, south, west} to previously placed cells is computed by a function D_v, which depends on all previously placed cells. Each cell is placed according to the BL-strategy and as close to the previously placed cells as allowed by D_v. Fig. 3.12 illustrates how D_v is computed. When testing if cell b_i can be placed at some given position (b_i^{xl}, b_i^{yl}), the four areas indicated by

3.2. Macro Cell Placement

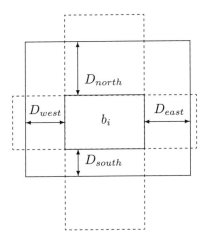

Figure 3.12. Estimation of routing area

dashed squares are considered. D_v depends on all terminals at side v of b_i and on all terminals in previously placed cells, which are (1) inside the square region at side v, (2) placed at some side parallel to side v of b_i, and (3) not shadowed by an intervening cell. Given this set of terminals, the channel density d_v is computed as if the square region were the routing channel. In order to account for global routing, D_v is then computed as

$$D_v = \begin{cases} \mu[d_v + \text{round}(a\sqrt{\frac{h_v}{\mu}} + b)] & \text{if } d_v > 0 \\ 0 & \text{if } d_v = 0 \end{cases}$$

where h_v is the length of side v of b_i, μ is the spacing in the routing grid, the rounding function returns the nearest integer value of its argument, and a and b are user defined parameters.

The area inside the solid rectangle shown in Fig. 3.12 is uniquely determined by D_v. Cell b_i can be placed at the given position if and only if this area contains no (parts of) cells apart from b_i itself. When $a = b = 0$, the estimated routing area is a lower limit of the area needed by any router regardless of the channel definition. If $a > 0$, the corresponding term of D_v increases with h_v. The argument for this definition is that the longer the channel, the more likely nets will pass through it [230]. Note that moving b_i in any direction may affect the value of D_v for all four values of v.

In summary, given a set of cells V, the genotype of an individual consists of the relations E and the functions π (priority) and o (transformation: reflection, rotation). The genotype (and the decoder) has the important property of implicitly representing most constraints of the problem. This simplifies the design of genetic operators, which assures that at any point in time, every individual satisfies all constraints.

Fitness Measure

Given (the phenotype of) an individual, its fitness is defined by the positive function F. Since the objective is to minimize layout area, initially the auxiliary function F_1 is defined as

$$F_1(p) = \frac{1}{A(B_p) - \sum_{i=1}^{n} A(b_i)}$$

where n is the number of cells of the placement problem, A is the area of a rectangular cell, and B_p is the rectangle B of the individual p. That is, $F_1(p)$ is the inverse of the total estimated routing area in p. All individuals having equal area will now have equal fitness. However, when fixing the total area of a placement, the probability of 100% routing completion within the estimated area is likely to increase as the total interconnect length decreases. Minimization of the total interconnect length is therefore introduced as a secondary optimization criterion. All individuals having the same area will have their fitness values adjusted so that fitness increases as the estimated interconnect length decreases.

The total interconnect length of an individual is estimated as in [93]: Let N denote the number of nets, and let n_k denote the total number of terminals of the kth net. Let $t_{ki} = (x_{ki}, y_{ki})$ denote the position of terminal i in net k. The *center of gravity* of the kth net is then defined by

$$T_k = (\overline{x}_k, \overline{y}_k) = \frac{1}{n_k} \sum_{i=1}^{n_k} t_{ki},$$

and the estimated total interconnect length L is defined as

$$L(p) = \sum_{k=1}^{N} \sum_{i=1}^{n_k} \|t_{ki} - T_k\|,$$

where $\|x\|$ denotes the usual Euclidean vector norm. Now suppose that the population is enumerated in ascending order according to F_1, and that

$$F_1(p_i) = F_1(p_{i+1}) = \ldots = F_1(p_j),$$

$i \leq j$. Thus, the fitness of $p_i, \ldots, p_j$ must be adjusted according to interconnect length. In order to assure that area always predominates interconnect length, this is done as follows. Sort $p_i, \ldots, p_j$ into decreasing order according to interconnect length; i.e., let us assume $L(p_i) \geq L(p_{i+1}) \geq \ldots \geq L(p_j)$. Define δA_{ij} as

$$\delta A_{ij} = \frac{F_1(p_{j+1}) - F_1(p_j)}{j - i + 1}.$$

A new fitness value F_2 is then computed as

$$F_2(p_k) = F_1(p_i) + (k - i)\delta A_{ij} \, , \, k = i, \ldots, j.$$

3.2. Macro Cell Placement

Following the normalization

$$F_3(p_i) = \frac{F_2(p_i)M}{\sum_{j=1}^{M} F_2(p_j)} \quad , i = 1, \ldots, M,$$

the final fitness values F are computed by linear scaling as in [79]:

$$F(p_i) = \xi F_3(p_i) + \eta \quad , i = 1, \ldots, M,$$

where the scaling coefficients ξ and η are determined so that three properties are obtained:

- The average value is unchanged; i.e., $\frac{1}{M} \sum_i F(p_i) = \frac{1}{M} \sum_i F_3(p_i)$.

- All new fitness values are positive.

- Maximum value is τ times the average; i.e., $\max_i\{F(p_i)\} = \frac{\tau}{M} \sum_i F_3(p_i)$ for the largest possible τ satisfying $1 < \tau \leq 2$.

The purpose of scaling is twofold. In the initial phase of the process, a few individuals having very high fitness compared to the average will be very dominating and may limit the search to a small region of the space too early. Scaling counteracts this effect by reducing the standard deviation of the fitness in the early phase of the optimization. In the final phase of the process, the difference between the best and the average individuals tends to be small due to the convergence. Selection then becomes almost random, thereby reducing the chances of further improvement. At this stage, scaling counteracts the problem by increasing the standard deviation.

Crossover Operator

Given two individuals ϕ and ψ, the crossover operator generates a feasible offspring θ. This operation is illustrated in Fig. 3.13. Throughout this section, a subscript specifies which individual the marked property is a part of.

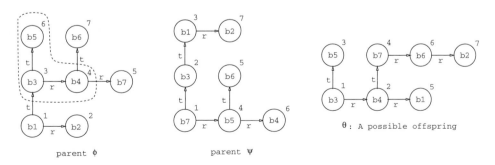

Figure 3.13. Crossover of ϕ and ψ

E_θ is constructed as follows. From the cell tree of ϕ, a connected subset $T' = (V', E')$, $V' \subset V$, $E' \subset E_\phi$ is chosen. T' is chosen at random but subject to the constraint that decoding T' in the order defined by π_ϕ, i.e., using $b \in V' \mid \forall b' \in V' \setminus \{b\} : \pi_\phi(b) < \pi_\phi(b')$ as root causes no constraint violations. The size of V' is determined by a normal distributed stochastic variable having mean $n/2$ and standard deviation 1. In Fig. 3.13, the chosen T' is indicated by the dashed line. Initially E_θ is defined to be E'. Hence, θ has inherited all cells in V' from ϕ. The remaining cells $V - V'$ are then inherited from ψ by extension of E_θ. The cell tree of ψ is traversed in ascending order according to π_ψ. At any node, it is checked if the corresponding cell b belongs to V', i.e., whether it has been placed in θ already. If so, the cell is skipped. Otherwise, b is added to the cell tree of θ by extending E_θ. The position at which to add b is randomly chosen among all free and feasible positions. The transformation of any cell is inherited unaltered together with the cell itself. π_θ is uniquely defined so that it corresponds to the order in which the cells were placed when creating E_θ.

Crossover is asymmetric in the sense that θ explicitly inherits a subtree from ϕ only. However, since the remaining cells from ψ are added to θ in the order defined by π_ψ, some relative cell positions from ψ will also be transferred to θ (at phenotype level), and in any case, the cell transformations are always inherited unaltered.

Mutation Operator and Hillclimber

The implementation of the operator $SAmutate$ of Fig. 3.10 performs four different types of random changes on the given genotype. Let b_i and b_j denote two randomly chosen cells, $i \neq j$. The four types of mutation are

1. Alter the set of edges E by moving a leaf b_i to another free and randomly chosen position. The type of edge going into the leaf may be changed as part of the move.

2. Alter the set of edges E by exchanging b_i and b_j. The priorities of the cells are exchanged simultaneously so that no pair of cells are prevented a priori from being exchanged due to the constraint that any node always has a higher priority than its predecessor. An example is shown in Fig. 3.14.

3. Alter π by exchanging $\pi(b_i)$ and $\pi(b_j)$.

4. Change the transformation of a cell by altering the value of $o(b_i)$.

When performing each of these mutations, a part of the genotype has to be decoded to check if the mutated individual satisfies all constraints. Mutations 1 and 4 require that all cells having priority $\pi(b_i)$ or higher are decoded, while mutations 2 and 3 require decoding from priority $\min(\pi(b_i), \pi(b_j))$. A mutation is only performed if it does not cause any constraint violations.

Because of the dual optimization criteria described in Section 3.2.2, the cost of an individual s cannot be suitably expressed in a single number $\mathcal{C}(s)$. Therefore,

3.2. Macro Cell Placement

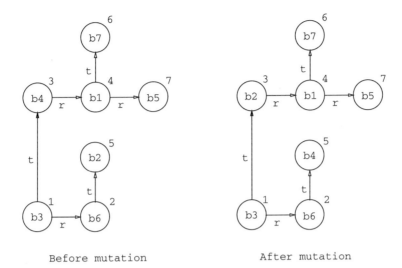

Figure 3.14. Mutation of type 2: Cells b_2 and b_4 are exchanged, while the priorities are still attached to the same positions in the tree

the acceptance criterion for mutations shown in Fig. 3.10 has to be modified slightly for this application. Let $\mathcal{C}_a(s)$ and $\mathcal{C}_w(s)$ denote the estimated area and total interconnect length of s, respectively. Each individual has two separate temperatures, T_s^a corresponding to area and T_s^w corresponding to interconnect length. These are defined and updated using the corresponding standard deviations of area and interconnect length, $\hat{\sigma}^a$ and $\hat{\sigma}^w$, respectively. The same parameter P_s (and $\mathcal{P}$) is used for both temperature computations.

The probability $P_{acc}(s,r)$ of accepting a mutation of s giving the result r is then calculated as

$$P_{acc}(s,r) = \begin{cases} \exp(\frac{\mathcal{C}_a(r)-\mathcal{C}_a(s)}{T_s^a}) & \text{if } \mathcal{C}_a(r) > \mathcal{C}_a(s) \\ \exp(\frac{\mathcal{C}_w(r)-\mathcal{C}_w(s)}{T_s^w}) & \text{if } \mathcal{C}_a(r) = \mathcal{C}_a(s) \text{ and } \mathcal{C}_w(r) > \mathcal{C}_w(s) \\ 1 & \text{otherwise.} \end{cases}$$

The implementation of the local hillclimbing performed by routine *optimize(t)* of Fig. 3.9 is very simple. It performs a sequence of mutations, each of which improves the fitness of t. An exhaustive strategy is used so that when *optimize(t)* has been executed, no single mutation exists that can improve t further.

Inversion Operator

Since several genotypes usually exist for the same phenotype, an inversion operator is desirable. The inversion operator selects a subtree at random and moves it to another free position in such a way that no constraints are violated and so that the corresponding phenotype is still the same. An example of this is shown in Fig. 3.15. The new genotype tree is generated by moving the subtree rooted at b_2 in the original genotype tree.

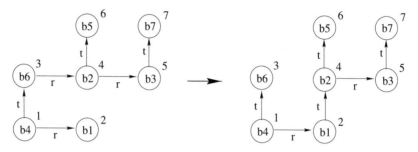

Figure 3.15. Genotype inversion

3.2.3 Experimental Results

In this subsection, experimental results obtained with an implementation of the application of SAGA to macro cell placement are reported. The implementation is written in the C programming language and consists of about 14,000 lines of source code. All experiments were performed on a DEC MIPS 5000-240 workstation. Performance is measured using three benchmarks from the 1992 MCNC International Workshop on Placement and Routing. Table 3.5 lists the main characteristics of these examples.

Table 3.5. Benchmark Characteristics

Benchmark	Cells	Nets	Terminals
Apte	9	97	287
Xerox	10	203	698
Hp	11	83	309

Behavior in Mixed Mode

The most interesting execution mode of SAGA is the mixed GA/SA mode, which also causes the most complex behavior. Figs. 3.16 through 3.19 illustrate a typical optimization process in the mixed mode. These graphs were extracted from the same sample execution of the algorithm using the parameter values $M_0 = 25$, $S = 200$, $P_{M,0} = 0.025$, $P_I = 0.05$, $\beta = 0.7$, $\gamma = 1.4$, $R = 80$, $\mathcal{P} = 0.99$, $\alpha = 0.6$, and $\lambda = 1$.

Fig. 3.16 shows for each generation the average of the estimated area of all individuals and the estimated area of the best individual. Both quantities improve very rapidly within the first 100 generations. Between generations 100 and 800, the best individual improves only very slowly. Up to this point, the observed

3.2. Macro Cell Placement

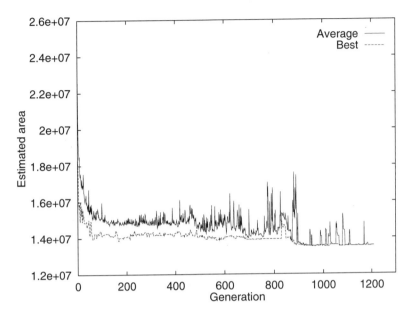

Figure 3.16. Estimated areas of average and best individuals as functions of generation number

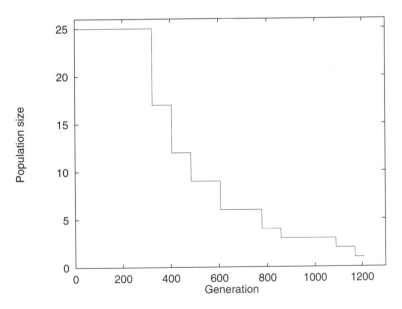

Figure 3.17. Population size as a function of generation number

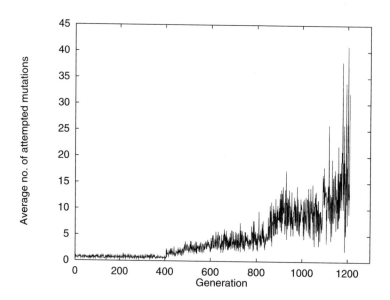

Figure 3.18. Average number of attempted mutations on each individual as a function of generation number

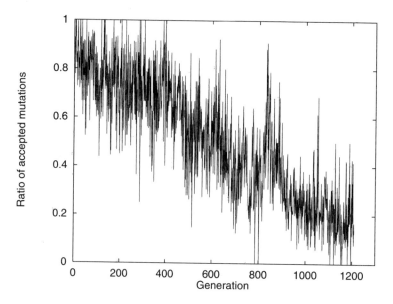

Figure 3.19. Ratio of attempted mutations which are actually accepted

3.2. Macro Cell Placement

behavior is typical of a pure GA. Due to the SA component of this algorithm, significant improvements are obtained between generations 800 and 1000. The very best individual emerges in generation 1009, and the process terminates after 1209 generations.

The population size M decreases as shown in Fig. 3.17. At generation 1170, it reaches 1, and the process becomes pure SA. Of course, this does not always happen. In many executions, the final population size is greater than 1.

For each generation, the average number of attempted mutations per individual is shown in Fig. 3.18. Offspring generated later in the optimization process are subjected to more mutations. However, the probability of accepting a cost-increasing mutation of an individual is decreased according to the number of mutations performed on it. The ratio of attempted mutations which are accepted is shown in Fig. 3.19 as a function of generation number. In the first phase of the optimization process, few mutations are attempted, and almost all of them are accepted. This resembles a pure GA process. In the final phase of the optimization, the probability of accepting cost-increasing mutations becomes small, and the ratio of accepted mutations decreases to about 20%.

The decrease of the population size and the increase of the number of SA-controlled mutations are both important components of the optimization process. Reducing the population size in an otherwise pure GA process will cause divergence, as illustrated in Fig. 3.20 for a sample execution of SAGA with the parameters $\mathcal{P} = 1 - 10^{-7}$, $\alpha = \gamma = 1.0$, and the remaining parameters as before. In other words, it is the SA-controlled mutations that allows the population size to be reduced while at the same time maintaining convergence. On the other hand, fixing the population size ($\beta = 1.0$) while performing an increasing number of SA-controlled mutations leads to an ineffective process. In the late phase, many mutations will be attempted on a large number of individuals, but only few mutations will actually be accepted and performed.

Comparing the GA with a Mixed Strategy

Using the benchmarks, the performance of a mixed GA/SA strategy has been compared to that of a pure GA. As for the previous experiments, the parameters $M_0 = 25$, $S = 200$, $P_{M,0} = 0.025$, and $P_I = 0.05$ were used for both strategies. For the GA, $\mathcal{P} = 1 - 10^{-7}$, while for the GA/SA, $\beta = 0.7$, $\gamma = 1.4$, $R = 80$, $\mathcal{P} = 0.99$, $\alpha = 0.6$ and $\lambda = 1$. These settings of the control parameters have been used for all three benchmarks; i.e., no problem-specific tuning has been made. Different sets of values were tried out, and the values listed were selected as they consistently yield good results. Suitable values for the parameters a and b used in the routing area estimate depend on the characteristics of the interconnections to be made; i.e., these parameters are problem dependent. Table 3.6 shows the values used.

For each benchmark and each set of control parameters, SAGA was executed 40 times. Table 3.7 summarizes the results. A_{best} and A_{avg} denote the best and average estimated areas in mm^2, respectively. A_σ denotes the standard deviation. Since

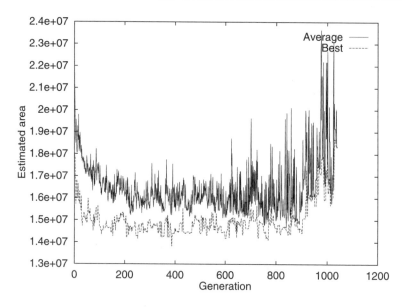

Figure 3.20. Effect of reducing population size in an otherwise pure GA process

Table 3.6. Values of Parameters for Routing Area Estimate

Parameter	Apte	Xerox	Hp
a	0.0	0.2	0.0
b	12.0	10.0	7.0

SAGA minimizes estimated area as opposed to area after routing and compaction, the best comparison of the two optimization approaches is obtained by comparing estimated areas. T_{avg} denotes the average CPU time in seconds, and T_σ is the standard deviation of the CPU time.

The results reported in earlier work [65] approximately correspond to what is here called the pure GA, and they are comparable to the best results published. It is therefore likely that the areas shown in Table 3.7 obtained by the GA are quite close to the optimum, suggesting that the room for improvement over the GA is small. However, for the Apte and Xerox benchmarks, the values of A_{best}, A_{avg}, and A_σ are significantly improved by the mixed strategy, while for the Hp benchmark, no significant improvement can be observed.

Comparison with Other Systems

Ten of the 40 placements generated for each benchmark by SAGA using the mixed GA/SA strategy have been routed and compacted using Mosaico [16], which is part of the Octtools CAD framework. In Table 3.8, the quality of the best completed

3.2. Macro Cell Placement

Table 3.7. Comparison of the Pure GA with a GA/SA Mixture

Benchmark	Quantity	GA	GA/SA
Apte	$A_{best}(mm^2)$	57.599	57.477
	$A_{avg}(mm^2)$	58.648	58.250
	A_σ	0.816	0.727
	$T_{avg}(s)$	3,134	3,328
	T_σ	1,563	2,259
Xerox	$A_{best}(mm^2)$	27.891	27.554
	$A_{avg}(mm^2)$	28.961	28.521
	A_σ	0.533	0.486
	$T_{avg}(s)$	9,354	13,192
	T_σ	3,908	2,568
Hp	$A_{best}(mm^2)$	12.805	12.803
	$A_{avg}(mm^2)$	13.310	13.294
	A_σ	0.357	0.369
	$T_{avg}(s)$	3,311	3,070
	T_σ	1,353	1,499

layouts are compared to the best published results. The absolute area is core area in mm^2. To ease comparison, relative areas have also been computed by assigning the best result for each benchmark the relative area 1. The total interconnect length in mm, and the total number of vias is also given.

BB [166] is a branch-and-bound approach, while Seattle Silicon [230] is based on SA. The results listed should be compared with some caution due to the impact of the different routing and compaction tools used, as well as variations in the exact definitions of the macro cell placement problem. Nevertheless, it is clear that SAGA is highly competitive with respect to layout area, the main optimization criterion. The SAGA layouts are routed using the global router of TimberWolfMC, which minimizes interconnect length rather than area. In [68], a GA-based global router is presented which explicitly minimizes area and obtains smaller layouts than TimberWolfMC. Therefore, it is most likely that the SAGA results listed could be improved even further by applying the global router of [68].

The total wire lengths of the completed SAGA layouts are 13% to 40% higher than those of Seattle Silicon, as shown in Table 3.8. This is due to the fitness computation, which always gives higher priority to area than wire length. Consequently, if solution A has a much larger estimated total wire length than B but just a slightly smaller estimated area, then A is always considered better than B. Since both area

Table 3.8. Comparison of SAGA (Mixed GA/SA Mode) with Other Systems

Bench-mark	System	Rel. Area	Abs. Area	Wire Length	Vias
Apte	SAGA	1.000	53.58	489	647
	BB [166]	1.009	54.05	460	-
	Seattle Silicon [230]	1.022	54.77	350	-
Xerox	Seattle Silicon [230]	1.000	25.79	601	1104
	BB [166]	1.015	26.17	628	-
	SAGA	1.053	27.15	679	1379
	BEAR [47]	1.104	28.47	633	897
	MOSAICO[2]	1.125	29.01	650	1173
	VITAL[2]	1.230	31.17	866	1029
Hp	SAGA	1.000	11.81	261	675
	Seattle Silicon [230]	1.003	11.85	200	-
	BB [166]	1.029	12.15	278	-

A hyphen indicates that the value is not available.

and wire length are estimated rather than exact quantities, the strict priority may not always correctly reflect the relative solution qualities of the completed layouts.

The current implementation of SAGA requires significantly more computation time than the other systems listed in Table 3.8. The BB approach [166] is about 2–3 times faster, and the Seattle Silicon approach [230] is about 10–20 times faster.[3] To some extent, this situation may simply reflect the nature of the problems considered. Many researchers have worked with the MCNC benchmarks, and as the gap to the global optimum is narrowed, it becomes increasingly hard to obtain even small improvements. Nevertheless, the current run time should and can be improved.

3.2.4 Limitations and Enhancements

The main limitation of SAGA is the run time requirement, which currently prevents larger circuits, such as Ami33 and Ami49, from being considered. However, there are a number of reasons why we expect that significant improvements can be obtained. First of all, in the current implementation, the majority of the run time is spent on repeated computation of channel densities. As various positions for a block are tried out, channel densities involving the same sets of cells are computed over and over again. The only difference from one computation to the next is that the block to be placed has been moved slightly. Currently, the channel density is computed from scratch whenever a new position is tested. Significant amounts of computation

---

[2]Referenced here as found in [47], [230].
[3]These figures do not account for the different machines used.

could thus be eliminated by using a data structure which allows the channel density to be dynamically updated as a block is shifted along one dimension.

Currently, a pool of M offspring is generated in each generation, as shown in Fig. 3.9. Many of these will probably never be used since they are worse than existing individuals. These computations can be eliminated by adapting the idea of the steady-state GA in which the crossover operator is applied only once per generation, and the offspring is inserted immediately into the population, for example by replacement of the current poorest solution. Although the total number of generations needed will then increase, a net gain is expected.

Interrupting decodings of solutions when an upper bound on cost is exceeded could also improve run time. Assume for example a steady-state GA, in which the generated offspring always replaces the current worst solution. If, during a decoding process initiated by the crossover operator, it can be determined that the cost of the partly generated offspring will exceed the cost of the current worst solution, then the decoding can be interrupted, since the offspring will not be inserted in the population anyway.

Another approach for handling larger circuits, which is independent of the above improvements, is to combine SAGA with a kind of hierarchical partitioning as used by, e.g., the BB algorithm [166] and the BEAR system [47]. Finally, in addition to or instead of the above improvements of the sequential algorithm, a parallel version of SAGA could be implemented. Due to the inherent parallelism of the algorithm, a good speedup can be expected on any MIMD architecture. A vast literature on parallel GAs exists, and almost linear speedups are reported [79], [148], [150].

SUMMARY

TOPICS STUDIED	KEY OBSERVATIONS AND POINTS OF INTEREST
Genetic Algorithm for Standard Cell Placement (GASP) • Representation • Crossover • Parameter Selection • Results	• Cell ID, X-Y coordinates, slot ID number
	• PMX crossover • Cycle crossover • Order crossover
	• Meta-genetic optimization • Cycle crossover with deterministic selection is best
	• Space and time complexity comparable to TimberWolf
Macro Cell Placement (SAGA) • Representation • Crossover • Results	• Two-dimensional B tree packing • Priority tree encoding
	• Subtree placement & encoding
	• Performance comparable with BB, Bear, Mosaico, etc.

CHAPTER 4
at a glance

- Steiner Problem in a Graph (SPG)
  - DNH genetic encoding
  - Results for B, C, D, and E graphs
  - Macro cell placement

- Macro cell global routing
  - Phase-I global router
  - Phase-II global router
  - Comparison with TimberWolfMC

Chapter 4

MACRO CELL ROUTING

H. Esbensen and P. Mazumder

After circuit components are placed, the interconnections between them satisfying the netlist must be routed to complete the layout synthesis process. The routing problem involves connecting the component terminals (pins) according to the circuit description such that the layout design rules are met. Additional objectives are to minimize the overall area of the layout and minimize the lengths of critical nets, not necessarily in that order. Since abstractions of the actual components and metal interconnection lines are used when solving a routing problem, only the component terminal locations are of concern, and the problem can be further simplified by assigning a grid to the routing surface. Grid lines are assumed to be spaced sufficiently far apart to meet the design rule requirements of the particular fabrication technology used. Following placement and routing, photolithography masks can be derived for use in chip fabrication.

After placement is completed, the routing surface is divided into smaller routing regions. Then a global router is used to assign each net to a few routing regions to accomplish the required interconnections. Finally, a detailed router is used to assign the nets to specific locations within the routing regions. Examples of detailed routers are channel routers, maze-type routers, and switchbox routers, and effective heuristics have been developed for such routers. Specialized routers are used for power and clock routing, and we will not consider them here. Instead, we will focus on the problem of global routing, and in particular, global routing for macro cells. Macro cells are large, irregularly sized parameterized circuit modules that are typically generated by a silicon compiler as per a designer's selected parameters. Global routers can be divided into four classes: (1) sequential (Steiner routing), (2) maze routing, (3) integer programming, and (4) stochastic techniques. Stochastic techniques include both simulated annealing and genetic algorithms and involve ripping up and rerouting various nets in an existing route. Before discussing the application of genetic algorithms to macro cell global routing, we address the Steiner problem in a graph, which is a subproblem of global routing.

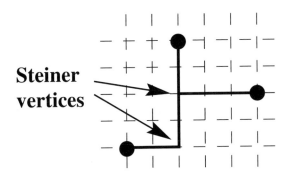

Figure 4.1. Rectilinear Steiner tree

4.1 The Steiner Problem in a Graph

The Steiner Problem in a Graph (SPG) is one of the classic problems of combinatorial optimization. Given a graph and a designated subset of the vertices, the task is to find a minimum cost subgraph spanning the designated vertices. The subgraph may also contain some graph vertices not in the designated subset. The SPG arises in a large variety of optimization problems such as network design, multiprocessor scheduling, and integrated circuit design [71].

Numerous algorithms of various kinds have been developed for the SPG. Exact algorithms can be found in [11], [14], [40], [60], [90], [140], [216]. However, since the SPG is NP-complete [111], these algorithms have exponential worst case time complexities. Therefore, a significant research effort has been directed toward polynomial time heuristics [11], [121], [172], [184], [222], [239]. Simulated annealing has also been applied to SPG [59].

The Rectilinear Steiner Problem (RSP) is an important special case of SPG [91], which is still NP-complete [74]. An example of RSP is shown in Fig. 4.1. The vertices of the graph are the grid points, and the designated subset of vertices, V', is indicated with filled circles. The Steiner tree contains V' and two additional grid points called Steiner vertices that lie at the intersections of the horizontal and vertical segments connecting the vertices in V'. The cost of the Steiner tree is equal to the sum of the lengths of the segments, i.e., 9 in this case. The general Steiner problem is not restricted to a rectilinear surface.

While at least two genetic algorithms for RSP have been published [94], [109], we are aware of only one previous GA for the SPG, developed by Kapsalis, Rayward-Smith, and Smith [110]. We present a GA for the SPG which differs significantly from the approach of Kapsalis et al. in a number of ways. While invalid solutions are allowed but penalized in [110], our approach is to enforce constraint satisfaction at all times, thereby eliminating the need for penalty terms in the cost function. Another major difference is our use of an inversion operator. The performance evaluation strategies also differ significantly. While the parameter settings used in [110] vary from problem to problem, a fixed set of parameter values has been used for all results reported here. From a practitioner's point of view, a stochastic algorithm is of limited use if it requires its parameters to be tuned every time a new problem instance is presented. Therefore, we consider a fixed parameter setting to be of major importance.

The algorithm that will be presented is tested on all SPG instances from the OR-Library [15]. This test suite consists of randomly generated graphs with up to 2500 vertices and 62,500 edges. The performance obtained is compared to that of the GA by Kapsalis et al. [110], an iterated version of the Shortest Path Heuristic called SPH-I, which is one of the very best deterministic heuristics [239], and two recent branch-and-cut algorithms by Lucena and Beasley [140] and Chopra, Gorres and Rao [40]. The experimental results show the following:

- The GA presented here clearly outperforms the GA in [110] with respect to solution quality and run time.

- The solution quality obtained by our GA is always at least as good as that obtained by SPH-I, and often the error ratio is an order of magnitude better. Depending on the problem, the two algorithms either require similar amounts of run time, or the GA is significantly faster.

- As opposed to the branch-and-cut algorithms, the GA is not guaranteed to find a global optimal solution. However, the experiments reveal that the GA does find the global optimum in more than 77% of all runs and is within 1% of the optimum in more than 92% of all runs. While the GA is capable of finding near-optimal solutions for *all* test examples in a moderate amount of time, the run time of the branch-and-cut algorithms varies widely and even prevents some of the largest problem instances from being solved.

4.1.1 Problem Definition

The graph terminology used is the same as in [7]. For a given graph $G = (V, E)$ and a subset $V' \subseteq V$, the *subgraph of G induced by V'* is a graph $G = (V', E')$ such that (1) $E' \subseteq E$, (2) $(v_i, v_j) \in E' \Rightarrow v_i, v_j \in V'$, and (3) $[\ v_i, v_j \in V' \wedge (v_i, v_j) \in E\ ] \Rightarrow (v_i, v_j) \in E'$. A graph is *complete* if it has an edge between every pair of vertices. The *distance graph* of G, denoted $D(G)$, is the complete graph having the same set V of vertices and in which the cost of each edge (v_i, v_j) equals the cost of the shortest path in G from v_i to v_j. For a given edge cost function $c : E \mapsto \Re$, the *cost of a graph G* is the sum of the costs of all edges of G and is denoted by $c(G)$. The problem considered can now be defined:

The Steiner Problem in a Graph (SPG): Given a connected, undirected graph $G = (V, E)$, a positive edge cost function $c : E \mapsto \Re_+$, and a subset $W \subseteq V$, compute a connected subgraph $G' = (V', E')$ of G, such that $W \subseteq V'$ and such that $c(G')$ is minimal.

4.1. The Steiner Problem in a Graph

Any acyclic subgraph G' of G such that $W \subseteq V'$ is called a *Steiner Tree for W in G*. A solution G' with minimal cost is called a *Minimal Steiner Tree (MStT) for W in G*. Vertices in the set $S \subseteq V \setminus W$ such that $V' = W \cup S$ are called the *Steiner vertices* of G'. Note the generality of this problem formulation. We do not require G to be planar, and we do not require c to satisfy the triangle inequality.

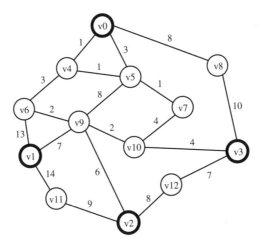

Figure 4.2. Example instance of the SPG. Highlighted vertices constitute W.

Throughout this section, let $n = |V|$, $m = |W|$ and $r = n - m$. If $m = 2$, SPG reduces to the shortest path problem, which can be solved by, e.g., Dijkstra's algorithm [137], in time $O(|E| \log n)$. If $m = n$, SPG is the Minimum Spanning Tree problem (MSpT), which can be solved in $O(n^2)$ time by, e.g., Prim's algorithm [7]. However, if $2 < m < n$, SPG is in general NP-complete [111].[1]

4.1.2 Description of the Algorithm

Fig. 4.3 shows a template for the GA considered here. Before the GA itself is executed, routine *graphReductions* tries to reduce the size of the given problem by applying some graph reduction techniques. Then the initial population is constructed from randomly generated individuals. Routine *evaluate* computes the fitness of each of the given individuals, while *bestOf* finds the individual with the highest fitness. One execution of the outer **While** loop corresponds to the simulation of one generation. Throughout the simulation, the number of individuals M is kept constant. We keep track of the best individual ever found. Routine *stopCriteria* terminates the simulation when no improvement of the best or average fitness has been observed for S consecutive generations, or when the algorithm has converged so that all individuals have the same fitness. In each generation, a set of offspring of size M is

---
[1]Some special graph topologies do exist for which SPG can still be solved in polynomial time [238].

evolved. The two mates p_1 and p_2 are selected from the population independently of each other, and each mate is selected with a probability proportional to its fitness. The *crossover* routine generates two offspring c_1 and c_2. Routine *reduce* returns the M fittest of the old and new populations, thereby keeping the population size constant. With a small probability, P_M, the *mutation* operator randomly changes each of the components, or *genes*, of each individual in the population, and with a small probability, P_I, inversion is performed on each individual. Routine *optimize* performs simple local hill climbing by executing a sequence of mutations on the best individual, each of which improves the fitness. An exhaustive strategy is used so that when the routine has been executed, no single mutation exists which can provide further improvements.

```
graphReductions();
generate initial Population randomly;
evaluate (Population);
bestIndividual = bestOf (Population);
While not stopCriteria() Do
    NewPopulation = ∅;
    For j = 1 to M/2 Do
        select p₁ and p₂ from Population;
        {c₁, c₂} = crossover (p₁, p₂);
        Add c₁ and c₂ to NewPopulation;
    evaluate (Population ∪ NewPopulation);
    Population = reduce (Population ∪ NewPopulation);
    For j = 1 to M Do
        mutate (Population[j], P_M);
        invert (Population[j], P_I);
    evaluate (Population);
    bestIndividual = bestOf (Population ∪ bestIndividual);
optimize (bestIndividual);
solution = bestIndividual;
```

Figure 4.3. Outline of the algorithm

Graph Reductions

Before the GA itself is executed, an attempt to reduce the size of the given problem is performed using standard graph reduction techniques. Routine *graphReductions* of Fig. 4.3 performs four kinds of simple reductions, all of which are described in [238], [239]. More elaborate reductions, as well as proofs of the correctness of the reductions used here, can be found in [61]. Let e_{vw} denote the edge between vertices

4.1. The Steiner Problem in a Graph

v and w, and let $\text{sp}(v,w) \subseteq E$ denote the shortest path between v and w. The four reductions used are

1. Assume $\text{degree}(v) = 1$ and $e_{vw} \in E$. If $v \in W$, any MStT must include e_{vw}. Hence, v and e_{vw} are removed from G, and w is added to V' if it is not already there. If $v \in V \setminus W$, no MStT can include e_{vw}; i.e., v and e_{vw} can also be deleted.

2. If $v \in V \setminus W$, $\text{degree}(v) = 2$, and $e_{uv}, e_{vw} \in E$, then v, e_{uv}, and e_{vw} can be deleted from G and replaced by a new edge between u and w of equivalent cost. More specifically, if $e_{uw} \notin E$, then $E = E \cup \{e_{uw}\} \setminus \{e_{uv}, e_{vw}\}$ and $c(e_{uw}) = c(e_{uv}) + c(e_{vw})$, where $c(e_{uw})$ is the cost of e_{uw}. If an edge from u to w exists, i.e., $e_{uw} \in E$, then $c(e_{uw}) = \min\{c(e_{uw}), c(e_{uv}) + c(e_{vw})\}$.

3. If $e_{vw} \in E$ and $c(e_{vw}) > c(\text{sp}(v,w))$, then no MStT can include e_{vw}, which therefore can be deleted.

4. Assume that $v \in V'$, and denote the closest neighbor to v as $u \in V$ and the second-closest neighbor as $w \in V$. Since G is connected, u always exists. If w does not exist, assume $c(e_{vw}) = \infty$. Let z be a vertex in $V' \setminus \{v\}$ which is closest to u. If $c(e_{vu}) + c(\text{sp}(u,z)) \leq c(e_{vw})$, then any MStT must include e_{vu}. Therefore, G can be contracted along this edge. Note that $u \in V' \Rightarrow z = u \Rightarrow c(\text{sp}(u,z)) = 0$; i.e., contraction can always be performed in this case.

To obtain the largest possible overall reduction of G, the above reductions are performed repeatedly as described below. Knowledge of the cost of a shortest path is required whenever a reduction of Type 3 or 4 is performed. Shortest paths are also repeatedly needed by the GA. Therefore, the distance graph $D(G)$ is computed initially using Floyd's algorithm [7], which requires time $O(n^3)$. Whenever one of the above reductions is performed, $D(G)$ has to be dynamically updated. When $D(G)$ is represented as an adjacency matrix, the update is trivial for reductions of Type 1 or 2: It simply consists of deleting the row and column corresponding to the deleted vertex. Reductions of Type 3 leave $D(G)$ unchanged. However, for reductions of Type 4, the update is slightly more involved. Whenever a contraction is performed, $D(G)$ is updated using an $O(n^2)$ algorithm by Dionne and Florian [57].

In [238], [239], the following reduction is also suggested along with the reductions described above: If $\max\{c(\text{sp}(v,u)), c(\text{sp}(v,w))\} < c(e_{uw})$, $e_{uw} \in E$, and $v \in W$, then no MStT can include e_{uw}, which therefore can be deleted. However, in this case, the required update of $D(G)$ has a worst case complexity of $O(n^3)$ using Dionne and Florian's algorithm [57]. The update, therefore, could be as expensive as recomputing the entire distance graph, and for this reason, this reduction is omitted.

When a sequence of reductions of the same type is performed, the overall result depends on the chosen traversal of the graph, i.e., the order in which reductions are tried out. Furthermore, reductions of distinct types are mutually dependent in the sense that performing all possible reductions of some type may allow new subsequent reductions of another type. It is not clear in which order reductions should be performed to obtain the overall best reduction of a given graph [239]. The arbitrarily chosen scheme for performing reductions in routine *graphReductions* is shown in Fig. 4.4. Routine *reductions* performs a single traversal of all vertices (or edges in the case of Type 3 reductions) of G in an unspecified order and carries out a reduction of a given type whenever possible. Routine *graphReductions* terminates when no reduction of any type succeeds for a complete iteration, i.e., when no reduction can reduce G further.

>
> compute $D(G)$;
> **While** improvements made **Do**
>   reductions(3);
>   reductions(2);
>   reductions(4);
>   reductions(1);

Figure 4.4. Outline of routine graphReductions.

To deduce the worst case time complexity of *graphReductions*, we start by considering the maximum total time spent on reductions of Type 4. Due to the required update of $D(G)$, a single reduction requires time $O(n^2)$. Since vertices can be added to V' when reductions of Type 1 are performed, $O(n)$ Type 4 reductions are possible. Hence, the total time spent on Type 4 reductions is $O(n^3)$. One execution of *reductions* requires at most time $O(n^2)$ when either $x \neq 4$ or $x = 4$ but no contraction is performed. Each of the reductions 1, 2, and 4 decreases the number of vertices in G by one, and Type 3 reductions are performed exhaustively in the sense that after executing *reductions(3)*, no edge exists which can be removed by a Type 3 reduction. Therefore, at least one vertex must be removed in every second iteration of the **While** loop in *graphReductions*. Hence, there can be no more than $O(n)$ iterations. In total, this gives routine *graphReductions* the time complexity $O(n^3)$.

Although it is not difficult to construct a graph for which none of the reductions performed by *graphReductions* applies, the routine has been observed to be very effective on many graphs, as will be seen in the experimental results. When applied to the graph of Fig. 4.2, the result is the degenerate graph consisting of one vertex only, implying that a MStT has been found. In general, reductions of Type 4 in particular have been observed to be very powerful when m, the number of vertices in W, is relatively large, which coincides with the results reported in [239].

4.1. The Steiner Problem in a Graph

Distance Network Heuristic (DNH)

The genetic encoding developed here is based on use of the Distance Network Heuristic (DNH), a deterministic heuristic for the SPG developed by Kou et al. [121]. Before proceeding with the genotype and decoder, we describe the DNH. Given a graph $G = (V, E)$, a cost function c, and a subset of vertices W in accordance with the definition of SPG in Section 4.1.1, the DNH computes an approximation T_{DNH} to the MStT for W in G in five steps:

1. Construct the subgraph G_1 of $D(G)$ induced by W.

2. Compute a MSpT T_1 of G_1.

3. Construct from T_1 the subgraph G_2 of G by substituting each edge in T_1 with the corresponding shortest path in G.

4. Compute a MSpT T_2 of G_2.

5. Compute T_{DNH} from T_2 by repeatedly deleting all vertices $v \in V \setminus W$ having degree equal to 1.

Any ties in Steps 2, 3, or 4 are broken arbitrarily. An example of how the DNH works is shown in Fig. 4.5, given as input the graph G of Fig. 4.2 and the subset $W = \{v_0, v_1, v_2, v_3\}$.

If $D(G)$ is not known, Step 1 of DNH requires time $O(mn^2)$ to compute shortest paths from each of the m vertices. Since G_1 is complete, the MSpT in Step 2 is computed using Prim's algorithm, requiring time $O(m^2)$. Each of the $m-1$ edges of T_1 may correspond to a path in G of up to $n-1$ edges. Hence, Step 3 requires time $O(mn)$, and Step 4 requires time $O(mn \log(nm))$ using Kruskal's algorithm [7]. The final step is done in time $O(n)$. Hence, if $D(G)$ is not known, Step 1 is the most expensive and gives the DNH a time complexity of $O(mn^2)$.

Genotype and Decoder

The basic idea of the genotype and the associated decoder is the following: The genotype specifies a set of selected Steiner vertices. The decoder computes the corresponding phenotype by executing the DNH using the union of the selected Steiner vertices and W as the set of vertices to be spanned. The selected Steiner vertices are specified by a bit string in which each bit corresponds to a specific vertex. If the bit is set, the vertex is selected. To allow inversion, we need the genotype to be independent of the ordering of the bits in the string. This is obtained by associating with each bit a tag which identifies the vertex specified by that bit.

Specifically, the genotype and decoder can be described as follows. For a given instance of SPG, assume a fixed numbering $0, 1, \ldots, r-1$ of the vertices in $V \setminus W$. Let $\pi : \{0, 1, \ldots, r-1\} \mapsto \{0, 1, \ldots, r-1\}$ be a bijective mapping. A genotype is then a set of r tuples:

$$\{(\pi(0), i_{\pi(0)}), (\pi(1), i_{\pi(1)}), \ldots, (\pi(r-1), i_{\pi(r-1)})\}$$

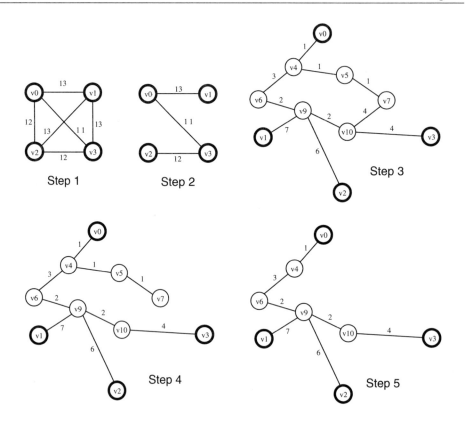

Figure 4.5. Steps of the DNH

where $i_k \in \{0, 1\}$, $k = 0, 1, \ldots, r-1$. The Steiner vertices $S \subseteq V \setminus W$ specified by the genotype are the set $S = \{v_k \in V \mid i_k = 1\}$. The Steiner tree in G corresponding to the genotype is the tree computed by DNH using the set $S \cup W$ as the vertices to be connected. In Step 5 of DNH, every vertex $v \notin W$ of degree 1 is deleted.

Any set of values of the i_k's in a genotype correspond to a valid phenotype. However, Lawler [133] has shown that a MStT in $D(G)$ exists which has at most $m - 2$ Steiner vertices. This result relies on the fact that regardless of the edge cost function c, the edge costs in $D(G)$ always satisfy the triangle inequality. Hence, it is sufficient to consider only the subset of genotypes which satisfies $|S| \leq \min(m-2, r)$. To take advantage of this reduction of the search space, a routine *filter* has been defined which, given any genotype g, enforces the satisfaction of $|S| \leq \min(m-2, r)$ by randomly selecting and clearing the necessary number of set bits.

When the initial random population has been generated, the filter is applied to each of the individuals. From then on, the search is limited to the restricted region by applying the filter to every new individual generated by one of the genetic operators.

It is important to note that the DNH is *not* chosen for use as the decoder because it is an especially good heuristic in terms of result quality. In [239], the performance of DNH is compared to that of two other well-known polynomial time heuristics for the SPG: the Shortest Path Heuristic (SPH) by Takahashi and Matsuyama [222] and the Average Distance Heuristic (ADH) by Rayward-Smith and Clare [184]. With respect to result quality, the DNH is clearly outperformed by both these heuristics. The reasons to use DNH for decoding are first, that it provides a way to interpret *any* set of selected vertices as a *valid* Steiner tree, and secondly, that it is relatively fast. The important advantage of considering valid Steiner trees only is that it eliminates the need for penalty terms in the cost measure and thus avoids potential problems of assigning a suitable cost value to an invalid or incomplete solution.

Fitness Measure

Given a population $P = \{p_0, p_1, \ldots, p_{M-1}\}$, the routine *evaluate* of Fig. 4.3 computes the fitness of each individual as follows. Let $C(p)$ be the cost of individual p, i.e., the cost of the Steiner tree represented by p, and assume that P is sorted so that $C(p_0) \geq C(p_1) \geq \ldots \geq C(p_{M-1})$. The fitness F of p_i is then computed as

$$F(p_i) = \frac{2i}{M-1}; \qquad i = 0, 1, \ldots, M-1.$$

This fitness computation scheme is called *ranking* and is discussed in [236]. Controlling the variance of the fitness values is one of the frequent problems of GAs [79]. Ranking assures that the variance is constant throughout the optimization process. The specific scheme chosen here constantly gives the best individual twice the probability of being selected for crossover as the median individual.

Crossover, Mutation, and Inversion

Given two individuals, the crossover operator generates two offspring in three steps. First, a copy of one of the parents is reordered so that it becomes homologous to the other parent. Then standard one-point crossover is performed, and finally, the offspring produced are subjected to the filter routine if necessary.

The mutation operator performs standard pointwise mutation; i.e., each bit in a given genotype is inverted with a given small probability P_M. If necessary, the genotype is then passed through the filter routine.

For the inversion operator to obtain a uniform probability of movement of all tuples, we consider the genotype to form a ring. A part of the ring is then selected at random and reversed. More specifically, two points $x, y \in \{0, 1, \ldots, r-1\}, x \neq y$, are selected at random. The operator then defines the new ordering π' of g as[2]

$$\pi'((x+i) \bmod r) = \begin{cases} \pi((y-i) \bmod r) & \text{if } 0 \leq i \leq (y-x) \bmod r \\ \pi((x+i) \bmod r) & \text{otherwise} \end{cases}$$

for all $i = 0, 1, \ldots, r-1$. The inversion operator is illustrated in Fig. 4.6.

Before inversion:

$\alpha : \{(2,0), (3,1), (0,1), (4,0), (1,0)\}$

After inversion with $x = 2$, $y = 0$:

$\alpha : \{(0,1), (3,1), (2,0), (1,0), (4,0)\}$

Figure 4.6. Illustration of the inversion operator with $r = 5$

Time Complexity

The filter routine, generation of each of the initial individuals, and the genetic operators *crossover*, *mutate*, and *invert* each require time $O(r) = O(n - m)$. The repeated decoding using DNH is the most expensive operation. Since knowledge of shortest paths is also required when performing some of the initial graph reductions, $D(G)$ is precomputed. This reduces the time of Step 1 of DNH to $O(1)$, and as a consequence, one decoding can now be performed in time $O(mn \log(nm))$. Fitness computation requires $O(M \log M)$ to sort the M individuals. In total, the GAs setup time is $O(n^3)$, and each generation requires time $O(M[nm \log(nm) + \log M])$.

Measurements reveal that the vast majority of the total run time is spent on decoding. It also turns out that in practice, the graph formed in Step 3 of the decoding process is almost always a tree, and as a consequence, Step 4 is rarely executed. Therefore, the true bottleneck of the algorithm is the MSpT computation performed in Step 2 of the decoding, which requires time $O(m^2)$.

4.1.3 Experimental Method

The algorithm was tested on all 78 SPG instances from the OR-Library [15]. These graphs are divided into four classes, denoted by B, C, D, and E, according to their sizes. All graphs were generated at random subject only to the connectivity constraint; i.e., the topology is random and the vertices to be spanned were selected

---

[2] The definition of π' relies on the mathematical definition of modulo, in which the remainder is always nonnegative.

4.1. The Steiner Problem in a Graph

at random. Every edge cost is a random integer in the interval [1,10]. In Class B, each graph has n equal to 50, 75, or 100. The value of m is either $n/6$, $n/4$, or $n/2$, and the average vertex degree is either 2.5 or 4. Since all combinations exist, Class B consists of 18 graphs. Classes C, D, and E consist of graphs with n equal to 500, 1,000, and 2,500, respectively; m equals 5, 10, $n/6$, $n/4$, or $n/2$, and the average vertex degree is 2.5, 4, 10, or 50. Thus, each of the Classes C, D, and E consists of 20 graphs.

One of the main advantages of using this test suite is that it facilitates comparison with the global optimal solutions. The global optima were first computed by J. E. Beasley, who developed a branch-and-cut algorithm which was executed on a Cray X-MP/48 supercomputer [14].

The iterated shortest path heuristic is also shown for comparison. As mentioned previously, a comparative study of the deterministic heuristics SPH, DNH, and ADH has been made by Winter and Smith [239]. Several variants of these heuristics, especially a number of repetitive variants of SPH, were also considered in the study. The ADH is in general considered to be one of the best deterministic heuristics, which is also confirmed by the investigation in [239]. However, the results also reveal that some of the repetitive variants of SPH consistently outperform ADH with respect to result quality. Furthermore, by applying initial graph reductions, the run time of the repetitive SPH variants can be made comparable to that of the other heuristics. One of the specific conclusions in [239] is that on the largest random graphs considered, the repetitive SPH variant, denoted SPH-ZZ, outperforms all other heuristics. Therefore, this heuristic has been chosen for comparison with the GA.

Fig. 4.7 outlines our implementation of SPH-ZZ, denoted by SPH-I. It starts by computing $D(G)$ and performing graph reductions. For given vertices x and y, $G_{xy} = (V_{xy}, E_{xy})$ denotes the subgraph of G corresponding to the shortest path between x and y. In each iteration of the outer loop, a tree T is built which spans all vertices in W. T is initialized with a shortest path between two of the vertices to be spanned, and T is then extended by repeated addition of a shortest path to a closest, not-yet-connected vertex. This scheme is tried for all possible initializations of T, and the algorithm outputs the best such tree obtained.

Routine *graphReductions* requires time $O(n^3)$. Construction of each candidate solution T takes time $O(m^2n)$, since the **While** loop is iterated $O(m)$ times, and it takes time $O(mn)$ to find each z vertex and extend T with a shortest path to it. This is due to the fact that all distances have been precomputed. Since $O(m^2)$ candidate solution trees T are computed, the total run time of SPH-I is $O(n^3 + m^4n)$.

For a given graph, the size of the search space $\mathcal{S}(n,m)$ to be explored by the GA is

$$\mathcal{S}(n,m) = \sum_{i=0}^{k} \binom{n-m}{i}$$

where $k = \min(m-2, n-m)$, since this is the number of possible distinct choices of the Steiner vertices. Some of the problem instances considered represent extremely

```
graphReductions();
$c(T_{SPH-I}) = \infty$;
$\forall\ x, y \in W, x \neq y$ do
    $T = G_{xy}$;
    $Q = W \cap V_{xy}$;
    **While** $W \setminus Q \neq \emptyset$ **Do**
        find a vertex $z \in W \setminus Q$ closest to a vertex in $T$;
        add to T a shortest path from $T$ to $z$;
        $Q = Q \cup \{z\}$;
    **If** $c(T) < c(T_{SPH-I})$ **then** $T_{SPH-I} = T$;
output $T_{SPH-I}$;
```

Figure 4.7. Outline of SPH-I

large search spaces. However, the corresponding phenotype spaces are smaller.

The GA was evaluated using four kinds of comparisons:

- The solution quality obtained was compared to the global optimum.

- The absolute run time was compared to that of two distinct branch-and-cut algorithms by Lucena and Beasley [140] and Chopra, Gorres, and Rao [40].

- Solution quality and absolute run time were compared to those of SPH-I.

- Solution quality and run time were compared to the available results for the GA by Kapsalis et al. [110].

The branch-and-cut algorithms are guaranteed to find the global optimum. However, run time may be unacceptable for some problem instances or may even prevent some problems from being solved. It is therefore of interest to investigate whether a near-optimal solution can be found for *all* problems by using a moderate amount of time.

The GA was executed 10 times for each example in the B, C, and D classes. Solution quality was then evaluated in terms of best, average, and worst results produced. However, due to run-time requirements, the GA was only executed once for each of the examples in Class E. The parameter settings were $M = 40$, $S = 50$, $P_M = 0.005$, and $P_I = 0.1$. These values were used for all executions; i.e., no problem-specific tuning has been made.

The GA as well as SPH-I were implemented in the C programming language. For both algorithms, examples from Classes B, C, and D were executed on a Sun Sparc IPX workstation having 32-Mb RAM. These examples require at most 10 Mb of memory. For the Class E examples, the memory requirement is about

4.1. The Steiner Problem in a Graph

58 Mb. Therefore, for these examples, the GA and SPH-I were executed on a DEC MIPS 5000-240 workstation having 128-Mb RAM.

The branch-and-cut algorithm by Lucena and Beasley [140] is a further development of the algorithm presented in [14], but instead of using a Cray, it was executed on a Sun Sparc 2 workstation. This machine is roughly as fast as the Sun Sparc IPX but probably somewhat slower than the DEC MIPS 5000-240. Chopra's algorithm [40] was executed on a VAX 8700, which is at most as fast as the other machines. When comparing absolute run times, the reader should keep these differences regarding the hardware used in mind. However, the run-time variations caused by the different machines are insignificant compared to the variations caused by different problem instances when considering a specific algorithm.

4.1.4 Results

In the following subsections, the detailed experimental results for all four problem classes are discussed.

The B Graphs

Table 4.1 lists the characteristics of the problems in Class B before and after the graph reductions are performed. The reductions significantly impact all graphs. In particular, graphs B-1, B-3, and B-9 are reduced to the degenerate graph consisting of a single vertex only, which means that the optimal solution is found solely by performing graph reductions.

Table 4.2 compares the solution quality obtained by the GA to the globally optimal solutions as well as to the solutions found by SPH-I. C_{opt} is the global optimum, and C_{sph} is the cost of the solution found by SPH-I. C_{best}, C_{avg}, and C_{worst} are for the best, average, and worst results produced by the GA in the 10 runs, while C_σ denotes the standard deviation of the 10 cost values. $\Delta C_{sph} = 100(C_{sph}/C_{opt} - 1)$ is the relative error in percent of the solution found by SPH-I compared to the optimum solution. Similarly, $\Delta C_{avg} = 100(C_{avg}/C_{opt} - 1)$ denotes the average error of the solutions found by the GA, and $\Delta C_{worst} = 100(C_{worst}/C_{opt} - 1)$ is the worst error produced by the GA. Finally, N_{ga} denotes the number of the 10 runs which did not find the global optimum. This notation is also used in the following sections. As can be seen, the GA finds the global optimum for all examples in every execution. SPH-I performs similarly for all graphs except B-13, for which it has a 1.82% overhead.

Table 4.3 compares the run time of the GA with that of SPH-I and the branch-and-cut algorithm by Lucena and Beasley [140]. T_{bc2} denotes the run time of the latter algorithm, and T_{sph} is the time of SPH-I. The average time spent by the GA is denoted T_{avg}, while T_σ denotes the standard deviation of the time for the 10 runs. Chopra et al. [40] give no computational results for these graphs. It can be seen that all run times are very small, and within the accuracy of these measurements, it is difficult to draw any conclusions regarding differences in speed for the different algorithms.

Table 4.1. Characteristics of Class B Graphs before and after Reductions

Graph	Problem size			Reduced size		
	n	m	$|E|$	n	m	$|E|$
B-1	50	9	63	1	1	0
B-2	50	13	63	7	4	12
B-3	50	25	63	1	1	0
B-4	50	9	100	34	7	72
B-5	50	13	100	35	10	76
B-6	50	25	100	25	10	60
B-7	75	13	94	16	6	26
B-8	75	19	94	16	7	25
B-9	75	38	94	1	1	0
B-10	75	13	150	50	10	115
B-11	75	19	150	47	8	108
B-12	75	38	150	31	11	74
B-13	100	17	125	28	9	47
B-14	100	25	125	22	8	42
B-15	100	50	125	16	9	28
B-16	100	17	200	63	9	148
B-17	100	25	200	51	12	113
B-18	100	50	200	35	12	84

Table 4.2. Comparison of Solution Quality for Graphs in Class B

Graph	C_{opt}	C_{sph}	ΔC_{sph}	C_{best}	C_{avg}	C_{worst}	C_σ	ΔC_{avg}	ΔC_{worst}	N_{ga}
B-1	82	82	0	82	82	82	0	0	0	0
B-2	83	83	0	83	83	83	0	0	0	0
B-3	138	138	0	138	138	138	0	0	0	0
B-4	59	59	0	59	59	59	0	0	0	0
B-5	61	61	0	61	61	61	0	0	0	0
B-6	122	122	0	122	122	122	0	0	0	0
B-7	111	111	0	111	111	111	0	0	0	0
B-8	104	104	0	104	104	104	0	0	0	0
B-9	220	220	0	220	220	220	0	0	0	0
B-10	86	86	0	86	86	86	0	0	0	0
B-11	88	88	0	88	88	88	0	0	0	0
B-12	174	174	0	174	174	174	0	0	0	0
B-13	165	168	1.82	165	165	165	0	0	0	0
B-14	235	235	0	235	235	235	0	0	0	0
B-15	318	318	0	318	318	318	0	0	0	0
B-16	127	127	0	127	127	127	0	0	0	0
B-17	131	131	0	131	131	131	0	0	0	0
B-18	218	218	0	218	218	218	0	0	0	0

Table 4.3. Comparison of CPU Time (in seconds) for Graphs in Class B

Graph	T_{bc2}	T_{sph}	T_{avg}	T_σ
B-1	0.1	0.1	0.1	0.0
B-2	0.1	0.1	0.2	0.0
B-3	0.1	0.1	0.1	0.0
B-4	0.6	0.1	1.2	0.6
B-5	1.9	0.1	0.7	0.2
B-6	0.6	0.1	0.7	0.1
B-7	0.2	0.2	0.5	0.1
B-8	0.1	0.2	0.5	0.1
B-9	0.1	0.2	0.2	0.0
B-10	3.1	0.3	1.7	0.5
B-11	1.4	0.3	1.4	0.6
B-12	0.6	0.3	0.6	0.1
B-13	0.7	0.4	1.4	0.4
B-14	1.2	0.5	0.9	0.3
B-15	0.3	0.5	0.8	0.1
B-16	18.4	0.6	4.4	1.9
B-17	3.3	0.6	2.3	0.6
B-18	1.0	0.6	1.5	0.3

The fact that all three algorithms find optimal solutions (except for SPH-I on B-13) in a very short time suggests that these examples are simply too small to facilitate any distinction of performance of the algorithms. For several of the graphs, the search spaces after graph reductions are indeed very small, and the largest search space is that of B-17 with less than 10^9 points, which is not large for a combinatorial optimization problem.

The C Graphs

From Table 4.4, it can be seen that the graph reductions are also very effective on most graphs in the C class. Note especially graph C-5, which after reductions has a search space size of only approximately 10^6 points. However, as the average vertex degree increases, the effect of reductions of Types 1 and 2 decreases significantly. When m is small, the effect of reductions of Type 4 is also very limited, as can be seen by the results for C-11, C-12, C-16, and C-17. The reductions in search space sizes obtained for these problems are negligible. The effect of reductions of Type 3 increases with the number of edges. For C-16 through C-20, about two-thirds of all edges are eliminated by graph reductions, mainly of Type 3. However, since the GA operates in terms of shortest paths, minimum spanning trees, etc., the number of edges is not very important for the performance of the algorithm.

Table 4.4. Characteristics of Class C Graphs before and after Reductions

Graph	Problem size			Reduced size		
	n	m	$|E|$	n	m	$|E|$
C-1	500	5	625	145	5	263
C-2	500	10	625	130	8	239
C-3	500	83	625	120	35	232
C-4	500	125	625	109	38	221
C-5	500	250	625	37	17	91
C-6	500	5	1,000	369	5	847
C-7	500	10	1,000	382	9	869
C-8	500	83	1,000	336	54	818
C-9	500	125	1,000	349	78	832
C-10	500	250	1,000	213	76	624
C-11	500	5	2,500	499	5	2,184
C-12	500	10	2,500	498	9	2,236
C-13	500	83	2,500	463	65	2,108
C-14	500	125	2,500	427	81	1,961
C-15	500	250	2,500	299	92	1,471
C-16	500	5	12,500	500	5	4,740
C-17	500	10	12,500	499	9	4,698
C-18	500	83	12,500	486	70	4,668
C-19	500	125	12,500	473	98	4,490
C-20	500	250	12,500	386	143	3,850

Table 4.5 shows that the GA finds the global optimum at least once for all examples and every time for 12 of the graphs, while SPH-I finds the optimum for 10 of the graphs. When neither the average GA run nor SPH-I finds the global optimum, ΔC_{avg} is often an order of magnitude better than ΔC_{sph}. This is the case for C-3, C-4, C-9, C-14, C-18, and C-19. For C-18 and C-19, the solutions produced by SPH-I are very poor, with errors in the 6–7% range. The results for C-16 are in direct contrast to all other results. While SPH-I finds the optimum, the GA encounters severe problems. In 7 of 10 runs, it misses the global optimum value of 11 and outputs a tree of cost 12. This corresponds to a huge relative error ΔC_{worst} of 9.09%.

In Table 4.6 and subsequent tables, T_{bc1} denotes the run time of the branch-and-cut algorithm by Chopra et al. [40]. Depending on the problem, the run time for both branch-and-cut algorithms varies significantly. Chopra's algorithm spans from 10 seconds for C-16 to more than 45,000 seconds for C-18, while Lucena's algorithm varies from 5 seconds for C-5 to more than 20,000 seconds for C-18. As a consequence, the branch-and-cut algorithms are significantly faster than both the GA and SPH-I for some graphs and significantly slower for others. The run times

4.1. The Steiner Problem in a Graph

Table 4.5. Comparison of Solution Quality for Graphs in Class C

Graph	C_{opt}	C_{sph}	ΔC_{sph}	C_{best}	C_{avg}	C_{worst}	C_σ	ΔC_{avg}	ΔC_{worst}	N_{ga}
C-1	85	85	0	85	85	85	0	0	0	0
C-2	144	144	0	144	144	144	0	0	0	0
C-3	754	757	0.40	754	754.2	755	0.4	0.03	0.13	2
C-4	1,079	1,081	0.19	1,079	1,079.1	1,080	0.3	0.01	0.09	1
C-5	1,579	1,579	0	1,579	1,579	1,579	0	0	0	0
C-6	55	55	0	55	55	55	0	0	0	0
C-7	102	102	0	102	102	102	0	0	0	0
C-8	509	512	0.59	509	509	509	0	0	0	0
C-9	707	714	0.99	707	707.4	708	0.5	0.06	0.14	4
C-10	1,093	1,098	0.46	1,093	1,093	1,093	0	0	0	0
C-11	32	32	0	32	32	32	0	0	0	0
C-12	46	46	0	46	46	46	0	0	0	0
C-13	258	263	1.94	258	259.7	260	0.6	0.66	0.78	9
C-14	323	327	1.24	323	323.4	324	0.5	0.12	0.31	4
C-15	556	558	0.36	556	556	556	0	0	0	0
C-16	11	11	0	11	11.7	12	0.5	6.36	9.09	7
C-17	18	18	0	18	18	18	0	0	0	0
C-18	113	121	7.08	113	114.3	115	0.8	1.15	1.77	8
C-19	146	155	6.16	146	147	148	0.4	0.68	1.37	9
C-20	267	267	0	267	267	267	0	0	0	0

Table 4.6. Comparison of CPU Time (in seconds) for Graphs in Class C

Graph	T_{bc1}	T_{bc2}	T_{sph}	T_{avg}	T_σ	Graph	T_{bc1}	T_{bc2}	T_{sph}	T_{avg}	T_σ
C-1	27	25	61	79	6	C-11	333	2769	119	187	20
C-2	812	45	61	79	3	C-12	120	1175	119	224	19
C-3	543	25	72	104	19	C-13	9170	9895	646	544	91
C-4	510	23	75	83	10	C-14	212	1150	1316	547	130
C-5	474	5	61	63	0	C-15	211	913	1544	262	56
C-6	49	561	83	130	11	C-16	10	877	119	180	22
C-7	83	522	86	153	24	C-17	98	14,557	119	203	26
C-8	674	1106	260	263	39	C-18	45,848	20,726	873	563	102
C-9	1866	5813	966	425	93	C-19	117	1689	3050	601	136
C-10	246	32	544	181	49	C-20	15	225	11,374	334	57

of the GA and SPH-I are similar for most graphs, although the GA is significantly faster for graphs C-15, C-19, and C-20. The time variation T_σ of the GA is relatively small.

The D Graphs

Graph reductions on the Class D graphs show a pattern similar to that observed for the Class C graphs, although now the pattern is even clearer. Most graphs are reduced significantly, especially D-5, as is apparent in Table 4.7. The effect of reductions decreases as m decreases and as the average vertex degree increases.

Table 4.7. Characteristics of Class D Graphs before and after Reductions

Graph	Problem size			Reduced size		
	n	m	$|E|$	n	m	$|E|$
D-1	1,000	5	1,250	274	5	510
D-2	1,000	10	1,250	285	10	523
D-3	1,000	167	1,250	224	67	441
D-4	1,000	250	1,250	159	66	339
D-5	1,000	500	1,250	97	48	246
D-6	1,000	5	2,000	761	5	1,741
D-7	1,000	10	2,000	754	10	1,735
D-8	1,000	167	2,000	731	124	1,708
D-9	1,000	250	2,000	654	155	1,613
D-10	1,000	500	2,000	418	146	1,317
D-11	1,000	5	5,000	993	5	4,674
D-12	1,000	10	5,000	1,000	10	4,671
D-13	1,000	167	5,000	922	122	4,433
D-14	1,000	250	5,000	853	160	4,173
D-15	1,000	500	5,000	550	157	2,925
D-16	1,000	5	25,000	1,000	5	10,595
D-17	1,000	10	25,000	999	9	10,531
D-18	1,000	167	25,000	978	145	10,140
D-19	1,000	250	25,000	938	193	9,676
D-20	1,000	500	25,000	814	324	8,907

Statistics on solution quality are given in Table 4.8 for the Class D graphs. SPH-I finds the optimum for 7 of the graphs, while the GA finds the optimum at least once for 17 graphs and every time for 13 graphs. SPH-I has relative errors exceeding 2% for 5 graphs, while that only happens for the GA on the graph labeled D-18. For all graphs, we have $C_{worst} \leq C_{sph}$, and $C_{worst} < C_{sph}$ holds for 13 graphs.

CPU times for graphs in Class D are shown in Table 4.9. On this class of problems, the run times for both branch-and-cut algorithms vary by three orders of magnitude and are as high as 200,000 seconds, corresponding to two to three days of computation. The run time of SPH-I also varies significantly. For practical reasons, it became necessary to introduce a CPU-time limit of 50,000 seconds for this algorithm on graphs from Classes D and E. When SPH-I did not complete its computation within this limit, it was terminated, and the best solution found so far

Table 4.8. Comparison of Solution Quality for Graphs in Class D

Graph	C_{opt}	C_{sph}	ΔC_{sph}	C_{best}	C_{avg}	C_{worst}	C_σ	ΔC_{avg}	ΔC_{worst}	N_{ga}
D-1	106	106	0	106	106	106	0	0	0	0
D-2	220	220	0	220	220	220	0	0	0	0
D-3	1,565	1,570	0.32	1,565	1,565	1,565	0	0	0	0
D-4	1,935	1,940	0.26	1,935	1,935	1,935	0	0	0	0
D-5	3,250	3,254	0.12	3,250	3,250	3,250	0	0	0	0
D-6	67	71	5.97	67	67.1	68	0.3	0.15	1.49	1
D-7	103	103	0	103	103	103	0	0	0	0
D-8	1,072	1,095	2.15	1,072	1,072.7	1,074	0.6	0.07	0.19	6
D-9	1,448	1,471	1.59	1,448	1,448.4	1,450	0.7	0.03	0.14	3
D-10	2,110	2,120	0.47	2,110	2,110	2,110	0	0	0	0
D-11	29	29	0	29	29	29	0	0	0	0
D-12	42	42	0	42	42	42	0	0	0	0
D-13	500	514	2.80	500	500.6	502	0.7	0.12	0.40	5
D-14	667	675	1.20	668	669.7	671	0.9	0.40	0.60	10
D-15	1,116	1,121	0.45	1,116	1,116	1,116	0	0	0	0
D-16	13	13	0	13	13	13	0	0	0	0
D-17	23	23	0	23	23	23	0	0	0	0
D-18	223	239	7.17	226	227.7	230	1.2	2.11	3.14	10
D-19	310	335	8.06	312	313.3	315	0.9	1.06	1.61	10
D-20	537	539	0.37	537	537	537	0	0	0	0

Table 4.9. Comparison of CPU Time (in seconds) for Graphs in Class D

Graph	T_{bc1}	T_{bc2}	T_{sph}	T_{avg}	T_σ	Graph	T_{bc1}	T_{bc2}	T_{sph}	T_{avg}	T_σ
D-1	476	200	486	523	8	D-11	1374	24,609	949	1070	59
D-2	284	148	488	537	13	D-12	305	5843	961	1085	20
D-3	2290	106	785	650	39	D-13	1864	91,718	15,187	2357	245
D-4	3529	41	689	554	21	D-14	3538	61,335	41,237	2601	393
D-5	811	37	522	504	8	D-15	1410	16,889	24,828	1302	102
D-6	2340	4148	687	788	44	D-16	871	9721	956	1047	21
D-7	100	1037	681	795	29	D-17	6965	147,598	950	1068	26
D-8	6985	17,858	13,237	2101	381	D-18	245,192	227,841	31,015	2536	491
D-9	4630	16,458	29,354	2744	624	D-19	878	304,380	50,003	3441	580
D-10	1312	1678	14,780	1100	169	D-20	47	1276	50,010	2638	658

was used. This happened for both graphs, D-19 and D-20. For these graphs, the total times needed by SPH-I are estimated to be 95,000 seconds and 679,000 seconds, respectively. These estimates can be considered to be quite accurate since they are based on measurements of the CPU time spent for each pair of vertices $x, y \in V$, cf., Fig. 4.7, which is then scaled with the relative number of vertex pairs not yet considered at the time the CPU limit is exceeded. The average run time of the GA varies from 504 seconds for D-5 to 3441 seconds for D-19, i.e., by a factor of 7. This variation is small compared to the variation of the other algorithms considered. For graphs labeled D-8, D-9, D-10, D-13, D-14, D-15, D-18, D-19, and D-20, the GA is on average an order of magnitude faster than SPH-I, while for the remaining graphs, the run times of these algorithms are comparable.

The E Graphs

For the graphs from Class E, the effect of graph reductions follows a pattern which coincides perfectly with the patterns observed for Classes C and D, as illustrated in Table 4.10. Even after reductions, the search space sizes for the Class E graphs are enormous. Using the bound

$$\mathcal{S}(n,m) > \binom{n-m}{k} \geq \left(\frac{n-m-k+1}{k}\right)^k$$

where $k = \min(m-2, n-m)$ reveals that a number of graphs in this class have search spaces exceeding 10^{100} points. In particular, the search space for E-13 exceeds 10^{231} points, and for E-18, it exceeds 10^{242} points. These bounds are computed after graph reductions have been performed.

Table 4.10. Characteristics of Class E Graphs before and after Reductions

Graph	Problem size			Reduced size		
	n	m	$\|E\|$	n	m	$\|E\|$
E-1	2,500	5	3,125	680	5	1,286
E-2	2,500	10	3,125	710	9	1,328
E-3	2,500	417	3,125	637	199	1,233
E-4	2,500	625	3,125	435	164	964
E-5	2,500	1,250	3,125	222	108	649
E-6	2,500	5	5,000	1,845	5	4,318
E-7	2,500	10	5,000	1,891	10	4,372
E-8	2,500	417	5,000	1,723	286	4,193
E-9	2,500	625	5,000	1,608	358	4,069
E-10	2,500	1,250	5,000	1,046	366	3,388
E-11	2,500	5	12,500	2,498	5	12,093
E-12	2,500	10	12,500	2,500	10	12,123
E-13	2,500	417	12,500	2,341	321	11,760
E-14	2,500	625	12,500	2,139	388	11,325
E-15	2,500	1,250	12,500	1,461	443	8,514
E-16	2,500	5	62,500	2,500	5	29,332
E-17	2,500	10	62,500	2,500	10	29,090
E-18	2,500	417	62,500	2,429	355	28,437
E-19	2,500	625	62,500	2,351	485	27,779
E-20	2,500	1,250	62,500	1,988	758	24,423

4.1. The Steiner Problem in a Graph

Table 4.11 lists the solution qualities obtained by the GA and SPH-I together with the run times of all algorithms considered. Due to the extensive run times required for the graphs in this class, the GA was executed only once for each example. C_{ga} denotes the cost obtained by the GA, ΔC_{ga} is the relative error of the solution found by the GA, and the time spent by the algorithm is denoted by T_{ga}. Hence, C_{ga} and T_{ga} can be considered to be estimates of C_{avg} and T_{avg}, respectively.

Table 4.11. Comparison of Solution Quality and CPU Time for Class E Graphs

Graph	Cost					CPU-time (secs)			
	C_{opt}	C_{sph}	ΔC_{sph}	C_{ga}	ΔC_{ga}	T_{bc1}	T_{bc2}	T_{sph}	T_{ga}
E-1	111	111	0	111	0	1,150	1,394	7,334	7,395
E-2	214	216	0.93	216	0.93	6,251	1,993	7,355	7,444
E-3	4,013	4,060	1.17	4,013	0	26,468	15,782	50,004	9,449
E-4	5,101	5,113	0.24	5,102	0.02	46,008	1,660	29,921	7,763
E-5	8,128	8,134	0.07	8,128	0	12,564	411	9,318	7,474
E-6	73	76	4.11	73	0	678	-	10,060	10,148
E-7	145	149	2.76	145	0	27,124	-	10,306	10,458
E-8	2,640	2,690	1.89	2,646	0.23	118,618	-	50,013	12,896
E-9	3,604	3,671	1.86	3,611	0.19	24,528	-	50,014	14,933
E-10	5,600	5,624	0.43	5,600	0	39,261	-	50,014	12,976
E-11	34	34	0	34	0	1,901	-	14,472	14,559
E-12	67	68	1.49	68	1.49	7,200	-	14,497	14,588
E-13	1,280	1,317	2.89	1,289	0.70	207,059	-	50,003	21,787
E-14	1,732	1,767	2.02	1,736	0.23	29,263	-	50,030	23,022
E-15	2,784	2,795	0.40	2,784	0	7,666	-	50,020	18,424
E-16	15	15	0	15	0	179	-	14,425	14,586
E-17	25	25	0	25	0	36,040	-	14,458	14,619
E-18	572	625	9.27	583	1.92	-	-	50,017	29,105
E-19	758	802	5.80	766	1.06	6,372	-	50,037	27,319
E-20	1,342	1,357	1.12	1,342	0	272	-	50,055	25,107

It should be noted that the listed value of C_{opt} for E-18 may not be the global optimum, but according to the information in OR-Library, it is the best known solution as found by Beasley's algorithm [14]. The optimum for this graph was not found within a CPU time limit of 21,600 seconds on the Cray X-MP/48. Chopra et al. [40] also encountered problems with E-18. No run time is listed for this graph since the algorithm did not terminate within a CPU time limit of 10 days on the VAX 8700 [40]. Lucena and Beasley [140] do not report any results for graphs labeled E-6 through E-20, and a reason is not given. However, considering the progression of run times for the graphs in Classes C and D, it is reasonable to assume that the algorithm is unable to solve some of these problems in a reasonable amount of time.

SPH-I exceeds the CPU time limit of 50,000 seconds for graphs labeled E-3, E-8, E-9, E-10, E-13, E-14, E-15, E-18, E-19, and E-20. The estimated total time required by SPH-I for these graphs varies from 81,000 seconds for E-3 to 4.3×10^7 seconds, or more than 16 months, for E-20. Compared to the branch-and-cut algorithms and SPH-I, the run times of the GA are very moderate for all graphs, with a maximum run time of 29,105 seconds for E-18. For most of the graphs for which SPH-I terminates within the CPU-time limit, the run times of the GA and SPH-I are very similar. Regarding solution quality, SPH-I finds the global optimum for 4 of the graphs and has a worst relative error ratio exceeding 9% for E-18. The GA finds the optimum for 11 graphs and has a worst relative error ratio less than 2%.

Summary of Results

This subsection summarizes the experimental results with respect to solution quality and run time. When comparing the solution quality obtained by the GA to that obtained by SPH-I for all graphs in Classes B, C, and D, the following can be observed: of a total of 58 graphs, SPH-I finds the global optimal solution for 34 graphs, while the GA finds the optimum 10 times out of 10 for 43 graphs and at least one time of 10 for 55 graphs. For the Class E examples, SPH-I finds the optimum for 4 of the 20 graphs, while the GA finds the optimum for 11 of these graphs. $\Delta C_{worst} \leq \Delta C_{sph}$ holds for all but one graph in Classes B, C, and D, and in Class E, we have $\Delta C_{ga} \leq \Delta C_{sph}$ for all graphs. In other words, with a single exception, even the worst results generated by the GA are equal to or better than the results generated by SPH-I. Furthermore, for the graphs where both SPH-I and the average execution of the GA fails to find the global optimum, the expected relative error ratio ΔC_{avg} of the GA is often an order of magnitude better than the error ratio ΔC_{sph} of SPH-I.

Table 4.12 summarizes the solution qualities obtained by the GA and SPH-I. These figures are based on the results of all 600 executions of the GA and all 78 executions of SPH-I performed in total. For each algorithm, Table 4.12 gives the accumulated percentage of runs which gave a result within the stated relative error from the optimum. For example, 66.7% of all executions of SPH-I gave a result which was less than 0.5% from the optimum solution. When computing the values listed for the GA, the results for the Class E examples have been weighted by a factor of 10 to compensate for the imbalance in the number of executions for each graph.

Table 4.12. Summary of Solution Qualities Obtained by the GA and SPH-I

Algorithm	Error Ratio		
	$= 0\%$	$< 0.5\%$	$< 1.0\%$
SPH-I	48.7	66.7	70.5
GA	77.1	86.7	92.6

4.1. The Steiner Problem in a Graph

The results regarding run times can be summarized in three main points:

- The GA is capable of finding a high-quality solution for *all* graphs considered in a moderate amount of time. This is not the case for any of the two branch-and-cut algorithms or for SPH-I.

- In most cases, the run time of the GA is very similar to that of SPH-I. In a few cases, the GA is significantly faster than SPH-I, while the opposite is never the case.

- The variation of GA run time is very small compared to the variation observed for the branch-and-cut algorithms as well as SPH-I. As a consequence, the branch-and-cut algorithms are significantly faster than both the GA and SPH-I for some examples, while they are significantly slower on other examples.

As problem size increases through the Classes B, C, D, and E, the above observations become increasingly pronounced. If only Class B graphs are considered, it is difficult to make any distinctions regarding performance of the algorithms. These examples appear to be too simple.

Comparison with Kapsalis Algorithm

We denote the GA developed by Kapsalis, Rayward-Smith, and Smith [110] as GA-KRSS. GA-KRSS differs from the GA presented here in a number of ways. Among other things, neither an inversion operator nor a hill climber is applied in GA-KRSS. However, the most significant differences concern the decoder and the cost computation. In GA-KRSS, a genotype is a bit string of length n in which the ith bit indicates whether the ith vertex is part of the phenotype tree. To ensure that every tree spans W, each genotype is ex-OR'ed with the fixed string specifying W. Hence, the encoding is very similar to our encoding. However, the interpretation of a genotype is very different. Assume that a genotype specifies the vertex set Z, $W \subseteq Z \subseteq V$. The corresponding graph is then computed as the subgraph G_Z of G induced by Z. In general, G_Z is not connected. Assume it consists of $k \geq 1$ components. The cost of a solution is defined as the sum of the costs of a minimum spanning tree for each component plus a penalty term which grows linearly with k.

Computational results were given only for the Class B graphs from the OR-Library. The solution quality obtained for each graph was reported as the best result of five runs. For each graph, *some* parameter setting of GA-KRSS has been found with which the global optimum is found in five runs. However, the parameter setting varies with the problem given. When fixing the parameter setting for all graphs, GA-KRSS finds the global optimum in approximately 70% of all runs, and the worst result generated is 7.3% above the global optimum.

All experiments with GA-KRSS were run on an Apple Mac IIfx. No total run times were given. Instead, the time spent until the best solution first appears, referred to as Last Improvement Time (LIT), was measured. It is not clear exactly which stop criterion was used, i.e., how long the algorithm takes to terminate beyond

LIT. For many of the graphs, the average LIT is in the range 200 to 2000 seconds. There is a time limit of 4000 seconds for a complete execution.

GA-KRSS is clearly inferior to each of the other algorithms considered here, both with respect to solution quality and run time. We believe that the main reason for the performance gap between GA-KRSS and the GA presented here is the different decoding strategy and consequently, the different cost evaluation strategy.

Typical Behavior

The progress of the typical optimization process is illustrated by Figs. 4.8, 4.9, and 4.10, which stem from a sample execution of the GA with graph D-15 as input. It should be emphasized that although the graphs are for a single run, the picture they give is very typical.

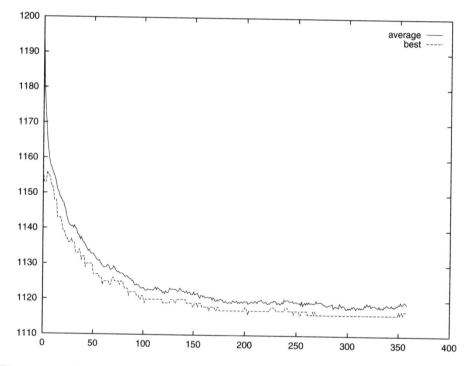

Figure 4.8. Cost of average and best individual as functions of generation number

For each generation, the top curve of Fig. 4.8 indicates the average cost of the individuals in the population at that time, while the bottom curve indicates the cost of the current best individual. Initially, the average cost is 1197, and the best is 1156. The global optimum of 1116 is first obtained in generation 203, and the algorithm terminates after 358 generations. Note that improvement is very rapid during the first part of the process. Then it levels out, and further improvements are obtained only slowly. As mentioned previously, the best and average costs are parts

4.1. The Steiner Problem in a Graph

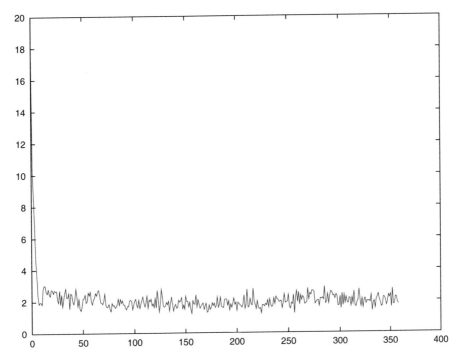

Figure 4.9. Standard deviation of cost as a function of generation number

of the stop criteria. If only the cost of the best solution were considered, the process would have terminated after generation 253, corresponding to a 29% reduction in the run time. However, the stop criteria used reflect a priority of solution quality as being more important than run time.

Fig. 4.9 shows the standard deviation of cost in the population for each generation. From an initial value of 19.2 in generation 0, the standard deviation decreases within 10 generations to about 2.0 and then stays at that level throughout the optimization process.

Each generation is initiated by the generation of M offspring individuals. From the total of $2M$ individuals, the best M individuals are then kept as members of the new population, while the rest are discarded. Fig. 4.10 shows for each generation, the percentage of individuals in the newly created population which have just emerged as a result of crossover. The percentage of newly generated individuals is very stable around 50. The important thing to note is that the fraction of new individuals does not decrease with time but is constant into the late phase of the process. In other words, throughout the process, half the individuals generated by the crossover operator are better than some other individuals already in the population. This confirms the role of crossover as the most important of the genetic operators.

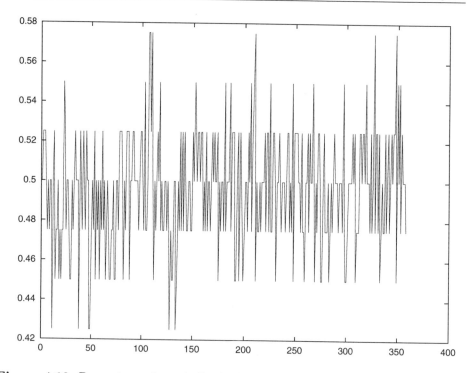

Figure 4.10. Percentage of new individuals in population as a function of generation number

4.1.5 Summary and Conclusion

In this section, a Genetic Algorithm (GA) for the Steiner Problem in a Graph (SPG) has been presented. The main idea behind the algorithm is the application of the distance network heuristic for interpretation of bit strings specifying selected Steiner vertices. This scheme ensures that every bit string corresponds to a valid solution and eliminates the need for penalty terms in the cost measure, thereby avoiding potential problems of assigning a suitable cost value to an incomplete or invalid solution.

The performance of the algorithm has been tested on random graphs with up to 2500 vertices and 62,500 edges. The experimental results show that in more than 92% of all executions, the GA finds a solution which is within 1% of the global optimum. This performance compares favorably with one of the very best deterministic heuristics for SPG as well as with an earlier GA by Kapsalis et al. Performance is also compared to that of branch-and-cut algorithms by Lucena and Beasley and by Chopra et al. While the run times of these algorithms vary significantly and prevent the solution of some of the problem instances considered, the GA is capable of generating a near-optimal solution for all problems within a moderate amount of time.

We therefore conclude the following: In cases where a globally optimal solution is absolutely required, the size of the given problem is not too big, and run time is not important, one of the branch-and-cut algorithms is preferable. On the other hand, if a near-optimal solution is sufficient or the problem is very large or a moderate run time limit is needed, the GA presented here is the best choice of the possibilities considered.

4.2 Macro Cell Global Routing

A common strategy for global routing of macro cell layouts which is used by, e.g. TimberWolfMC [206], consists of two phases. In the first phase, a number of alternative routes are generated for each net. The nets are treated independently, one at a time, and the objective is to minimize the length of each net. In the second phase, a specific route is selected for each net, subject to channel capacity constraints, so that some overall criterion, typically area or total interconnect length, is minimized.

We present a global router which minimizes area and, secondarily, total interconnect length. While also being based on the two-phase strategy, this router differs significantly from previous approaches in that each phase is based on a GA. The GA used in the first phase provides several high-quality routes for each net. In the second phase, another GA minimizes the dual optimization criteria by appropriately selecting a specific set of routing channels for each net. The router was first presented in [66].

4.2.1 Routing Phase I : Applying the GA for the SPG

Before the global routing process itself is initiated, a rectilinear *routing graph* is extracted from the given placement. Routing is then performed in terms of this graph; i.e., computing a global route for a net is done by computing a corresponding path in the routing graph. Roughly speaking, each edge of the graph corresponds to a routing channel, and each vertex corresponds to the intersection of two channels.

Before finding routes for a given net, vertices representing the terminals of the net are added to the routing graph at appropriate locations, as illustrated in Fig. 4.11. Finding the shortest route for the net is then equivalent to finding a minimum cost subtree in the graph which spans all of the added terminal vertices, assuming that the cost of an edge is defined as its length. This problem is known as the Steiner Problem in a Graph (SPG) and has been addressed earlier in this chapter. When a net has been treated, its terminal vertices are removed from the routing graph before considering the next net, thereby significantly reducing the size of the SPG instances to be solved.

In the first phase of the global router the nets are routed independently, one net at a time. Several alternative routes are generated for each net. Two-terminal nets are handled by an algorithm of Lawler [133], while nets with three or more terminals are handled by the GA for the SPG presented in Section 4.1.

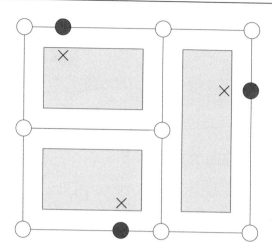

Figure 4.11. A placement and its corresponding routing graph. Addition of terminal vertices (shaded) for a net with three terminals (marked with crosses) is shown.

4.2.2 Routing Phase II : Minimizing Layout Area

In the second phase of the router, a specific set of routing channels is assigned to each net. Area estimation is based on the formation of horizontal and vertical polar graphs, and estimating the width and the height of the layout, respectively. The formation of such polar graphs is explained in detail in [164].

In the GA used in the second phase of the router, a global routing solution is represented by specifying for each net, which of its possible routes is used. For N nets, an individual is a list of N tuples. For example, the tuple (3,7) specifies that the third net uses its seventh route.

Fig. 4.12 outlines Phase II of the algorithm. Initially, the current population of size M contains $M-1$ randomly generated individuals and a single individual consisting of the shortest route found for each net. Routine *evaluate* computes the fitness of each of the given individuals. Fitness is rank-based, following a lexicographical sorting of the individuals using area as the primary criterion and wire length as the secondary criterion. Routine *bestOf* keeps track of the best individual ever found. Routine *stopCriteria* terminates the simulation when no improvement has been observed for S consecutive generations. Each generation is initiated by the formation of a set of offspring of size M. Using standard one-point crossover, two offspring c_1 and c_2 are generated from parents p_1 and p_2. Routine *reduce* returns the M fittest of the given individuals, and standard pointwise mutation and standard inversion are then performed on each offspring. Routine *optimize* performs simple hill climbing by executing a sequence of mutations on the best individual, each of which improves the fitness.

4.2. Macro Cell Global Routing

```
generate initial Population;
evaluate (Population);
bestIndividual = bestOf (Population);
While not stopCriteria() Do
   NewPopulation = ∅;
   For j = 1 to M/2 Do
      select p₁ and p₂ from Population;
      crossover(p₁, p₂, c₁, c₂);
      Add c₁ and c₂ to NewPopulation;
   evaluate (Population ∪ NewPopulation);
   Population = reduce (Population ∪ NewPopulation);
   For j = 1 to M Do
      mutate (Population[j], P_M);
      invert (Population[j], P_I);
   evaluate (Population);
   bestIndividual = bestOf (Population ∪ bestIndividual);
optimize (bestIndividual);
solution = bestIndividual;
```

Figure 4.12. Outline of Phase II

4.2.3 Experimental Results

The router was implemented in the C programming language, and all experiments were performed on a Sun Sparc IPX workstation. The router was interfaced with the macro cell layout system Mosaico, which is a part of the Octtools CAD framework developed at the University of California, Berkeley. This integration allows for comparison of the router's performance to that of TimberWolfMC [206], a well-known global router also interfaced to Mosaico.

Slightly modified versions of three of the MCNC macro cell benchmarks were used for the experiments: xerox, ami33, and ami49. The modified versions are referenced using an "-X" suffix. Examples xerox-X, ami33-X, and ami49-X were placed by Puppy, a placement tool based on simulated annealing, also included in Octtools; ami33-2-X and ami49-2-X are alternative placements of ami33-X and ami49-X, respectively.

For each of the placed examples, Mosaico was executed to generate a complete layout, using either TimberWolfMC or the GA-based router for the global routing task. All other steps of the layout process were performed by the same tools. The same set of GA control parameters was used for all program executions; i.e., no problem-specific tuning was performed.

Table 4.13 summarizes the impact on the completed layouts of using the GA-based router instead of TimberWolfMC. The best and average of five runs are given. A_{tot} denotes total area, A_{route} denotes routing area, i.e., the part of the total area not occupied by cells, and WL denotes total wire length. Each entry is computed as $100 \times (\frac{GA-result}{TimberWolfMC-result} - 1)$. Hence, a negative value indicates a reduction in percentage obtained by the GA-based router, while a positive value indicates a percentage overhead as compared to TimberWolfMC.

Table 4.13. Relative Improvements Obtained by the GA-Based Router

Problem	Solution	A_{tot}	A_{route}	WL
xerox-X	best	−1.9	−4.7	+0.0
	average	−1.4	−3.5	+0.8
ami33-X	best	−3.2	−5.1	−3.2
	average	+1.6	+2.5	−0.2
ami33-2-X	best	−3.0	−4.7	−1.5
	average	−1.1	−1.7	−0.2
ami49-X	best	−1.9	−3.3	−1.5
	average	−0.5	−0.8	+0.3
ami49-2-X	best	−4.2	−7.3	−4.0
	average	−3.7	−6.3	−2.9

On average, the router requires about 22, 12, and 130 minutes to route examples based on xerox, ami33, and ami49, respectively. TimberWolfMC spends about 30 seconds for examples based on xerox and ami33, and about 5 minutes for ami49-based examples. Hence, the GA-based router is clearly inferior to TimberWolfMC with respect to run time. However, the run time of the current implementation can be improved significantly, since the vast majority of the run time is spent (re-)computing channel densities in an inefficient nonincremental fashion.

4.2.4 Summary and Conclusion

In this chapter, a novel approach to global routing of macro cell layouts based on genetic algorithms has been presented. The performance of the router was compared to that of TimberWolfMC on MCNC benchmarks. Experimental results show that the quality of completed layouts generally improves when using the GA-based router instead of TimberWolfMC. The router is inferior to TimberWolfMC with respect to run time, but major improvements are possible. We conclude that the genetic algorithm is well suited as the basic algorithm of a global router.

4.2. Macro Cell Global Routing

SUMMARY

TOPICS STUDIED

- Steiner Problem in a Graph (SPG)
- Distance Neighbor Heuristic (DNH)

KEY OBSERVATIONS AND POINTS OF INTEREST

- SPG is an NP-complete problem
- GA finds global optimal solution (within 1%) more than 77% of time
- Graph reduction is applied to expedite GA
- GA requires $O(n^2)$ setup time and $O(M(mn \log(mn)) + \log(M))$ time for each generation, where M is the GA population size, n is the number of vertices in the graph, and m is the subset of vertices to be connected

- Macro Cell Global Routing

- Phase 1 router applies Lawler's algorithm for 2-terminal nets and applies GA for SPG
- GA-based router gives better layout quality than TimberWolfMC in MCNC benchmark circuits

CHAPTER 5
at a glance

- FPGA technology mapping via circuit segmentation
  - Table Lookup (TLU) FPGA architecture
  - Genetic algorithm for circuit segmentation

- Pseudo-exhaustive testing via circuit segmentation

Chapter 5

FPGA TECHNOLOGY MAPPING

Venkataramana Kommu and Irith Pomeranz

Field Programmable Gate Arrays (FPGAs) have become very popular for prototyping new designs of digital logic circuits. This is because the hardware implementation of the design is relatively easy to implement using FPGAs, thus allowing logic verification to be performed early in the design process and reducing the turnaround time [26]. This has further ramifications on the manufacturing costs. In implementing a design in FPGAs, the optimized logic description obtained during logic synthesis must be mapped to the modules and routing resources available in the FPGAs. The objective is to find the best mapping of logic to FPGA modules, in terms of number of modules required. Other factors, such as performance, may also be considered. In this chapter, we present a genetic algorithm for FPGA technology mapping.

The chapter is organized as follows. An overview of the problem of FPGA technology mapping is given in Section 5.1. In Section 5.2, the proposed solution to the FPGA mapping problem is described. The proposed approach is based on circuit segmentation, i.e., partitioning the circuit into a number of subcircuits that are not necessarily disjoint. In Section 5.3, another application of segmentation, to divide a circuit into smaller blocks in order to pseudo-exhaustively test the circuit, is briefly presented. In Section 5.4, experimental results for both problems are presented to demonstrate the effectiveness of the GA solution proposed. Concluding remarks are given in Section 5.5.

5.1 Problem Description

Synthesis and mapping of circuits to FPGA architectures has received wide attention [12], [26], [33], [72], [73], [152], [153], [157]. We look into the mapping of circuits onto a class of FPGAs that use Configurable Logic Blocks (CLBs) containing Table Lookups (TLUs). The CLBs can implement any Boolean function with k inputs, where k is a given constant. In the TLU architecture, the CLBs contain RAM/ROM based lookup tables which are user programmable. Technology mapping for a TLU-based FPGA architecture maps a circuit, optimized according to criteria such as area and delay, into an interconnection of CLBs that implement the same function. We concentrate on the area of the resulting circuit, as measured by the number of CLBs required to implement it. The FPGA mappers in [72], [152] use the bin-packing approach, where the CLBs are treated as bins and the circuit parts are packed into them. This approach has fast execution times but suffers from limitations in exploring a large number of solutions. This problem is magnified by the large number of options that FPGA mapping allows, e.g., decomposition of gates and replication of gates. Our approach using the GA considers many of these aspects implicitly and thus performs global optimization.

Our approach to technology mapping is based on circuit segmentation: Given a combinational logic circuit and a positive integer k, partition the circuit into (not necessarily disjoint) subcircuits, such that the number of inputs to each subcircuit does not exceed k and the number of lines connecting different subcircuits is minimized. Fig. 5.1 shows a circuit partitioned into subcircuits with $k = 5$. The lines

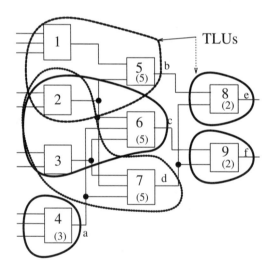

Figure 5.1. Circuit with segmentation cells ($k = 5$)

connecting different subcircuits are a, b, c, and d. The subcircuit with line a as the output has inputs $\{p, q, r\}$. Similarly, the subcircuit with line c as output has $\{l, m, n, o, a\}$ as its inputs. We will refer to the points where the subcircuits meet as *segmentation points*.

The problem of circuit segmentation is highly constrained. In constrained search spaces, the genetic operators may produce infeasible solutions (which do not satisfy the constraints). Some of the effort may thus be spent in exploring solutions that do not contribute toward the final, feasible result. For example, during crossover, we may produce a solution which does not map a function to a CLB and is therefore an invalid solution. In our approach, a randomized validation procedure is used to convert an infeasible solution into a feasible one. The validation procedure introduces information about the problem constraints into the GA and as a result, the validation procedure allows an efficient genetic search using extremely small population sizes.

5.2 Circuit Segmentation and FPGA Mapping

In this section, we describe the general genetic solution to the segmentation problem. We then describe additional features of the solution, that are unique to the FPGA problem. We start by introducing the terminology used by referring to Fig. 5.1. A line on which a segmentation point is placed becomes an output of the subcircuit driving it and an input of the subcircuits it drives. It is referred to as a *pseudo-output* of the subcircuit driving it, and a *pseudo-input* of the subcircuits it drives. Thus, we refer to lines a, b, c, and d in Fig. 5.1 as pseudo-outputs. Line a is also a pseudo-input to gates 6 and 7. The input cone of a primary or pseudo-output consists of all the gates driving it; however, a cone does not contain other pseudo-outputs. Thus, the cone of primary output e consists only of gate 8. The number of inputs that a primary output or pseudo-output depends on is found by traversing the circuit backwards from the primary output or pseudo-output, until a primary or pseudo-input is reached. For example, the input cone of line c in Fig. 5.1 consists of gates 2, 3, and 6, and its inputs are the primary inputs l, m, n, o, and the pseudo-input a. In the figure, all the pseudo-outputs and primary outputs have a number in parentheses inside the gates driving them, indicating the number of inputs that they depend on.

The following sections describe the FPGA mapping problem and our enhanced genetic solution to the circuit segmentation problem with specific application to FPGA mapping.

5.2.1 TLU-Based FPGA Architecture

An FPGA consists of an array of logic blocks (CLBs) whose interconnections are user programmable. The CLBs can implement any Boolean function with k inputs, where k is a given constant. In the TLU architecture, the CLBs contain RAM/ROM based lookup tables which are also user programmable. The CLBs in the Xilinx 3000 chip can implement

- any single function with $k = 5$ inputs, or

- two functions if the total number of inputs to both functions is at most 5 and each function depends on at most 4 inputs.

We use the following terminology. To map a circuit onto an FPGA, the circuit is partitioned into k-input subcircuits. Each subcircuit corresponds to a TLU; thus, each TLU has a single output. Two TLUs can be combined and mapped onto a single CLB, under the constraints above. Our goal is to minimize the number of CLBs, thus potentially minimizing the area of the circuit.

Next, we review issues in FPGA mapping, including gate decomposition and logic replication.

Gate Decomposition: Gate decomposition is the operation of replacing an m-input gate by several gates with smaller numbers of inputs. Correct decomposition of a gate within a circuit may result in a reduction in the number of TLUs required to implement the circuit, as demonstrated below. Chortle-crf [72] solves a series of bin-packing problems to find the decomposition of a gate. This may lead to a suboptimal decomposition, since it depends on the order in which the gates are decomposed. MIS-pga [152] uses Roth-Karp decomposition, Shannon decomposition, AND-OR decomposition, and disjoint decomposition.

We use two types of gate decompositions. One is the Complete Binary Tree Decomposition (CBTD) [96], and the other is what we refer to as the Long Decomposition (LD). These are shown in Fig. 5.2. The way in which the inputs of the decomposed gate are arranged also has an effect on the number of TLUs. For example, in Fig. 5.3(a), with the original order of the inputs of the CBT decomposed gate

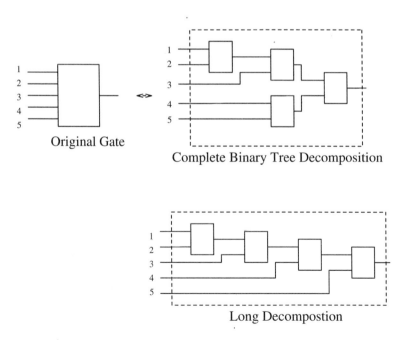

Figure 5.2. Gate-level decomposition

5.2. Circuit Segmentation and FPGA Mapping

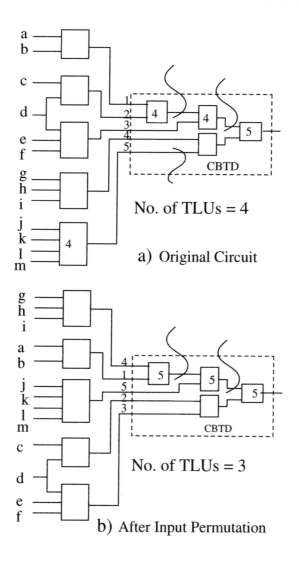

Figure 5.3. Effect of input permutations (CBT decomposition)

and assuming $k = 5$, we require four TLUs, while under a different permutation shown in Fig. 5.3(b), we use only three TLUs. With no decomposition, six TLUs are required. For a gate having five inputs or less, all different decompositions can be obtained using LD, CBTD, and input permutation [119]. In our solution, both decomposition and input permutation are done automatically as part of the genetic search, as will be explained later.

Replication of Gates: Replication of gates may also reduce the number of TLUs. A replicated gate is included in several TLUs, avoiding the need to define a separate TLU containing that gate. Chortle-crf [72] checks whether replicating a gate would result in a reduction in the TLU-count. If it does, the gate is replicated. In our mapping method, replication is done automatically by allowing partitioning into overlapping subcircuits. All the circuitry in the intersection of the subcircuits is replicated. An example is given in Fig. 5.1. Here, the subcircuits present in the overlap of the subcircuits having outputs b, c and c, d are replicated.

5.2.2 Mapping Using Circuit Segmentation

Given a circuit in which every gate has at most k inputs, we can find a mapping onto the TLU blocks by finding an appropriate circuit segmentation. For example, in Fig. 5.1, by inserting segmentation points on lines a, b, c, d, e, and f, we have six TLU blocks. We use a simple GA to solve the circuit segmentation problem. Roulette wheel selection and one-point crossover are used. The best solution found in any generation is stored and returned as the final solution. In the following paragraphs, we describe the string representation, fitness function, and GA parameters used for FPGA mapping.

Representation: The GA requires a string representation of the solution. In our implementation, each solution is represented by two strings. The first string stores the segmentation point placement in binary form. For a circuit with N gates, we represent the solution by a binary string of length N, denoted $x_1, x_2, \ldots, x_N$ (0 stands for no segmentation point on the output of a gate, and 1 indicates that a segmentation point exists on the output of a gate). To simplify the solution, segmentation points are not placed on fanout branches (to allow segmentation points on fanout branches, fanout branches can be treated as single-input single-output gates).

The second string represents the decomposition state of a gate, where a value of 0 implies that the gate is not decomposed, 1 stands for CBTD, and 2 represents LD. Since the decomposition of a gate changes the number of gates in the circuit, we take the following approach. An n-input gate can be either undecomposed or decomposed into at most $n - 1$ 2-input gates. A gate can thus be replaced by at most $n - 1$ gates when it is decomposed. To keep the length of the strings fixed, we define $n-1$ positions for each gate in every string and save the last of the $n-1$ positions for the original gate (in the undecomposed case) or for the output gate of the CBTD or LD. The status of the original gate (CBTD, LD, or no decomposition) as recorded in the second string, indicates how the other $n-2$ gates defined for this gate should be considered: as part of a CBTD, as part of an LD, or ignored if the gate is not decomposed.

Every gate has a $\{0, 1\}$ value in the first string. Only the original gates have an entry in the second string. The other gates have don't-care entries in their positions in the second string. The GA, while trying to optimize the number of TLUs in the circuit, also selects at the same time a decomposition for every gate that minimizes

5.2. Circuit Segmentation and FPGA Mapping

the number of TLUs. In Fig. 5.1, gates 1, 4, 6, and 7 have more than 2 inputs, and are therefore decomposed into two input nodes. The string representation for the circuit is 00 | 0 | 0 | 01 | 1 | 01 | 01 | 0 | 0, where the first 2 bits represent node 1 decomposed into 2 nodes, the next bit represents node 2, and so on. For gate 4, we have a 01 entry, implying there is a segmentation point on its output. If in the solution, only node 1 is decomposed (say, into an LD), the string representation for the decomposition status would be 02 | 0 | 0 | 00 | 0 | 00 | 00 | 0 | 0. The entry for gate 1, 02, implies that the original gate is decomposed into an LD gate.

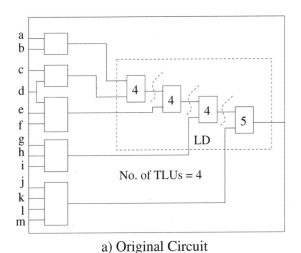

a) Original Circuit

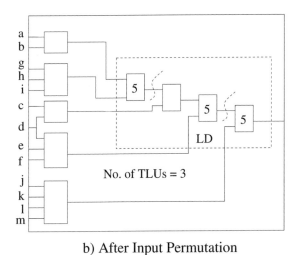

b) After Input Permutation

Figure 5.4. Effect of input permutations (LD decomposition)

In the above representation, the segmentation points are captured by a simple binary string, and a simple crossover operator can be used.

Crossover and Mutation: During crossover, the offspring receives segmentation point positions from both parents. A crossover probability of 0.9 is used. The mutation operator randomly selects positions on the string and complements them, with a mutation probability of $1/N$, where N is the number of gates in the circuit. Thus, it may introduce a segmentation point in a position where it is not present and may remove it from a position where it is present. Both crossover and mutation may transform parents which are valid solutions (the number of inputs of a subcircuit is within the bound) into an invalid offspring. For example, assuming that no gate is decomposed, for the circuit of Fig. 5.1, 0000011010111 and 0111010000000 represent valid segmentations for $k = 5$. A crossover between them with a crossover point after position 6 produces an offspring 0000010000000. Since node 8 in the offspring depends on 8 inputs, it is an invalid solution. Our solution involves correcting invalid solutions and transforming them into valid ones. This is discussed in detail below.

Validation: A subcircuit is *valid* if it depends on at most k inputs. Similarly, a solution is valid if all the subcircuits in the circuit are valid. As pointed out above, invalid individuals may result from crossover and mutation operations on valid solutions. In the literature on the GA, two main alternatives have been proposed to deal with invalid solutions obtained due to crossover and mutation [51]. The first alternative is to keep the invalid solution in the population but appropriately reduce its fitness to ensure that its contribution to following generations is minimized. This may also cause it to disappear in later iterations. The second alternative is to devise operators that will only produce valid solutions.

The selection of the most appropriate approach depends on the problem being solved. Since design automation problems are usually highly constrained, the first approach may result in the population containing too many invalid solutions. This may lead to the formation of an individual that may be the only valid individual in the population, and thus the population may converge to this individual without exploring other valid solutions. The second approach is not applicable to all problems and significantly complicates the crossover operation when it is applicable. Our approach is to alter every invalid solution and transform it into a valid one. This approach proved to be successful for other problems [119].

In the circuit segmentation problem, if an invalid solution is obtained after a crossover operation, we insert segmentation points at various positions in the invalid subcircuit, until the subcircuit is partitioned into valid subcircuits. The segmentation points are inserted

1. at randomly chosen inputs of the gate at the output of the subcircuit.

2. at selected inputs of the gate at the output of the subcircuit. The input that belongs to the largest number of other subcircuits is chosen to be the site of the segmentation point.

5.2. Circuit Segmentation and FPGA Mapping

3. anywhere in the subcircuit.

4. at the gate that belongs to the largest number of subcircuits.

For the purpose of illustration, consider Fig. 5.1 again. Assume that crossover produced an offspring 0000010000000, in which output e is invalid. Using the first method, a segmentation point is inserted at the output of gate 5 or 7. Using the second method, the segmentation point is inserted at the output of gate 7 first, because it occurs in the subcircuits of both outputs. Using the third method, the choice for segmentation point insertion falls at the outputs of 1, 2, 3, 5, or 7. Using the fourth method, the segmentation point is inserted at the outputs of gates 2, 3, or 7 since they occur in two subcircuits (of e and f), while 1 and 5 occur in one subcircuit only (of e). If new invalid subcircuits are created, they are considered as well, until a valid solution is obtained.

The validation method to use on an infeasible solution is chosen randomly. The motivation for the second and fourth methods is to insert the segmentation points at the output of the gate which appears in the largest number of subcircuits. This may cause these subcircuits to depend on less than k inputs and in subsequent iterations of the algorithm may lead to the removal of other segmentation points that they depend on. The first and the third methods introduce randomness into the choice of positions for segmentation points required to get out of local optima. The randomness introduces diversity into the population and enables us to use small population sizes without a loss in the solution quality.

Population Size: The population size used by the GA has been a subject of research and controversy. A larger population size may reduce the errors associated with the selection of the parents for propagation. A smaller population size increases the efficiency of the GA and makes it competitive with other algorithms in terms of execution time. We use a population size of 4, since we found that the population maintains its diversity, in spite of its small size, due to the randomness of the validation scheme.

Fitness Function: The fitness function is an important component of the GA and has an important role in directing the search toward high-quality solutions. An individual's fitness reflects its quality as a solution to the problem, relative to the other individuals. We use two different fitness functions.

The first fitness function is used to minimize the number of TLUs. We use the following inverse quadratic fitness function to evaluate the solutions:

$$\sigma = \frac{1.0}{(z)^2 + \alpha \times y} \quad (5.2.1)$$

where z is the number of TLUs,

$$y = \sum_i x_i \times (\text{number of inputs gate } i \text{ depends on}),$$

α is a small constant such that $\alpha \times y < 1$, and the term $\alpha \times y$ contributes less than an additional segmentation point. Recall that x_i is equal to 1 if gate i has a segmentation point at its output and is equal to 0 otherwise. The number of TLUs is equal to the sum of the number of segmentation points in the circuit and the number of primary outputs.

The above objective function ensures that solutions having lower numbers of TLUs would have a greater fitness. This objective function was chosen from a variety of other formulations (linear, inverse functions $(1/x)$). The choice was based on the experimental results obtained for these formulations. The term $\alpha \times y$ distinguishes among solutions having the same number of segmentation points but having different numbers of inputs driving the output gates of the resulting subcircuits. Thus, it may cause the best solution to be changed and reduces the probability of premature convergence.

The second fitness function uses the number of Xilinx CLBs which can implement 2-output TLUs based on the conditions discussed in Section 5.2.1. This approach directly computes the number of CLBs using a heuristic procedure that merges pairs of TLUs into single CLBs (the merging problem can be transformed into a Maximum Cardinality Matching problem [152]). The following fitness function is then used in the second approach to qualify the fitness of a solution:

$$\sigma = \frac{1.0}{\text{Number of CLBs required by the solution}}. \quad (5.2.2)$$

Termination: If 300 consecutive iterations do not show any improvement in the solution quality, the circuit is perturbed locally by randomly choosing a decomposed gate and changing its input order. If there is still no improvement for 100 iterations, then the circuit is perturbed again. If there is an improvement, then the circuit is perturbed only after 300 iterations without improvement, and 100 iterations thereafter, until an improvement is noticed again. The above process continues until there is no improvement in the solution quality for 3000 iterations.

5.3 Circuit Segmentation for Pseudo-Exhaustive Testing

Exhaustive testing, which consists of applying all possible input combinations to a circuit-under test, is a method to test modeled as well as unmodeled faults in digital logic circuits. For circuits whose outputs depend on a large number of primary inputs, exhaustive testing is prohibitive in time. Pseudo-Exhaustive Testing (PET) [147] was proposed to reduce the complexity of exhaustive testing. In PET, the circuit is partitioned into possibly overlapping subcircuits. Partitioning is done such that the output of each subcircuit depends on at most k inputs (hence the relationship to the circuit segmentation problem). These subcircuits are exhaustively tested by applying all 2^k input combinations to each set of inputs. Thus, all the multiple stuck-at faults within every subcircuit are detected, and all the single stuck-at faults in the whole circuit are detected.

Many approaches for finding the locations of segmentation points for pseudo-exhaustive testing have been reported in the literature. Most approaches use greedy heuristics which explore relatively small parts of the solution spaces [186], [18], [229], [92]. The CMP algorithm [186] is a greedy heuristic for finding segmentation points. The DC algorithm [92] is based on a divide-and-conquer approach that is augmented by hill-climbing procedures. Here the circuit is initially divided into subcircuits (at fanout points), and a locally optimum segmentation is obtained on each subcircuit. The GL algorithm [92] is a faster heuristic which uses global information to choose the segmentation points. The PEST algorithm [241] chooses the segmentation point which reduces the largest number of subcircuits having greater than k inputs. The H-heuristic [220] was used to find segmentation points in two phases. In the first phase, a greedy procedure determines an initial set of candidate segmentation points, and in the next phase, the necessity of placing a segmentation point at each candidate position is checked. This heuristic performs well for smaller circuits.

In contrast to the greedy heuristic procedures, genetic search allows various solution subspaces to be efficiently explored in parallel, thus improving the quality of the final solution obtained.

5.4 Experimental Results

In this section, we will first present experimental results for the FPGA technology mapping problem. We will then show the results for the segmentation of the circuits in order to pseudo-exhaustively test them.

5.4.1 Results for FPGA Technology Mapping

Table 5.1 shows results of mapping circuits to TLU blocks and to CLBs using the GA technology mapper, GAFPGA, described above. The circuits were obtained by running a MIS standard script for area optimization on the MCNC logic synthesis benchmark circuits. OBJ1 refers to the results obtained using the first objective function (Equation 5.2.1), which minimizes the number of TLUs. The CLBF header refers to the second objective function (Equation 5.2.2), which uses the number of CLBs to compute the fitness of the solution. t_{total} indicates the total time in seconds taken for the GA when optimizing for the number of CLBs, and it includes the time to find the best solution and the additional time during which no improvement was observed until the termination condition was satisfied. The GA optimizing for TLUs always took considerably less time than the GA optimizing the number of CLBs. The experiments were run on a Sun Ultra Sparc machine. For the purpose of comparison, we executed Chortle-crf on the same circuits using the recommended script for area optimization:

Chortle -i -A -r -Z 12 -f -a -K 5 -e in.eqn.

Table 5.1. Results for FPGA Technology Mapping

Circuit	Chortle-crf		GAFPGA		
			OBJ1	CLBF	
	TLU	CLB	TLU	CLB	t_{total}
z4ml	6	3	5	3	6.88
misex1	19	13	16	12	13.55
vg2	23	18	22	16	26.07
5xp1	28	21	22	17	29.6
count	31	27	31	23	39.57
9symml	55	47	57	45	140.9
9sym	64	52	56	45	111.3
apex7	64	46	61	41	99.40
rd84	45	32	43	32	55.99
e64	81	55	81	47	114.3
c880	86	74	86	64	303.4
alu2	112	92	112	81	222.7
duke2	123	99	119	82	300.7
c499	70	50	64	49	336.1
rot	203	145	200	124	521.8
apex6	213	167	196	150	177.8
alu4	197	150	195	145	629.7
apex4	566	450	639	470	687.4
des	955	762	1010	740	14472

The performance of our mapper is tested by comparison with the results obtained by Chortle-crf. For most of the circuits, our mapper performs better than Chortle-crf, both in the number of TLUs and the number of CLBs. In fact, the results indicate that segmenting a circuit leads to a better packing into CLBs. Our CLB results are also better than those obtained by ASYL [12]. The performance for the ISCAS85 benchmarks is shown in Table 5.2. The GA mapper takes more time than Chortle-crf; however, in most cases, it yields better results. In some cases, significantly better results are obtained. Relatively long execution time is a known characteristic of the sequential GA. This problem has been addressed to a great degree by the use of parallel processing. (Theoretical and experimental research into parallel GAs have shown superlinear speedups.) Direct comparison with other mappers was not possible, since for a meaningful comparison, the same circuits, synthesized in the same way, must be used.

Though direct comparison with MIS-pga [152] is not possible since MIS-pga combines logic optimization and mapping, a comparison indicated that our mapper yielded less CLBs than MIS-pga for most circuits. However, for some circuits there

5.4. Experimental Results

Table 5.2. Results of Technology Mapping for ISCAS Circuits

Circuit	Chortle-crf		GAFPGA		
			OBJ1		CLBF
	TLU	CLB	TLU	CLB	CLB
c432	71	53	61	53	43
c1355	74	54	66	58	48
c1908	155	124	137	108	98
c2670	327	253	288	193	175
c3540	367	302	334	277	237
c5315	484	376	480	368	350
c6288	494	248	489	307	263
c7552	673	521	645	507	467

were large differences in circuit size (in terms of the number of gates) obtained by MIS-pga as compared to the MIS-standard area optimization script, causing large differences in the number of CLBs. Using the circuits synthesized by MIS-pga as the starting circuits, GAFPGA obtains the same number of CLBs as those obtained by MIS-pga.

For delay optimization, the length of the critical path should be as small as possible. There are many algorithms, e.g., Chortle-d [73], DAG-MAP [33], and MIS-pga [153], which reduce the number of levels in the mapped circuit at the cost of increasing the number of TLUs. For the purpose of comparison, we calculated the depth of the solutions obtained through the iterations of the GA and stored the best depth obtained. Some of the results are shown in Table 5.3. These results show that delay optimization must be explicitly targeted if small delays are desired. Note that here, the results for Chortle-d and DAG-MAP are taken from [73] and [33], respectively, and the circuits used may not have been synthesized in an identical way to the ones we used.

Another issue we investigated is the effect of increasing the number of inputs, k, that the TLU block is allowed to have, on the quality of the solution. With improvement in technology, the number of inputs to a TLU block is expected to increase, and the scalability of FPGA mapping to larger CLBs is important. We applied our approach to TLUs with larger numbers of inputs, and compared with Chortle-crf. Table 5.4 shows the results of our experiments for $k = 7$ and $k = 9$. Our results are in most cases superior to those obtained by Chortle-crf. Note especially the results for rd84 and $k = 9$, where the GA yields 4 TLUs, whereas Chortle-crf yields 27 TLUs.

Table 5.3. Delay Comparison

Circuit	Chortle-d		DAG-MAP		GAFPGA OBJ1	
	TLU	Depth	TLU	Depth	TLU	Depth
z4ml	25	3	5	2	5	2
vg2	55	4	29	3	24	4
count	91	4	31	5	31	5
9sym	63	5	60	5	61	6
rd84	61	4	46	4	44	5
c880	329	8	128	8	86	9
alu2	227	9	156	9	120	14
c499	382	6	68	4	64	6
apex6	308	4	246	5	206	7
des	2086	6	1423	5	1007	10

Table 5.4. Number of TLUs for Chortle vs. GA: Comparison for Higher k

Circuit	$k = 7$		$k = 9$	
	Chortle	GA	Chortle	GA
z4ml	5	4	4	4
vg2	18	13	10	8
count	23	23	19	19
9sym	17	17	3	1
rd84	34	24	27	4
c880	59	60	52	54
alu2	80	72	59	34
c499	71	54	48	48
apex6	156	148	140	134
apex4	435	467	366	379
des	742	641	592	365

5.4.2 Experimental Results of Segmentation for PET

Tables 5.5 and 5.6 compare the results obtained by the GA with other heuristics for the PET problem. Results for k having values of 16 and 20 are compared in the two tables; i.e., each subcircuit has at most k inputs.

The columns t_{best} and gen_{best} indicate the time and the number of generations, respectively, required to obtain the best solution. t_{total} indicates the total time taken (in seconds), which includes the time to find the best solution and the additional

5.4. Experimental Results

Table 5.5. Results for Pseudo-Exhaustive Testing

Ckt	DC	time *	CMP	PEST	H	H_{time}	GA	t_{best}	gen_{best}	t_{total}
c432	27	4013	56	–	27	–	27	0.18	3	177.6
c499	8	104	20	–	8	–	8	0.08	1	328.3
c1355	8	7455	20	–	8	–	8	0.7	2	1718
c1908	21	6326	51	–	18	–	17	47.7	133	1160
c2670	45	55267	108	–	33	–	32	223.2	461	1722
c3540	87	22438	138	–	87	–	77	736.8	908	3318
c5315	52	19637	96	–	52	–	48	840.1	905	3357
c6288	93	15390	115	–	–	–	95	63988	11672	81243
c7552	100	74316	118	–	–	–	94	3431	1857	8401

$k = 16$

* These times are on a Sun 3/280 Computer
– Results are not available

Table 5.6. Results for Pseudo-Exhaustive Testing

Ckt	DC	time *	CMP	PEST	H	H_{time}	GA	t_{best}	gen_{best}	t_{total}
c432	20	619	44	19	20	200.5	19	0.08	1	233.7
c499	9	6	32	8	8	96.5	8	0.08	1	350.3
c1355	9	7423	32	8	8	96.5	8	1.34	3	1784
c1908	18	6083	53	15	14	439.5	11	28.2	59	1509
c2670	37	3972	77	30	29	2440	28	325.0	551	2090
c3540	63	3661	105	–	58	7502	61	733.8	664	4020
c5315	42	27814	82	30	37	7907	36	3238	3236	6231
c6288	65	40916	74	–	67	48648	55	97973	10534	97973
c7552	79	29379	75	75	79	433613	65	4328	2544	9218

$k = 20$

* These times are on a SUN 3/280 Computer
† These times are on a SUN 4/370 Computer
– Results are not available

time during which no improvement was observed until the termination condition was satisfied. We do not have the results for $k = 16$ for the PEST algorithm [241]. In most cases, the GA gives results which are superior to those obtained by all other algorithms. The most noteworthy result is for c1908 and $k = 20$, where the GA obtains a reduction in the number of segmentation points by 38% over DC [92] and 22% over PEST [241]. By changing the parameters of the GA, e.g., the crossover and mutation probabilities, we have obtained better results than those shown for some circuits; e.g., for c5315 and $k = 20$, the GA obtains a circuit with

29 segmentation cells compared to 36 shown. The results of Tables 5.5 and 5.6 were obtained with a single set of crossover and mutation rates for all circuits.

5.5 Summary

In this chapter, we have presented a GA approach for circuit segmentation with applications in FPGA technology mapping. We also briefly described the application of segmentation to pseudo-exhaustive testing of the circuit. The experimental results have demonstrated the effectiveness of this approach in solving this problem. Design automation problems generally have highly constrained solution spaces. This has prevented the widespread use of the genetic optimization strategy in solving them. We presented an approach using randomized constraint-based validation that can also be applied to other problems with constrained solution spaces.

SUMMARY

TOPICS STUDIED

KEY OBSERVATIONS AND POINTS OF INTEREST

Genetic Algorithm for Circuit Segmentation in FPGA Technology Mapping and Pseudo-Exhaustive Testing
- Representation
- Validation
- Fitness Function
- Results

- Binary string indicating which gates are segmentation points
- Binary string representing decomposition state of each gate

- Invalid offsprings are altered to create valid solutions

- Minimize the number of TLU functions
- Minimize number of CLBs used in FPGA implementation

- Better results than heuristic approaches for FPGA mapping, but execution times are longer
- Better results than other algorithms for pseudo-exhaustive testing, and execution times are often shorter

CHAPTER 6
at a glance

- Automatic Test Generation (ATG) for sequential circuits
- GA framework for ATG
  - Encoding alphabet
  - GA parameters
  - Fitness function
- Test application time reduction
- Deterministic/genetic hybrid test generation
  - GA for state justification
  - Alternating deterministic/genetic phases
- Use of finite state machine sequences in fault-oriented ATG
- Dynamic test sequence compaction

Chapter 6

AUTOMATIC TEST GENERATION

E. M. Rudnick, M. S. Hsiao, J. H. Patel

The Integrated Circuit (IC) fabrication process is prone to random defects, which may affect the functionality of a device and cause erroneous outputs. Testing of finished circuits is essential in weeding out defective parts to ensure that electronic systems built using the ICs function correctly. Sets of test vectors applied to circuits by a tester must have high defect coverages if they are to be effective in identifying defective chips. Furthermore, since the cost of testing VLSI chips is a significant fraction of the overall manufacturing cost, the time required to test a chip should be minimized. Effective tools for Automatic Test Generation (ATG) are needed to obtain compact test sets with high defect coverages.

In this chapter, we will discuss how genetic algorithms can be used for automatic test generation. Genetic algorithms have been very effective for sequential circuit test generation, especially when combined with deterministic algorithms. We begin with an overview of the test generation problem and then discuss how tests can be generated in a GA framework. Details are given of an encoding alphabet, GA parameters, and fitness function, and an implementation is described. Results of an evaluation of various GA parameters are presented, and extensions to the GA framework for reducing test application time in circuits that use scan as a Design-For-Testability (DFT) methodology are described. Two approaches for integrating GAs with deterministic algorithms into a hybrid test generator are described next, followed by a discussion of how problem-specific knowledge in the form of finite state machine sequences can be incorporated into the GA in order to achieve better results. Finally, the use of GAs for dynamic test sequence compaction is described.

6.1 Problem Description

The objective of automatic test generation is to obtain a set of test vectors that will detect any defect that might occur in the manufacturing process. However, covering all potential defects would be very difficult and would require an inordinate number of test vectors. Therefore, automatic test generators operate on an abstract representation of defects referred to as *faults* and model a subset of the potential faults. A fault represents the logical effects of a defect on the circuit [2]. The single stuck-at fault model is used most often, with the assumption that many additional faults will be detected by tests generated using this fault model. With the single stuck-at fault model, faults that tie a circuit node to logic one or logic zero for every node in a circuit are considered. For example, Fig. 6.1 shows a circuit with a single stuck-at fault in which node D is tied to logic 0 (D is s-a-0). We assume that only a single fault is present in the circuit to simplify the problem. A logic one must be applied to node D if there is to be a difference between the faulty and fault-free circuits. Also, a logic zero must be applied to node C so that if the fault is present, it can be detected at the output E. In addition to the s-a-0 fault at node D, several other faults must be considered during the test generation process: s-a-1 at D, s-a-0 at A, s-a-1 at A, s-a-0 at B, etc. Some of these faults are logically equivalent, and no test can be obtained to distinguish between them. For example, in Fig. 6.1, A s-a-0, B s-a-0, and D s-a-0 are equivalent since they are detected by the same tests. Equivalent fault collapsing is often used by test generators to identify equivalent faults in order to reduce the number of faults that must be targeted [2].

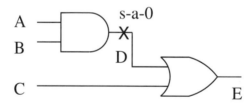

Figure 6.1. Single stuck-at fault

In a typical ATG tool, individual faults in a circuit are targeted, and after a test (one or a sequence of test vectors) is successfully generated to detect a fault, a fault simulator is used in order to identify additional faults that are covered by the test. These faults are removed from the fault list and need not be targeted by the more computation-intensive test generator.

Automatic test generation for combinational circuits is commonly done using variations of the PODEM algorithm [85]. Values required to excite a targeted fault are determined, and these values are backtraced through logic gates to the Primary Inputs (PIs) of the circuit using knowledge of the logic gate functions. Backtracing is a critical step and is used to determine the component input values required to obtain a particular output value. Handling components other than simple gates

6.1. Problem Description

is especially difficult because of the backtracing step. Decisions are made at the PIs based on the backtraced values, and alternative choices are placed on a decision stack, which is used during decision backtracking. Forward implication is performed when a value is assigned to a PI to determine the values of other nodes affected by the PI. After all values required for fault excitation are backtraced to the PIs and the corresponding forward implications performed, the fault effects are propagated to the Primary Outputs (POs), again using knowledge of the logic gate functions. During fault effect propagation, additional backtracing and forward implications are done as necessary. If conflicts are found, then the test generator backtracks to a previous decision point and makes an alternative choice. If all choices are exhausted, the fault is declared to be untestable.

Suppose we want to use the PODEM algorithm to generate a test for G s-a-1 in the circuit of Fig. 6.2. A logic zero is required at node G to excite the fault. Either node A or node F (or both) must therefore be logic zero. If F is chosen, then nodes D and C must be logic one. Since C is a PI, it is assigned a value of 1, and forward implications are performed. As a result, nodes D and E are assigned values of 1, node F is assigned a value of 0, and node G is assigned a value of 0/1, 0 for the fault-free circuit and 1 for the faulty circuit. Propagation of the fault effects to a PO requires a 0 on node H, and consequently a 0 on node B. Since B is a PI, it is assigned a value of 0, and H and Y are assigned values of 0 and 0/1, respectively, through logic implication. For this simple circuit, a test is easily obtained with no backtracking on decisions.

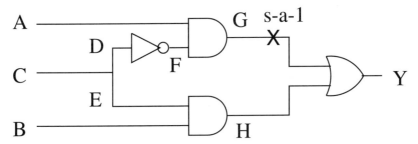

Figure 6.2. Combinational circuit test generation

Full scan is often used to convert sequential circuits into combinational circuits during testing [64], [237]. With this DFT approach, all flip-flops are arranged in one or more scan chains when a circuit is being tested, and flip-flop values are scanned in before each test vector is applied and scanned out after each test vector is applied. Hence, combinational circuit test generators can be used. However, full scan is not desirable for many circuits, due to area and performance overhead, and sequential circuit test generators are required. Partial scan requires less overhead since only a subset of the flip-flops are scanned. The complexity of test generation is reduced compared to the original nonscan circuit; however, the circuit is still of sequential nature, and a sequential circuit ATG is needed.

Sequential circuit test generation using deterministic algorithms is highly complex and time consuming [37], [76], [134], [141], [145], [160], [204]. A sequential circuit can be viewed as a combinational circuit expanded over several time frames, where the next state values in one time frame correspond to the present state values in the next time frame. During test generation, each target fault must be excited and the fault effects propagated to a PO. Fault effects may propagate directly to a PO in the same time frame in which the fault is excited or through flip-flops to the POs in subsequent time frames, as illustrated in Fig. 6.3(a). The required state must then be justified through reverse time processing. Values required at the flip-flops are backtraced time frame by time frame until a time frame is reached in which all flip-flops have don't-care values, as illustrated in Fig. 6.3(b). Test sets can be long, and development of a deterministic test generator is very time consuming. For these reasons, many researchers have suggested a simulation-based test generation approach. In a simulation-based approach, processing occurs in the forward direction only; i.e., no backtracing is required. Therefore, complex component types are handled more easily. As a result, the development time is greatly reduced.

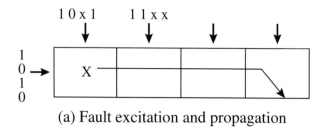

(a) Fault excitation and propagation

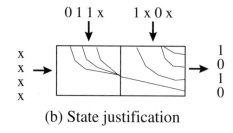

(b) State justification

Figure 6.3. Deterministic algorithm for sequential circuit ATG

Seshu and Freeman [207] first proposed simulation-based test generation, and several simulation-based test generators have since been developed [6], [20], [139], [203], [218], [242]. Breuer [20] used a fault simulator to evaluate sets of random vectors and to select the best vector to apply in each time frame. Weighted random pattern generators were interfaced with fault simulators in [139], [203], [242], and high fault coverages were obtained for combinational circuits. The test generators in [6], [218] were also built around fault simulators, but only candidate vectors of

Hamming distance one from the previous vector were considered; i.e., successive vectors differed in a single bit only. Specific faults were targeted in [218], with a backtrace step used to select the bit to be flipped. Cost functions calculated during concurrent fault simulation were used to evaluate candidate vectors in [6]. While development of these random and mutation-based test generators was simplified and test generation time was reduced, the test sets generated were typically much longer than those generated by deterministic test generators. One common attribute to all the work on simulation-based test generation is that it is by and large ad hoc, with numerous heuristics. The heuristics are often very clever, but there is no underlying well-defined algorithmic framework such as genetic algorithms or simulated annealing.

Genetic algorithms were first used as a framework for simulation-based test generation by Saab et al. in the test generator CRIS [195] and by Srinivas and Patnaik [219]. CRIS used a GA with overlapping populations modeled after Holland's GA [95] in which two new individuals were evolved in each generation. The new individuals replaced the least-fit individuals with some high probability. Individuals in the population represented test sequences, and each test sequence was evaluated by using a logic simulator and measuring the activity at internal nodes in a circuit. A circuit was divided into partitions using depth-first search from the POs, and the fitness function used favored test sequences that resulted in the same activity levels in all partitions of the circuit. Presumably, test sequences that result in similar activity levels at all nodes are expected to provide high fault coverage. However, this fitness function is only a rough approximation of the actual quality of a given test sequence. A ranking scheme was used for selection of individuals, and a heuristic crossover scheme was used to exploit problem-specific knowledge. The test sets generated tended to be longer and have lower fault coverage than test sets generated using a deterministic test generator because of the inaccuracies inherent in the fitness function used. Furthermore, the heuristic crossover scheme makes it difficult to separate the effects of the GA from the application-specific heuristics used. Nevertheless, significant speedups were obtained. In a more recent version of CRIS [197], fault simulation was used in the evaluation of candidate tests after the easy-to-test faults were detected. Fault coverages improved for many circuits, but execution times also increased.

In the GA-based test generator developed by Srinivas and Patnaik [219], only combinational circuits were handled, since a combinational circuit fault simulator was used in evaluating candidate tests. Each individual in the GA population represented a test vector, and uniform crossover was used. The selection scheme was not reported. An adaptive GA was used to vary the crossover and mutation rates depending on the fitness value of an individual. The fitness function included costs for activation and propagation of fault effects. The test sets generated were significantly larger than the CRIS test sets for the combinational benchmark circuits, although the fault coverages were higher.

The GATTO sequential circuit test generator used GAs in the fault propagation phase of a two-phase test generation strategy [46], [181]. Sequences of random vectors were fault simulated until one or more faults were excited. The 64 faults that propagated furthest were selected as target faults, and a GA was used to propagate the fault effects even further toward the flip-flops and primary outputs. If the fault effects reached the primary outputs, the corresponding test sequences were added to the test set; otherwise, the faults were aborted after several unsuccessful attempts. The GA used ranking for selection of individuals, and the best individuals were saved from one generation to the next. Fault coverages were sometimes greater than and sometimes less than those for CRIS, as were the test set sizes, but the implementation for which these results were obtained required the existence of a reset state.

A GA-based test generator was also developed by Pomeranz and Reddy [179] that used problem-specific knowledge particularly well suited for combinational circuit test generation. PIs that are in the same cones of logic for specific internal nodes are grouped together, and crossover is performed at group boundaries only. The idea is to link the PIs that affect fault excitation and fault propagation. The internal nodes are selected so that each PI is in at least one group, and the number of PIs in a group is not too large or too small. Uniform crossover is used, and since a PI can be assigned to more than one group, the order of copying values for groups of PIs from a parent to a child must be prioritized. This is done randomly. A steady-state GA is used, and only individuals that improve the fault coverage are added to the population. The population size increases throughout the process, and processing terminates when the number of iterations performed reaches a given limit or until all faults are detected.

The goal in our work has been to use simulation-based test generation implemented on the well-established framework of GAs. In our initial implementation, we used problem-specific knowledge in the fitness evaluation only, which allowed us to study the effectiveness of GAs alone for the test generation application. We applied the simple genetic algorithm described by Goldberg [79] to the generation of individual test vectors for combinational and sequential circuits [188]. A sequential circuit fault simulator was used in the fitness evaluation. Compact test sets with high fault coverages were obtained for most combinational circuits, and high fault coverages were obtained for many of the sequential circuits. However, fault coverages were lower for highly sequential circuits.

The GA-based test generator was extended to evolve test sequences in addition to individual test vectors [189], [194]. The GA generates candidate test vectors and sequences, and the fitness of each candidate test is computed by a sequential circuit fault simulator. One issue in evolving test sequences is the alphabet size, i.e., whether a binary or nonbinary coding should be used. In a binary coding, the individual vectors in a sequence are packed into a single string, and the GA operates on that string. In a nonbinary coding, each possible vector is a separate character in the alphabet, and the GA operates on the test sequence as a string of characters in the alphabet, with special operators developed for the nonbinary alphabet. Another concern is to achieve compact test sets in a reasonable execution time. An accurate fitness function is needed, while approximations are required to speed up the execution. We have chosen to use fault simulation of candidate tests, rather than the less accurate logic simulation used in CRIS [195], to provide a better quality test set. A small sample of faults can be used in the fitness computation to speed up the execution. Another technique to reduce execution time is to use overlapping populations in the GA in which only a fraction of candidate tests is replaced in each generation.

6.2 Test Generation in a GA Framework

Genetic algorithms can be used to generate populations of candidate test vectors and sequences and to select the best test to apply in a given time frame. This process is illustrated in Fig. 6.4. The test generator begins by generating individual

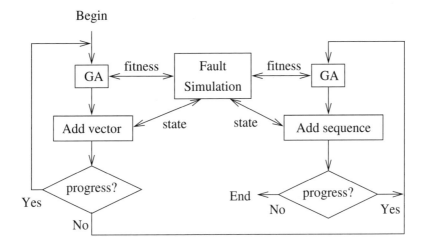

Figure 6.4. GA-based test generation

test vectors. When no more improvements in fault coverage can be made with the individual test vectors, the test generator proceeds to test sequence generation if the circuit is sequential. Test sequences are generated until no more progress is made, at which point test generation terminates. A GA having a random initial population is used to generate each test vector or sequence, as shown in Fig. 6.5, and a sequential circuit fault simulator is used to evaluate the fitness of each candidate test. The best test evolved in any generation is selected and added to the test set. Then the fault simulator is used to update the state of the circuit and to drop detected faults.

Generation of individual test vectors is repeated until a given number of vectors are successively generated that do not improve the fault coverage. This *progress limit* is set to ten for combinational circuits and a small multiple of the sequential

```
generate initial Population;
evaluate (Population);
bestIndividual = bestOf (Population);
For i = 1 to NumGenerations Do
    NewPopulation = ∅;
    For j = 1 to PopulationSize/2 Do
        select p₁ and p₂ from Population;
        crossover(p₁, p₂, c₁, c₂);
        mutate(c₁);
        mutate(c₂);
        Add c₁ and c₂ to NewPopulation;
    evaluate (NewPopulation);
    Population = NewPopulation;
    bestIndividual = bestOf (Population ∪ bestIndividual);
solution = bestIndividual;
```

Figure 6.5. Genetic algorithm for test generation

depth for sequential circuits. A new random initial population is used with each attempt; therefore, a useful vector may be found even after several unsuccessful attempts. For sequential circuits, several vectors may be required to change to a state in which additional faults may be detected. However, since the circuit is not guaranteed to go into a desirable state, a reasonable progress limit should be used to limit the execution time and test set size. For combinational circuits, only vectors that improve the fault coverage are added to the test set, and the second best test vector evolved is included in the initial GA population for the next time frame, since it may be useful.

Even when test sequences are being generated, a sequence may not be found to improve the fault coverage if the initial population does not contain the right combination of vectors. Therefore, the GA is reinitialized with a new random population for each attempt at generating a useful test sequence. Test generation for a given sequence length is terminated if four consecutive attempts fail to improve the fault coverage, and only sequences that improve the fault coverage are added to the test set. In addition, various sequence lengths are used in an attempt to achieve a high fault coverage in a reasonable time. In this work we use one, two, and four times the sequential depth for most circuits, beginning with the smallest sequence length. Many of the faults are detected by the individual test vectors and shorter sequences, and only the most difficult faults remain for the longest sequences. Execution time is thus reduced.

6.2.1 Encoding Alphabet

The mapping from a GA individual to a test vector or sequence is illustrated in Fig. 6.6. During generation of individual test vectors (Fig. 6.6(a)), each character

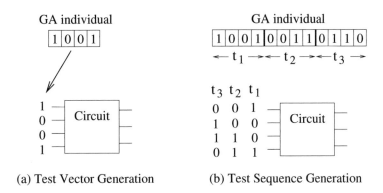

Figure 6.6. GA encoding

of a string in the population is mapped to a Primary Input (PI). A binary coding is used; therefore, the string represents a test vector. Thus, all PIs are set to known zero or one states. The fitness of each string is evaluated directly through fault simulation. During test sequence generation (Fig. 6.6(b)), the GA individual represents a sequence of vectors to be applied to the PIs in successive clock cycles (t_1, t_2, and t_3 in the figure). Either a binary or nonbinary coding can be used, and the genetic operators used depend on the encoding scheme. If a binary coding is used, the individual vectors in a sequence are placed in adjacent positions on a single string. Then the GA processes that string using the same selection, bitwise crossover, and bitwise mutation operators that are used in generating individual test vectors. Crossover can occur at test vector boundaries, but it can also occur at positions internal to a test vector. Mutation involves flipping of a single bit in a given string position.

In a nonbinary coding, all 2^L possible test vectors are separate characters in the alphabet, where L is the vector length, and the individual characters of a string represent separate test vectors in a sequence. In our implementation, the characters are mapped to their binary equivalents to enable simulation of candidate tests, but test vector boundaries are maintained to separate characters. The increased alphabet size has no effect on the selection operator, but the crossover and mutation operators must be modified. Crossover can occur at test vector boundaries only, and mutation involves replacing a given vector in a sequence with a randomly generated vector.

Whether a binary or nonbinary coding is preferable in general is open to debate. However, if a nonbinary coding is used, a larger population size and mutation rate may be required to ensure adequate diversity.

6.2.2 GA Parameters

Several GA parameters are important in achieving good results. A sufficient population size is needed to ensure adequate diversity. All characters in the alphabet should be present at every string position, and a sufficient population size is needed to provide good combinations of characters. For example, in test vector generation, a group of PIs may have specific required values to activate certain faults. If no string in the population has all the PIs set to the required values, the faults will not be activated. The number of combinations of bits increases with increasing test vector length. Similarly, the numbers of characters and combinations of characters increases with increasing vector and sequence length during test sequence generation. Therefore, the required population size increases with increasing test vector and test sequence length. At the same time, a reasonable limit on the population size is needed to reduce computations. A second key parameter is the number of generations. A sufficient number of generations is needed to allow useful combinations of characters from several strings to mix. In this work, we limit the number of generations to eight to reduce the run time. For combinational circuits, the number of generations is increased to twenty in the later part of the execution in order to expand the search space. Alternatively, we could have increased the population size. Typically, many faults are detected by each of the first several vectors, but the number of faults detected drops off with successive vectors. A relatively small population size and small number of generations can be used to find good vectors in a reasonable time, although better vectors might exist. Then, when a time frame is encountered in which no faults are detected, the search space is expanded. The impact of increasing the search space on execution time is smaller after many faults have been dropped, since less time is required to simulate each candidate vector.

Tournament selection without replacement and uniform crossover are used by default. The remaining parameters of interest are the crossover and mutation probabilities. We use a crossover probability of one; i.e., two individuals are always crossed in generating two new individuals. Mutation is used to prevent the loss of key characters at the various string positions. However, mutation also destroys good combinations of characters, so a balance must be found. The population sizes and mutation probabilities used in this work are shown in Table 6.1 for the generation of individual test vectors. During test sequence generation, a population size of 32 and a mutation rate of 1/64 are used for all circuits.

6.2.3 Fitness Function

An accurate fitness function is needed to achieve a high-quality test set. Several factors may be important in generating a test vector or sequence, depending on the phase of test generation. Furthermore, with a proportionate selection scheme, such as roulette wheel selection or stochastic universal selection, the combination of factors in the fitness function is critical. In CONTEST [6], test generation takes place in three phases, each having its own fitness measure. We use a similar strategy.

In the initial phase of test vector generation, test vectors are generated to ini-

Table 6.1. GA Parameter Values

Vector Length (L)	Population Size	Mutation Probability
< 4	8	1/8
4–16	16	1/16
17 - 49	16	1/L
50 - 63	24	1/L
64 - 99	24	1/64
> 99	32	1/64

tialize the flip-flops. Therefore, the fitness of a candidate vector is a measure of the number of flip-flops set to a known (zero or one) state. To differentiate vectors that cause the same number of flip-flops to be set, we also include the fraction of flip-flops changing values since the previous time frame. Only a good circuit simulation is required to obtain the flip-flop state information. When all flip-flops are set, the test generator switches to *Phase 2* in which test vectors are generated to maximize the number of faults detected. In this phase, the fitness of a candidate vector indicates the number of faults it detects. To differentiate vectors that detect the same number of faults, we include the number of fault effects propagated to flip-flops in the fitness function, since fault effects at the flip-flops may be propagated to the POs in the next time frame. However, the number of fault effects propagated is offset by the number of faults simulated and the number of flip-flops to ensure that the number of faults detected is the dominant factor in the fitness function. When a test vector is generated that detects no additional faults, the test generator enters *Phase 3* and begins counting the number of noncontributing test vectors. In order to encourage the evolution of useful vectors, we add the good and faulty circuit activity levels to the other two measures used in Phase 2. Vectors that activate more faults and propagate more fault effects will then have higher fitness values, and the GA will be more likely to evolve a vector that can propagate the effects of some fault to a PO. If a test vector is found that detects any faults before the number of noncontributing vectors generated reaches the progress limit, the test generator switches back to Phase 2, and the noncontributing vector count is reset to zero. The procedure for generation of individual test vectors is summarized in Fig. 6.7.

When the number of successive noncontributing vectors generated exceeds the progress limit, the test generator proceeds with test sequence generation. In this phase, the fitness function used is the same as that for the second phase of test vector generation except that the test sequence length is included in the metric for the number of fault effects propagated to flip-flops. In summary, the fitness of a

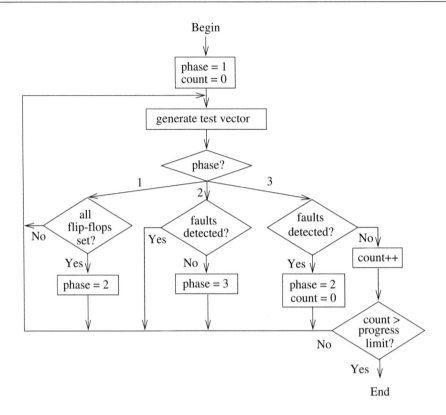

Figure 6.7. Generation of individual test vectors

candidate test is calculated as follows:

Phase 1: $fitness = total\ flip\ flops\ set + fraction\ of\ flip\ flops\ changed$

Phase 2: $fitness = \#\ faults\ detected + \dfrac{\#\ faults\ propagated\ to\ flip\ flops}{(\#\ faults)(\#\ flip\ flops)}$

Phase 3: $fitness = \#\ faults\ detected + \dfrac{\#\ faults\ propagated\ to\ flip\ flops}{(\#\ faults)(\#\ flip\ flops)}$
$+ \dfrac{2(\#\ good\ and\ faulty\ circuit\ events)}{(\#\ circuit\ nodes)(\#\ faults)}$

Phase 4 (Test Sequence Generation):

$fitness = \#\ faults\ detected + \dfrac{\#\ faults\ propagated\ to\ flip\ flops}{(\#\ faults)(\#\ flip\ flops)(sequence\ length)}$

Note that the computational cost of calculating a fitness function is the dominant cost in the execution of the GA. The string manipulations (selection, crossover, and mutation) are not very time consuming. While an accurate fitness function is essential in achieving a good solution, the high computational cost of fault simulation may be prohibitive, especially for large circuits. To avoid excessive computations, we can approximate the fitness of a candidate test by using a small sample of faults. In particular, we can use a small fraction of the remaining faults chosen at random, e.g., 1%–10%, or a set sample size, e.g., 100–300 faults. Another alternative is to avoid using the costly fault simulator altogether by choosing a fitness function based on good circuit simulation, as is done in CRIS [195]. Of course, using a less accurate fitness function has an adverse effect on the quality of the solution; i.e., in our case, it may result in more test vectors and lower fault coverage.

6.2.4 An Implementation

The GA-based test generator, GATEST, was implemented around the sequential circuit fault simulator PROOFS [161] in 3000 additional lines of C++ code. Results obtained for the ISCAS85 combinational benchmark circuits [22] are shown in Table 6.2 averaged over ten runs, with standard deviations shown in parentheses. Execution times are for a Sun SPARCstation II with 64-MB RAM. High fault coverages and very compact test sets were obtained, although better results can be obtained with deterministic algorithms [188]. Specifically, deterministic algorithms can achieve 100% coverage of testable faults in much less time. Test set sizes for deterministic algorithms that use compaction techniques are only marginally smaller, however.

One hundred percent coverage of testable faults in the ISCAS85 circuits was also achieved by the GA-based test generator developed by Pomeranz and Reddy [179],

Table 6.2. GATEST Results for Combinational Circuits

Circuit	Gates	PIs	Faults			Test Vectors	Exec Time
			Total	Testable	Detected		
c432	160	36	524	520	520.0 (0.0)	42.5 (1.2)	43.4s
c499	202	41	758	750	750.0 (0.0)	52.2 (0.4)	38.4s
c880	383	60	942	942	942.0 (0.0)	35.7 (1.8)	1.24m
c1355	546	41	1574	1566	1566.0 (0.0)	84.1 (0.3)	1.97m
c1908	880	33	1879	1870	1868.5 (0.5)	115.2 (2.1)	4.57m
c2670	1193	233	2747	2630	2526.3 (46.3)	62.4 (6.4)	16.1m
c3540	1669	50	3428	3291	3288.8 (1.6)	121.2 (3.9)	17.7m
c5315	2307	178	5350	5291	5291.0 (0.0)	83.6 (2.8)	21.4m
c6288	2416	32	7744	7710	7710.0 (0.0)	17.9 (0.7)	6.38m
c7552	3512	207	7550	7419	7356.0 (13.7)	133.2 (6.8)	55.5m

but test set sizes were larger. For the hard-to-test circuits (c2670 and c7552), many more candidate tests were evaluated than for GATEST, and execution times can be expected to be correspondingly longer. Nevertheless, the results indicate that GAs are effective for combinational circuit test generation when problem-specific knowledge is incorporated [179].

Table 6.3. GATEST Results for Sequential Circuits

Circuit	PIs	Seq Depth	Total Faults	HITEC Det	HITEC Vec	HITEC Time	GATEST Det	GATEST Vec	GATEST Time
s298	3	8	308	**265**	306	4.44h	**264.7**	161	6.05m
s344	9	6	342	328	142	1.33h	**329.0**	95	5.85m
s349	9	6	350	**335**	137	52.2m	**335.0**	95	5.83m
s382	3	11	399	**363**	4931	12.0h	347.0	281	8.91m
s386	7	5	384	**314**	311	1.03m	295.2	154	3.45m
s400	3	11	426	**383**	4309	12.1h	365.1	280	9.45m
s444	3	11	474	**414**	2240	16.1h	405.7	275	10.5m
s526	3	11	555	365	2232	46.8h	**416.7**	281	14.3m
s641	35	6	467	**404**	216	18.0m	**404.0**	139	8.24m
s713	35	6	581	**476**	194	1.52m	**476.0**	128	9.41m
s820	18	4	850	**813**	984	1.61m	516.5	146	13.4m
s832	18	4	870	**817**	981	1.76m	539.0	150	12.3m
s1196	14	4	1242	**1239**	453	1.53m	1232	347	11.6m
s1238	14	4	1355	**1283**	478	2.20m	1274	383	16.0m
s1423	17	10	1515	-	-	-	1222	663	2.83h
s1488	8	5	1486	**1444**	1294	3.60h	1392	243	25.2m
s1494	8	5	1506	**1453**	1407	1.91h	1416	245	23.2m
s5378	35	36	4603	-	-	-	3175	511	6.08h
s35932	35	35	39,094	34,902	240	3.80h	**35,003**	200	106.7h

Tests were generated for the ISCAS89 sequential benchmark circuits [23] using GATEST on a SUN SPARCstation II with 64-MB RAM. Circuit descriptions and test generation results are shown in Table 6.3. The structural sequential depth shown in the table is taken from [159]. It is defined as the maximum structural sequential depth of all flip-flops in the circuit, where the structural sequential depth of a flip-flop is defined as the minimum number of flip-flops that must be passed through to reach that flip-flop from a PI. The numbers of faults detected (**Det**), the numbers of test vectors generated (**Vec**), and the execution times are shown for tournament selection without replacement and uniform crossover. A binary coding was used. The progress limit for test vector generation was equal to the sequential depth for s5378 and s35932 and four times the sequential depth for all other circuits. Test sequence lengths were 1/4, 1/2, and 1 times the sequential depth for s5378 and s35932 and 1, 2, and 4 times the sequential depth for all other circuits. Each result is the average of ten runs, and a new random seed was used for each run. The numbers of faults detected, the numbers of vectors generated, and the execution times on a Sun SPARCstation SLC are shown for the HITEC deterministic, fault-oriented test generator [159] for comparison, and the highest fault coverages achieved are

6.2. Test Generation in a GA Framework

highlighted in bold. The number of faults detected was greater than or equal to that of HITEC for 7 of the 17 circuits for which fault coverages were available. The fault detection count was within 10 faults for another 3 circuits and within 20 faults for an additional 3 circuits. Test set length was 42% that of HITEC on average.

In most cases, test generation time for GATEST is a small fraction of the time required by HITEC. Thus, GATEST can be used as a first pass in test generation to screen out many of the faults before applying a deterministic test generator. Note that untestable faults cannot be identified by a simulation-based test generator, so the deterministic fault-oriented test generator is still needed for this purpose.

Test generation results for the CRIS GA-based test generator [195] are shown in Table 6.4 for comparison. Execution times are reported for a Sun SPARCstation SLC with 16-MB RAM. Recall that CRIS uses the less accurate logic simulation, rather than fault simulation, in the fitness evaluation. Fault coverages for GATEST were higher than those for CRIS for 17 of 18 circuits, although the execution time was between 6 and 40 times as long, depending on the circuit. Test set length was one-third that of CRIS on average.

Table 6.4. CRIS Results for Sequential Circuits

Circuit	Det	Vec	Time
s298	253	476	16.2s
s344	328	115	19.9s
s382	274	246	16.7s
s386	292	1230	16.9s
s400	357	758	35.3s
s444	397	519	34.2s
s526	428	692	48.4s
s641	398	628	1.45m
s713	475	1124	1.58m
s820	451	1381	2.26m
s832	370	1328	1.75m
s1196	1180	2744	1.62m
s1238	1229	4313	2.19m
s1423	1167	2696	9.68m
s1488	1355	1960	3.10m
s1494	1357	1928	2.85m
s5378	3029	1255	15.8m
s35932	34,481	1525	2.65h

6.2.5 Evaluation of GA Parameters

Results for various selection and crossover schemes are shown in Table 6.5. Again results are shown averaged over ten runs. Circuits that had about the same fault coverages for all selection and crossover schemes are omitted from the table. Significant differences in fault coverage were found for many of the larger circuits. The best selection scheme was tournament selection without replacement; both tournament selection schemes gave better results than either of the proportionate selection schemes. Uniform crossover gave consistently better results than 1-point or 2-point crossover.

Table 6.5. Selection and Crossover Scheme Comparison: Detected Faults

Cir-	Roulette Wheel Selection			Stochas. Univer. Selection			Tournament Selection					
							No Replacement			Replacement		
cuit	1-pt	2-pt	Unif	1-pt	2-pt	Unif	1-pt	2-pt	Unif	1-pt	2-pt	Unif
s386	294	293	296	297	296	**298**	295	297	295	297	296	296
s526	420	420	418	**422**	415	418	416	417	417	417	418	420
s820	501	478	514	503	497	524	520	520	517	**528**	528	505
s832	512	504	507	501	516	513	522	516	**539**	516	502	515
s1196	1228	1228	**1232**	1229	1228	1231	1227	1229	**1232**	1227	1225	1230
s1238	1270	1272	1274	1273	1271	**1275**	1269	1272	1274	1268	1272	**1275**
s1423	1243	1229	**1257**	1210	1243	1223	1242	1219	1222	1250	1227	1212
s1488	1363	1381	1352	1378	1360	1367	1392	1390	1392	1380	1388	**1395**
s1494	1357	1362	1361	1352	1401	1394	1412	1388	**1416**	1384	1391	1408
s5378	3169	3160	3216	3124	3183	3167	3175	3165	3175	3168	3150	**3180**

1-pt: one-point crossover, **2-pt**: two-point crossover, **Unif**: uniform crossover

The effects of mutation rate on fault coverage were also investigated. Results are shown in Table 6.6 averaged over ten runs for various mutation rates used during test sequence generation; mutation rates given in Table 6.1 were used while individual test vectors were being generated. Tournament selection without replacement and uniform crossover were used. Circuits having about the same fault coverage for all mutation rates are not included. The mutation rate had a much smaller effect on fault coverage than the selection and crossover schemes. Significant differences in fault coverage were obtained for only two of the larger circuits, s1423 and s5378, but for these circuits, the lowest mutation rates (1/128 and 1/256) tended to give the highest fault coverage.

Binary and nonbinary codings of test sequences are compared in Table 6.7. Results are averaged over ten runs. Population sizes of 16, 32, and 64 were used during test sequence generation; the population sizes given in Table 6.1 were used while individual test vectors were being generated. Circuits having about the same fault coverage for all experiments are not shown. Tournament selection without replacement and uniform crossover were used. Fault coverages tended to improve

6.2. Test Generation in a GA Framework

Table 6.6. Mutation Rate Comparison: Detected Faults

	Mutation Rate				
Circuit	1/16	1/32	1/64	1/128	1/256
s386	296	**297**	295	296	296
s820	511	509	**517**	510	510
s832	534	534	**539**	534	533
s1196	1231	1230	**1232**	1231	1230
s1238	1274	1275	1274	**1276**	1274
s1423	1216	1226	1222	1244	**1258**
s1488	**1394**	**1394**	1392	1393	1391
s1494	1416	1415	1416	**1418**	1417
s5378	**3204**	3159	3175	3175	3192

with increasing population size, as expected. The best results were obtained using the largest population size, but good results were also obtained with population sizes of 16 and 32. Therefore, we recommend using a population size of 16 or 32 to reduce the execution time. At the largest population size, a nonbinary coding gave better results, but significant differences were obtained for two circuits only. At the smaller population sizes, a binary coding usually gave better results.

Table 6.7. Binary and Nonbinary Coding Comparison: Detected Faults

Circuit	Pop Size = 16		Pop Size = 32		Pop Size = 64	
	Binary	Nonbinary	Binary	Nonbinary	Binary	Nonbinary
s386	294	294	295	295	**297**	296
s820	507	508	**517**	508	509	510
s832	533	535	**539**	534	533	534
s1196	1228	1223	1232	1228	**1233**	1229
s1238	1273	1262	1274	1267	**1277**	1273
s1423	1196	1202	1222	1219	1246	**1266**
s1488	1389	1386	1392	1387	**1396**	1395
s1494	1416	1413	1416	1416	**1417**	1415
s5378	3162	3165	3175	3190	3179	**3205**

Execution times may be reduced by using only a small fraction of the fault list in the fitness evaluation. Results of using this approach are shown in Table 6.8 averaged over ten runs for sample sizes of 100, 200, and 300 faults. Tournament selection without replacement, uniform crossover, and a binary coding were used. If the number of faults remaining in the fault list dropped below the fault sample size, then all remaining faults were simulated. For the smaller circuits, the differences in fault coverage were due to the nondeterminism of the algorithm used, since the undetected fault list size eventually dropped below the fault sample sizes. The highest fault coverages were typically obtained when the entire fault list was used. In general, the execution times were lower for the smaller fault sample sizes. The minimum size of the fault sample needed to obtain good results tends to increase with increasing circuit size. Note that the results for s35932 were influenced by the large number of combinationally redundant faults (3984). Untestable faults cannot be identified using a simulation-based approach, and they were not filtered out in our experiments. Thus, the untestable faults were heavily represented in the fault samples towards the end of the test generation process, providing no useful information to the GA.

Table 6.8. Fault Sampling

Circuit	Sample Size								
	100 Faults			200 Faults			300 Faults		
	Det	Vec	Spd	Det	Vec	Spd	Det	Vec	Spd
s298	264.5	161	1.05	264.7	168	0.99	265.0	179	0.95
s382	348.1	295	1.06	347.2	277	1.03	347.3	274	1.01
s386	286.8	128	1.16	297.3	133	1.11	295.3	143	1.07
s526	417.0	293	1.79	417.4	314	1.04	418.8	295	1.04
s820	494.7	144	2.75	536.8	157	1.77	532.2	155	1.45
s832	476.4	137	2.51	526.3	158	1.70	546.2	156	1.40
s1196	1230	373	1.55	1231	384	1.08	1230	348	1.12
s1238	1269	389	1.26	1274	375	1.19	1274	381	1.18
s1423	1245	619	3.28	1255	587	2.32	1287	778	1.11
s1488	1153	211	2.14	1394	272	1.03	1378	233	1.12
s1494	1303	267	1.65	1370	235	1.17	1400	242	1.10
s5378	3048	394	6.31	3095	409	5.24	3130	450	4.25
s35932	34854	282	3.53	34955	221	4.80	34937	203	4.55

$$\text{Spd} : \frac{Execution\ time\ using\ full\ fault\ list}{Execution\ time\ using\ fault\ sample}$$

Overlapping populations were also investigated as a means of reducing execution time. Generation gaps of $2/N$, $1/4$, $1/2$, and $3/4$ were tried, where N is the population size. Corresponding population sizes used were 3-, 2-, 1.5-, and 1-times the population size used for nonoverlapping populations. Larger populations were used for the smaller generation gaps to provide diversity, since overlapping populations sacrifice exploration in favor of exploitation. The number of generations was also adjusted for generation gaps of $2/N$ and $1/4$ to provide approximately the same number of evaluations for all experiments. Since the number of evaluations was about 81% of the number used for nonoverlapping populations, execution times were correspondingly smaller. Results for overlapping populations are given in Table 6.9 averaged over ten runs. Fault coverages for a generation gap of $3/4$ were

6.2. Test Generation in a GA Framework

only 0.4% lower on average than for nonoverlapping populations. Generation gaps of 1/2, 1/4, and 2/N resulted in successively lower fault coverages overall.

Table 6.9. Overlapping Populations

Cir-cuit	Generation Gap											
	2/N			1/4			1/2			3/4		
	Det	Vec	Spd	Det	Vec	Spd	Det	Vec	Spd	Det	Vec	Spd
s298	264	205	1.03	264	183	1.14	265	173	1.12	265	167	1.27
s382	348	270	1.24	348	277	1.23	347	283	1.17	347	270	1.28
s386	294	137	1.28	295	134	1.34	296	142	1.26	297	144	1.30
s526	417	306	1.20	420	299	1.21	417	298	1.13	418	301	1.25
s820	520	155	1.28	522	144	1.37	520	141	1.34	500	138	1.38
s832	512	140	1.22	508	154	1.14	522	151	1.14	501	142	1.21
s1196	1231	341	1.30	1231	374	1.20	1231	356	1.22	1230	385	1.20
s1238	1271	388	1.30	1274	393	1.31	1274	378	1.27	1273	394	1.36
s1423	1213	666	1.23	1216	677	1.20	1247	657	1.14	1239	669	1.16
s1488	1381	220	1.38	1410	252	1.33	1393	231	1.28	1404	247	1.35
s1494	1410	256	1.21	1402	236	1.28	1402	250	1.15	1408	239	1.32
s5378	3164	522	1.12	3170	560	1.09	3156	490	1.23	3193	500	1.33

$$\text{Spd}: \frac{\textit{Execution time using nonoverlapping populations}}{\textit{Execution time using overlapping populations}}$$

Several synthesized circuits were used to evaluate GATEST. Characteristics of these circuits are shown in Table 6.10, including sequential depth, numbers of flip-flops, PIs, and POs, and total number of collapsed faults. The am2910 12-bit microprogram sequencer is similar to the one described in [4]; div16 is a 16-bit divider that uses repeated subtraction to perform division; mult16 is a 16-bit two's complement multiplier that uses a shift-and-add algorithm; pcont2 is an 8-bit parallel controller used in DSP applications; and piir8o is an 8-point infinite impulse response filter for DSP applications. Results of running GATEST on the synthesized circuits are shown in Table 6.11. The number of faults detected, the number of test vectors generated, and the execution times are shown for each circuit. Tournament selection without replacement, uniform crossover, and a binary coding were used. Test sequence lengths were 6, 12, and 24, and the progress limit for generation of individual test vectors was 24. Fault samples of size 100 were used in the fitness evaluation. Results for HITEC [160] are shown for comparison. HITEC was run with time limits of 2, 20, and 200 seconds per fault for all circuits except pcont2, which was run with time limits of 2 and 20 seconds only. Times are shown for a Sun SPARCstation II with 32-MB RAM. GATEST gave higher fault coverages for four out of five circuits and lower execution times for all five circuits.

Table 6.10. Characteristics of Synthesized Circuits

Circuit	Seq Depth	FFs	PIs	POs	Total Faults
am2910	4	87	20	16	2391
div16	19	50	33	34	2147
mult16	9	55	18	33	1708
pcont2	3	24	9	8	11,300
piir8o	5	56	9	8	19,936

am2910: 12-bit microprogram sequencer
div16: 16-bit divider
mult16: 16-bit two's complement multiplier
pcont2: 8-bit parallel controller for DSP applications
piir8o: 8-point infinite impulse response filter for DSP applications

Table 6.11. Test Generation Results for Synthesized Circuits

Circuit	GATEST			HITEC		
	Det	Vec	Time	Det	Vec	Time
am2910	2163	745	1.47h	2164	874	5.19h
div16	1739	634	59.0m	1665	189	21.5h
mult16	1653	204	34.8m	1640	273	3.66h
pcont2	6826	272	1.08h	3354	7	38.1h
piir8o	15,013	531	2.71h	14,221	347	3.67h

6.3 Test Generation for Test Application Time Reduction

Typical Design-For-Testability (DFT) techniques involve improving the controllability and observability of internal circuit nodes. Full scan design provides complete controllability and observability at the flip-flops by placing the flip-flops in a scan chain so that values can be scanned in before each test vector is applied and scanned out after each test vector is applied. Thus, any state can be scanned in, which simplifies test generation, and fault effects need only to be propagated to the flip-flops. Combinational circuit test generators specifically targeted at generating compact test sets have been developed to limit the test application time for full scan designs [173]. However, test application time can still be long, especially for long scan chains, since flip-flop values must be scanned in and out before and after each test vector is applied. Multiple scan chains [158], parallel scan [185], [232], and selectable-length scan [149] have also been used to reduce test application time by reducing the number of clock cycles required for scan operations, but these approaches require greater hardware overhead. Selective scan [89], [129] has been used in a manner similar to selectable-length scan but without the hardware overhead; in this method, only required flip-flop values are shifted in and shifted out. This method can be combined with ordering of flip-flops in the scan chain to minimize the test application time [89], but extra routing overhead may be incurred. Partial scan design, in which only a subset of the flip-flops are placed in the scan chain, has been proposed as a means of reducing the area overhead and performance penalties of the DFT hardware [36], [38], [225]. Scan chain lengths are reduced, which may result in reductions in test application time. However, sequential circuit test generators are usually required since not all flip-flops are scanned, and the resulting test sets tend to be much less compact than those generated for full-scan implementations. Partial reset, in which reset signals are provided for a subset of the flip-flops, is also effective in reducing test application time, but fault coverage may also decrease [146].

Reductions in test application time have been made for some full-scan circuits by eliminating many of the scan operations [135], [180]. A hybrid approach to test generation was used in [180], in which sequences of parallel vectors were generated in addition to scan vectors. We refer to vectors applied without scan as *parallel vectors* and vectors applied with scan as *scan vectors*. The improvement in fault coverage obtainable by a sequence of parallel vectors was compared to the maximum fault coverage improvement for a set of scan vectors requiring an equivalent test application time in deciding whether to apply the parallel vectors or the scan vectors in a given time frame [180]. More significant reductions in test application time were obtained by targeting faults in reverse order of testability, i.e., by targeting the hardest faults first, and by including sequences in which flip-flop values are either scanned in or out, but not both [135]. Deterministic sequential circuit test generators were used to generate test sequences in both of these hybrid test generators, while combinational circuit test generators were used to generate the scan vectors. Since combinational circuit test generators are used, the same approach cannot be applied to partial scan circuits. Modifications to an existing sequential circuit test generator for reduction of test application time in partial scan circuits were reported in [136]. Again, faults are targeted in reverse order of testability, and for each fault, attempts are made to generate normal sequences of parallel vectors, scan-initiated sequences, or scan-terminated sequences to detect the fault. If these attempts are unsuccessful, the test generator resorts to using scan vectors.

Modifying a deterministic test generator to generate scan-terminated sequences is relatively easy, but handling scan-initiated sequences is much more difficult for partial scan circuits. The test generator must do reverse time processing to find a state in which only flip-flops in the scan chain have specified values. Backtracing decisions are typically made based upon node controllability values, but these values change depending on when the scan operation occurs. In a simulation-based approach to test generation, these problems are avoided. Processing occurs in the forward direction only, and values to be scanned into the flip-flops are evolved by a genetic algorithm. Vectors in the test sequences are also evolved by the GA, and a sequential circuit fault simulator is used to select the best normal or scan-initiated sequence to apply in a given time frame. Each test sequence is specifically targeted at detecting the maximum number of faults in the smallest test application time possible. Since the objective of the GA is to maximize improvement in fault coverage for each test generated, a very compact test set results. Furthermore, the same approach can be used for both full-scan and partial-scan circuits.

In our work, we use a GA-based test generator to generate test sequences. Both scan-initiated sequences and normal sequences are generated. In a *scan-initiated sequence*, the first vector is scanned, but all successive vectors are parallel vectors. Sequences of parallel vectors are called *normal sequences*. The ratio of the number of faults detected to test application time is compared for the scan-initiated and normal sequences, and the sequence with the highest ratio is added to the test set for a given time frame.

Flip-flop values may or may not be scanned in before a test sequence is applied, and they may or may not be scanned out after a test sequence is applied. Thus, a sequence may be normal, scan-initiated, scan-terminated, or both scan-initiated and scan-terminated. Faults whose effects propagate to the POs are detected, but faults whose effects propagate to flip-flops in the scan chain after the last vector in the sequence is applied are also detected if the sequence is scan-terminated. Ideally, we would like to evolve sequences of each of the four types and then choose the best sequence. In evaluating the fitness of a sequence that is not scan-terminated, we would include only faults detected at the POs. Sequences that are not scan-initiated would not be considered if the previous sequence selected was scan-terminated, and faults detected at the flip-flops in a scan-out operation at the end of the previous sequence would not be included in the fitness of the current scan-initiated sequence. However, the number of test sequences evaluated can be cut in half if the decisions on scan operations are restricted to whether a scan operation will occur before the current test sequence is applied. A test sequence becomes scan-terminated only if the next sequence is a scan-initiated sequence. The scan-out operation of one sequence is overlapped with the scan-in operation of the next sequence; therefore, no additional test application time is incurred due to the scan-out operation.

For a given time frame, we generate both scan-initiated and normal sequences. In computing the fitness of a candidate sequence, we assume that the sequence will be scan-terminated. This fitness function encourages evolution of sequences that can propagate fault effects to flip-flops in the scan chain. If the next sequence added to the test set is scan-initiated, the faults will be detected in the scan operation. Otherwise a better sequence can be found that does not require a scan operation, most likely because the fault effects at the flip-flops can be propagated to the POs by the sequence. In addition, if the current sequence is scan-initiated, then faults detected at the flip-flops in the scan operation are also included in the fitness function. Thus, the fitness computed is an approximation of the actual quality of a test sequence. This approximation may result in a small increase in test application time, but the execution will be speeded up significantly. The sequence having the

6.3. Test Generation for Test Application Time Reduction

best ratio of number of faults detected to test application time is selected and added to the test set.

In applying GAs to test generation, successive vectors in a test sequence are placed in adjacent positions along a string in the population. The string is then an encoding of several successive test vectors to be applied to the PIs of the circuit. We use a binary coding in which the two characters, 1 and 0, represent the values to be applied to the various PIs. For scan-initiated sequences, the state of the flip-flops to be scanned in is prefixed to each string, as illustrated in Fig. 6.8. Each character in the string now corresponds to a 1 or 0 value to be scanned into a flip-flop or applied

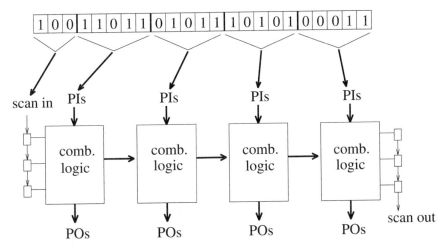

Figure 6.8. Generation of scan-initiated test sequences

to a PI. Since the state evolves along with the test vectors, and selection is biased toward better individuals, a scan-initiated sequence detecting additional faults is expected to be generated. Scan operations are useful for both controllability and observability enhancement; therefore, additional faults that can be detected at the flip-flops if they are scanned out are included in the fitness function computation.

The generation of scan-initiated sequences for partial scan circuits is the same as that for full scan circuits, except that only a subset of the flip-flops are scanned. Flip-flops not in the scan chain are simply loaded with the normal next state logic values from the previous time frame. Alternatively, *destructive scan* could be assumed in which flip-flops not in the scan chain are placed in an unknown state. In determining if additional faults are detected at the flip-flops in a scanout operation, only flip-flops in the scan chain are checked. The GA-based test generation framework for scan circuits is illustrated in Fig. 6.9. The simple GA for test generation outlined in Fig. 6.5 is again used. The GA generates populations of scan-initiated or normal sequences and selects the best test sequence to apply in the current time frame. The fitness of each candidate test is evaluated using the PROOFS sequential circuit fault simulator [161], and the fault simulator is also used to update

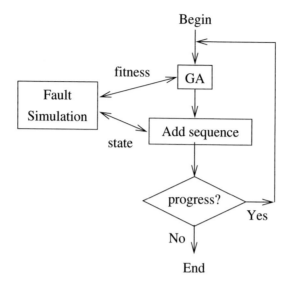

Figure 6.9. GA-based test generation for scan circuits

the state of the circuit and to drop detected faults once a test is generated. The GA is reinitialized with random sequences for each attempt at improving the fault coverage.

The overall objective of test generation is detection of all testable faults in the smallest test application time possible. In trying to achieve this objective, we attempt to find a test sequence that has a large ratio of number of faults detected to test application time. A normal sequence of moderate length typically has a lower test application time than a shorter scan-initiated sequence. However, the scan-initiated sequence has the potential to detect more faults. Therefore, we use a GA to generate both scan-initiated and normal sequences for the current time frame. We typically use a normal sequence length, L_N, of half the sequential depth of the circuit and a scan-initiated sequence length, L_S, of two. In addition we also generate a test sequence of length one, since the computation cost is relatively small, and the smaller sequence may reduce the test application time. To select the sequence to be applied at the current time frame, the test generator compares the fitness of each of the three tests, namely the scan-initiated sequences of lengths 1 and 2 and the normal sequence of length L_N, and selects the best among these three if it results in an improvement in fault coverage.

To speed up the execution, we use fault samples consisting of 200 faults in evaluating the fitness of each candidate test, although the full fault list could be used to obtain a more accurate result. The fitness values of scan-initiated and normal sequences are calculated as follows:

$$fitness(S) = \frac{C + \#\ faults\ scanned\ out}{L_S + 2F} \qquad (6.3.1)$$

$$fitness(N) = \frac{C + \#\ faults\ scanned\ out}{L_N + F}, \qquad (6.3.2)$$

where L_S and L_N are the lengths of the scan-initiated and normal sequences, F is the number of flip-flops, $L_S + 2F$ and $L_N + F$ are measures of the test application time, and C is a measure of the fault coverage:

$$C = \#\ faults\ detected + \frac{\#\ faults\ propagated\ to\ flip\ flops}{(\#\ faults\ simulated)(\#\ flip\ flops)}. \qquad (6.3.3)$$

Since detected faults are dropped, only undetected faults are targeted. The fraction of fault effects propagated to flip-flops is included in the fitness function since a fault whose effects propagate to the flip-flops in one time frame is very likely to be detected in the next time frame. This metric is useful in guiding the GA to generate sequences that detect a greater number of faults. The number of faults detected at the flip-flops if the sequence is scan-terminated is included separately. If the next sequence is scan-initiated, these faults will be detected. Otherwise, they are still more likely to be detected by the next sequence.

Alternatively, faults detected at the POs could be separated from faults that would be detected at the flip-flops if the sequence is scan-terminated. The cost of the faults detected in the scan-out operation is then higher in terms of test application time than faults detected at the POs, and the fitness functions are modified as follows:

$$fitness(S) = \frac{C}{L_S + F} + \frac{\#\ faults\ scanned\ out}{L_S + 2F} \qquad (6.3.4)$$

$$fitness(N) = \frac{C}{L_N} + \frac{\#\ faults\ scanned\ out}{L_N + F}. \qquad (6.3.5)$$

Test generation terminates after four consecutive attempts at improving the fault coverage fail. The GA-based test generator does not always achieve 100% fault coverage. For the remaining faults, we assume that each vector is scan-initiated and scan-terminated and use the deterministic test generator HITEC [160] after the GA-based test generator terminates.

A scan capability was added to GATEST in 400 additional lines of C++ code. Results for the ISCAS89 sequential benchmark circuits are shown in Table 6.12, including number of testable faults, total execution time on a Sun SPARCstation II with 64-MB RAM, and total test cycles required. Results are an average of ten runs. Results for TARF [135] and COMPACTEST (COMP) [173] are shown for comparison. COMPACTEST is a deterministic combinational circuit test generator that uses various compaction techniques. Reductions in test application time were

Table 6.12. Test Application Time Reduction for Full-Scan Circuits

Circuit	Testable Faults	Exec Time	Test Cycles		
			GATEST	COMP	TARF
s298	308	2.92m	**233**	404	376
s344	342	1.95m	**123**	139	166
s349	348	2.30m	**123**	239	125
s382	399	4.84m	**478**	593	680
s386	384	5.10m	**303**	496	376
s400	420	5.61m	**470**	-	-
s444	460	5.49m	**439**	615	788
s526	554	14.1m	**995**	1231	1551
s641	467	5.79m	**438**	-	-
s713	543	9.00m	**466**	-	-
s820	850	18.1m	**445**	641	617
s832	856	18.0m	**437**	623	589
s1196	1242	38.7m	854	2640	**465**
s1238	1286	46.0m	917	2773	**448**
s1423	1501	53.7m	**2450**	2624	3222
s1488	1486	22.5m	**396**	846	641
s1494	1494	23.6m	**399**	811	582
s5378	4563	7.56h	34,967	**19,979**	-
s35932	35,110	117.9h	38,200	**24,205**	-

observed for most of the smaller full-scan circuits as compared to the previous approaches. For the largest full-scan circuits, COMPACTEST is so effective in producing compact test sets that the genetic approach cannot compete.

However, combinational circuit test generators cannot be used for partial-scan circuits, and the genetic approach to test generation for test application time reduction provides significant improvements over the traditional approach [190], as shown in Table 6.13. For each circuit, the number of flip-flops placed in the scan chain, number of faults detected, and the number of test cycles required and execution time on a Sun SPARCstation II with 64-MB RAM for both the modified GATEST and the traditional approach using HITEC are given. Again, results for the GA-based approach are an average of ten runs. Further improvements in test application time can be obtained by combining this method with a multiple-scan-chain approach.

6.4 Deterministic/Genetic Test Generator Hybrids

A comparison of results for deterministic and GA-based test generators shows that each approach has its own merits. For some circuits, deterministic test generators

6.4. Deterministic/Genetic Test Generator Hybrids

Table 6.13. Results for Partial-Scan Circuits

Circuit	Flip-Flops	Faults Total	Det	HITEC Cycles	Time	GATEST Cycles	Time
s298	1	308	292	1517	1.21m	1039	6.98m
s344	5	342	340	581	6.37s	163	4.72m
s349	5	350	346	671	24.4s	125	4.13m
s382	9	399	394	1579	5.27s	1003	7.24m
s386	5	384	384	1007	3.55s	424	5.51m
s400	9	426	415	1689	5.65s	998	7.88m
s444	9	474	455	1699	9.11s	1009	10.2m
s526	6	555	544	14,216	6.30m	10,430	21.7m
s641	7	467	442	1031	17.4s	193	7.83m
s713	7	581	514	1015	48.4s	199	10.9m
s820	4	850	850	1789	11.7s	547	21.0m
s832	4	870	860	1949	13.5s	746	20.2m
s1196	0	1242	1239	460	22.7s	246	50.1m
s1238	0	1355	1283	469	34.6s	253	56.8m
s1423	37	1515	1462	13,907	2.70h	2387	1.32h
s1488	5	1486	1486	1679	21.8s	415	24.5m
s1494	5	1506	1498	1739	22.6s	399	31.4m
s5378	72	4603	4474	58,326	34.8m	21,042	6.90h
s35932	306	39,094	35,110	156,262	5.74h	114,364	35.0h

provide higher fault coverages, while for other circuits, GA-based test generators are better. The simulation-based approach is particularly well suited for data-dominant circuits, while deterministic test generators are more effective for control-dominant circuits. Untestable faults can be identified by using deterministic algorithms, but significant speedups can be obtained with the genetic approach. Hence, combining the two approaches could be beneficial. A straightforward solution would be to start with a fast run of the GA-based test generator and then to use a deterministic test generator to improve the fault coverage and to identify untestable faults. Saab's CRIS-hybrid test generator [196] switches from simulation-based to deterministic test generation when a fixed number of test vectors are generated without improving the fault coverage; several targeted faults may be identified as untestable during the deterministic phase, and simulation-based test generation resumes after a test sequence is obtained from the deterministic procedure. We present two different approaches to combining deterministic and genetic algorithms for test generation. The first approach, implemented in the GA-HITEC test generator [191], [192], uses deterministic algorithms for fault excitation and propagation, and a GA for state justification. Individual faults in a circuit are targeted, as is normally done in deterministic test generators. Deterministic procedures for state justification are used if the GA is unsuccessful, to allow for identification of untestable faults and to improve the fault coverage.

The second approach, implemented in the ALT-TEST test generator [97], alternates repeatedly between GA-based and deterministic test generation in a manner similar to the CRIS-hybrid but with a few notable differences. A fault simulator is used to evaluate candidate test sequences in ALT-TEST instead of a logic simulator, and small fault samples are used to reduce the execution time. New measures are included in the fitness function to increase the number of states visited, which has been shown to increase the fault coverage [144], [175]. HITEC [160] is used as the deterministic test generator in ALT-TEST. The number of calls to the deterministic test generator is on the order of hundreds in the CRIS-hybrid but an order of magnitude lower in ALT-TEST.

6.4.1 GA-HITEC Hybrid

Deterministic algorithms for combinational circuit test generation have proven to be more effective than genetic algorithms [188]. Higher fault coverages are obtained, and the execution time is significantly smaller. A hybrid test generator would then naturally include the deterministic algorithm for fault excitation and propagation within a single time frame. Since we have access to the HITEC [160] source code, we also chose to use the deterministic algorithms for fault propagation in successive time frames. State justification using deterministic algorithms is a much more difficult problem, however, and is prone to many backtracks, which can lead to high execution times. In the GA-HITEC hybrid test generator, we use a simulation-based approach to state justification in which candidate sequences evolve over several generations, as controlled by a GA. When a sequence that justifies the desired state is found, execution of the GA terminates. Deterministic procedures for state justification are used only if the genetic approach is unsuccessful.

Test generation using the GA-HITEC hybrid approach is illustrated in Fig. 6.10. An individual fault in the circuit is targeted. The fault is excited and required values are backtraced to the PIs and flip-flops. Next, the fault effects are propagated to a PO, either in the current time frame or in successive time frames. Again, required values are backtraced to the PIs and to flip-flops in time frame zero, in which the fault was excited. If any conflicts are found during fault excitation and propagation, the test generator backtracks to a decision point and makes an alternative choice. Finally the required state in time frame zero is justified by using GAs. Several candidate sequences are simulated, starting from the last state reached after any previous tests have been applied. If a sequence is found that justifies the state, then the sequence is added to the test set, along with the vectors required for fault excitation and propagation. If a sequence cannot be found to justify the desired state, then backtracks are made in the fault propagation phase, and attempts are made to justify any new state.

One drawback to this approach is that some untestable faults may not be identified. Even if a sequence exists that justifies a given state, the GA is not guaranteed to find it. Therefore, deterministic algorithms for state justification are still required in a complete test generator. Hence, our overall approach to test generation

6.4. Deterministic/Genetic Test Generator Hybrids

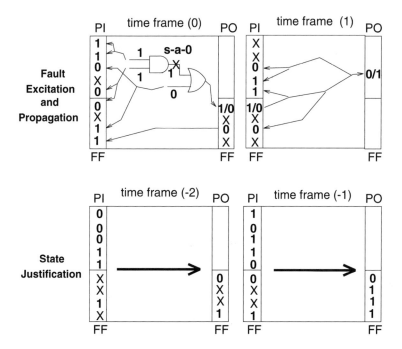

Figure 6.10. Test generation using GA for state justification.

includes both genetic and deterministic approaches for state justification, as indicated in Table 6.14. The test generator makes several passes through the fault list, with different conditions and time limits imposed in each pass. Faults are removed from the fault list once they are detected. After each pass, the user is prompted as to whether to continue with another pass, and execution terminates when the user responds negatively.

In the first pass through the fault list, state justification is performed using a GA. A time limit of one second per fault is imposed, but the time is checked at backtrack decision points only; i.e., the GA evolves over four full generations before the time is checked. Therefore, the actual time spent per fault could be greater than one second. A small population size of 64 is used, and the number of generations is limited to four to reduce the execution time. A sequence length of $\frac{1}{2}x$ is used, where x is supplied by the user. Many of the testable faults are detected in this pass, but untestable faults are identified only if conflicts are found without doing state justification. In the second pass through the fault list, GAs are again used for state justification, but the search space is expanded. In particular, the population size is increased to 128, the number of generations is increased to eight, and the sequence length is doubled. Also, the time limit is increased to 10 seconds per fault to enable more backtracking in the fault propagation phase. Finally, deterministic algorithms are used for state justification for any additional passes through the

Table 6.14. GA-HITEC Hybrid Test Generation

Pass	Test Generation Approach		Time/Fault
I	Fault Excitation and Propagation with HITEC	⇒ State Justification with GA Population: 64 Generations: 4 Seq. length: 1/2x	1 s
II	Fault Excitation and Propagation with HITEC	⇒ State Justification with GA Population: 128 Generations: 8 Seq. length: x	10 s
III	Fault Excitation, Propagation, and State Justification with HITEC		100 s and greater

fault list. Required values at the flip-flops are backtraced to the PIs in previous time frames through reverse time processing. An untestable fault is identified when all possible choices at decision points prove unsuccessful in generating a test to detect the fault. The time limit per fault is increased to 100 seconds in the third pass and multiplied by ten in successive passes to expand the search space. The time is checked before each new time frame is processed, as well as at backtrack decision points. In this manner, tests are generated for many of the testable faults by using the GA for state justification. The deterministic algorithms for state justification are used to identify untestable faults and to generate tests for hard-to-detect faults only.

In applying GAs to state justification, we use each string in the population to represent a candidate test sequence. A binary coding is used, and successive vectors in the sequence are placed in adjacent positions along the string. Tournament selection without replacement and uniform crossover are used. Sequences are evolved over several generations, with the fitness of each individual being a measure of how closely the final state reached matches the desired state. If any sequence is found that produces the desired state, the search is terminated, and the sequence is added to the test set, along with the fault excitation and propagation vectors. Otherwise, the GA runs to completion for a limited number of generations according to the procedure shown in Fig. 6.5. The test sequence length used is typically a multiple of the structural sequential depth of the circuit.

Since the fitness of an individual sequence indicates how closely the state it produces matches the desired state, simulation is required. The presence of a particular fault may affect the state; thus, both good and faulty circuit simulations are required for an accurate result. If tests have already been added to the test set, then

6.4. Deterministic/Genetic Test Generator Hybrids

the current good circuit flip-flop values may already be known. However, the state is not known for the faulty circuit unless a faulty circuit simulation is performed using all previously generated test vectors. Instead of simulating the faulty circuit, we initialize the faulty circuit flip-flops to unknown values. Before the search is begun for a sequence to justify a required state, the desired good circuit state is compared to the current good circuit state, and the desired faulty circuit state is compared to the all-unknown state. (Note that separate values are maintained for the good and faulty circuits during the fault excitation and propagation phases.) If the states match, no justification is required.

If the current state does not match the desired state, then several candidate sequences are simulated for both the good and faulty circuits. Fault injection is performed by modifying the circuit description, as is done in PROOFS [161]; e.g., an OR-gate is inserted to simulate a stuck-at-one fault, and the second input of the OR-gate is set to zero for the good circuit and one for the faulty circuit. The bitwise parallelism of the computer word is used, which allows 32 sequences to be simulated in parallel. Two bits are required to represent the three possible logic values: one, zero, and X (unknown). Thus, two computer words are used at each node to simulate the good circuit, and two computer words are used at each node to simulate the faulty circuit. PI values are mapped from the sequences in the GA to the respective bit positions at the PI nodes. Simulation is done in an event-driven manner, with good and faulty circuit simulations done together.

The test sequence length is set to a fixed value, but the state is checked after each test vector is simulated to determine whether it matches the desired state. If it does, the search is terminated. Therefore, the length of the actual test sequence used may be less than the given value. However, for the purposes of the GA, the fitness function measures how closely the final state matches the desired state:

$$fitness = \frac{9}{10}(\# \ matching \ flip \ flops \ in \ good \ circuit)$$

$$+ \frac{1}{10}(\# \ matching \ flip \ flops \ in \ faulty \ circuit).$$

A flip-flop is considered to match if it requires no particular value or if the desired and actual values are equal. If the states match in both the good and faulty circuits, then the fitness will equal the number of flip-flops in the circuit. The two terms in the fitness function correspond to the two goals of the GA: finding a state justification sequence for the good circuit and finding a state justification sequence for the faulty circuit. Unequal weights are used in order that the GA can be targeted at one goal at a time. When a GA has two or more goals, the optimum fitness function does not necessarily weight the goals equally. If equal weights are used, the GA jumps back and forth among the goals, and none of the problems gets solved quickly. A heavy weighting of one goal ensures that the strings evolve steadily in one direction. Experiments on several circuits confirmed that the weights chosen work better than equal weights of 1/2.

Since 32 sequences can be evaluated in parallel, the population size should be a multiple of 32. Initially, we use a small population size of 64 to limit the execution time. We increase it to 128 in the second pass through the fault list, expanding the search space. The number of generations is initially limited to four, again to reduce the execution time. We increase the number of generations to eight in the second pass, when expanding the search space. Tournament selection and uniform crossover are used, since these schemes worked well in simulation-based test generation [189]. Crossover and mutation probabilities of one and 1/64, respectively, are used. Nonoverlapping generations are used, since exploration of the search space is paramount.

The GA-HITEC hybrid test generator was implemented using the existing HITEC [160] source code and 2700 additional lines of C++ code. Results for the ISCAS89 sequential benchmark circuits demonstrate the effectiveness of GAs for state justification. Higher fault coverages are obtained for GA-HITEC as compared to HITEC for many circuits. About the same number of untestable faults are identified for the two test generators, and GA-HITEC executes more quickly for many of the circuits [191], [192].

Results of running GA-HITEC on several circuits synthesized from high-level descriptions are shown in Table 6.15. The circuits are described in Section 6.2.5. Test sequence lengths of 24 and 48 were used in the first two passes through the fault list. Results for HITEC are shown for comparison. An HP 9000 J200 with 256-MB RAM was used for both GA-HITEC and HITEC. The three lines of results for each circuit correspond to the individual passes through the fault list. GA-HITEC yielded higher fault coverages than HITEC for all five circuits, and the GA-HITEC execution times were also smallest.

6.4.2 ALT-TEST Hybrid

A fast run of a simulation-based test generator followed by a slow but more successful run of a deterministic test generator (on faults left untested by the first test generator) could provide better results in a shorter time than would be obtained by the deterministic test generator alone. For the case of combinational circuits, this approach was proposed as a random phase followed by a deterministic phase [2], [84]. This idea can be extended to multiple such tries, as is done in ALT-TEST. The simulation-based test generator in ALT-TEST is controlled by a GA, and alternation between the two algorithms is done in such a way that most processing is left for the GA. The deterministic test generator, HITEC, is used mainly to guide the GA to new state spaces of the circuit that have not been visited. By visiting new state spaces, the test generator maximizes the search space. The need for visiting a large number of states is apparent because the outputs of flip-flops can be considered to be pseudo-inputs of the circuit, and the more distinct states traversed, the more patterns the pseudo-inputs will take. Pomeranz and Reddy [175] showed that most test sets derived from deterministic test generators traverse a large portion of

6.4. Deterministic/Genetic Test Generator Hybrids

Table 6.15. GA-HITEC Test Generation Results: Synthesized Circuits

Circuit	Total Faults	GA-HITEC			HITEC		
		Det	Vec	Time	Det	Vec	Time
am2910	2391	2163	747	1.70m	1991	348	4.13m
		2175	880	6.90m	2130	585	12.8m
		2187	1002	34.3m	2171	871	56.4m
div16	2147	1722	229	4.93m	1664	224	6.62m
		1722	229	29.3m	1664	224	37.9m
		1723	251	4.39h	1667	228	5.35h
mult16	1708	1548	236	3.55m	1319	69	5.26m
		1550	285	22.4m	1487	90	23.6m
		1606	306	1.56h	1582	111	1.90h
pcont2	11,300	6748	174	48.1m	3514	7	2.25h
		6752	206	4.59h	3514	7	9.58h
		6752	206	29.3h	3514	7	79.5h
piir8	19,936	11,504	53	10.4h	9003	21	3.62h
		11,504	53	44.0h	9003	21	13.1h
		11,504	53	94.7h	9003	21	98.8h

Det: number of faults detected
Vec: number of test vectors generated
am2910: 12-bit microprogram sequencer
div16: 16-bit divider
mult16: 16-bit two's complement multiplier
pcont2: 8-bit parallel controller for DSP applications
piir8: 8-point infinite impulse response filter

the possible states in a circuit. A similar observation was made by Marchok et al. [144]. Furthermore, using a deterministic test generator at an early stage of test generation helps to identify untestable faults, thus saving simulation time in later GA runs. The GA is used to identify "hard" faults that HITEC should target for test generation, since targeting "hard" faults seems to reduce the total execution time.

The test generation process in ALT-TEST is divided into three stages, as illustrated in Fig. 6.11. Each of the three stages is composed of alternating phases of GA-based and HITEC test generation. The first stage attempts to detect as many remaining faults as possible from the fault list. The second stage tries to maximize the number of visited states and propagate fault effects to flip-flops. The third and final stage attempts to detect the final remaining faults and to visit new states.

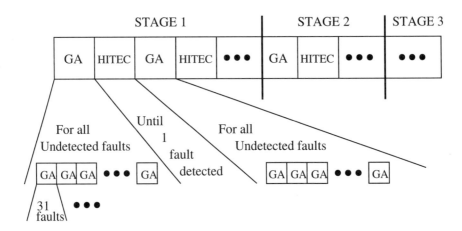

Figure 6.11. ALT-TEST structure

In each of the three stages, the GA is first run until little or no more improvement is obtained; then the deterministic approach is used to target undetected faults. The two phases are repeated many times in each stage. The GA targets all faults in the fault list in groups of 31 faults, traversing the fault list several times until little or no more improvement is made. Between consecutive GA runs, useful vectors from the best sequence are added to the test set, and sequences having fitness values greater than or equal to one half of the best fitness are used to seed the population in the next GA run. The remaining individuals in the population are initialized with random test sequences. After the GA phase, the deterministic test generator, HITEC, is activated in an attempt to bring the GA to a new circuit state region that has not yet been explored. HITEC takes the *hard-to-detect* faults provided by the GA as the fault list to work on. *Hard-to-detect* faults are defined as those faults that have never been detected by any of the individuals in the GA population. These faults are either hard to excite, hard to propagate, or both. HITEC imposes a time limit for each fault, initially using a minimal time limit of 0.4 seconds per fault; the time limit is increased only if no fault is detected. Alternating phases of GA-based and HITEC test generation are repeated several times in each stage of the test generation process. A stage is finished when no more improvements are made for the remaining undetected faults. The pseudo-code for test generation within a stage is given in Fig. 6.12.

The HITEC test sequence is likely to bring the circuit to a previously unvisited state. The number of faults detected as a function of time for alternating phases of GA-based and HITEC test generation is illustrated in Fig. 6.13 for a case in which HITEC successfully finds a sequence for one fault. The sequence found by HITEC is used as an initial seed for the next GA run. Details of GA-based test generation and the interaction with HITEC are given next, followed by experimental results.

6.4. Deterministic/Genetic Test Generator Hybrids

While there is improvement in this stage **Do**
      // GA-based test generation
  While there is improvement in the GA phase **Do**
    For all undetected faults, in groups of 31 faults **Do**
      target-faults = next 31 undetected faults;
      best-sequence = GA-test-generate (target-faults);
      fault-simulate (best-sequence);
      update test set;
      seed the next GA population (better sequences);
    compute improvement;
        // deterministic test generation, HITEC
  target-faults = hard-to-detect faults;
  test-sequence = HITEC-test-generate (target-faults);
  fault simulate (test-sequence);
  update test set;
  seed-GA (test-sequence);
  compute overall improvement;

Figure 6.12. ALT-TEST algorithm

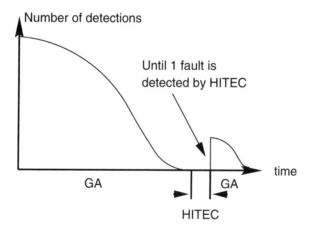

Figure 6.13. Number of detections in alternating phases

GA-Based Test Generator

The simple GA of Fig. 6.5 is used, and each individual represents a test sequence. The GA population evolves over several generations until no more improvements in fitness values are obtained or the maximum of eight generations is reached. A binary coding is used in which successive vectors in a sequence are placed in adjacent positions along a string, as for GATEST. The test sequence length is set equal to the sequential depth in the first stage, two times the sequential depth in the second stage, and four times the sequential depth in the third stage. In the first and second stages, the population size is set to $4 \times \sqrt{sequence\ length}$ when the number of PIs is less than 16 and $16 \times \sqrt{sequence\ length}$ when the number of PIs is greater than 16. The population size is doubled in the third stage to expand the search space. Tournament selection without replacement and uniform crossover are used; mutation is done by simply flipping a bit.

Each individual in the population is simulated using a group of 31 faults. The reason 31 faults are chosen is to maximize the execution speed. Fault simulation on the entire fault list is very costly, and results for GATEST have shown that small fault samples give good quality results as well. A new fault simulator was built and used for this purpose. It is based on an algorithm similar to that of PROOFS [161], but several features have been added. New features include the ability to unroll time frames in order to save storage and retrieval costs at the flip-flops and an ability to keep track of states that have been visited. Thus, the new fault simulator allows sequences of vectors to be evaluated without intermediate storage and retrieval of flip-flop state information. Furthermore, instead of two passes (fault-free circuit simulation followed by a 32-fault parallel simulation), a single pass using the fault-free circuit and 31 faulty circuits in parallel is performed to save simulation time.

Initially, the first 31 undetected faults in the fault list are targeted. All individuals in the population target the same group of 31 faults, and the individuals evolve in successive generations until no more improvements in fitness values are obtained. Then a full-scale fault simulation is performed, using the entire list of undetected faults and the best test sequence found; all detected faults are dropped from the fault list. For successive GA runs, faults should be chosen cleverly so that work done previously may assist in the test generation process. If any fault effects have been propagated to the flip-flops by the previous sequence generated, the respective faults should be placed in the next targeted fault group. This is done with the expectation that faults propagated to flip-flops have a greater chance of propagating to the POs. If the effects of more than 31 faults have propagated to the flip-flops, preference is given to those that have propagated to more *observable* flip-flops. Observability of a node is defined as the difficulty of propagating a fault effect from the given node to a PO. If the effects of fewer than 31 faults have propagated to the flip-flops, the remaining faults in the group are filled from the undetected fault list, starting at the position where the previous fault group left off.

Not all vectors in the best test sequence are necessarily added to the test set. Only the good portion of the sequence is added to keep the test set compact.

6.4. Deterministic/Genetic Test Generator Hybrids

Since the three stages of ALT-TEST target different goals, their corresponding fitness functions are different. The parameters that affect the fitness of an individual in the population are

P_1: *Number of faults in the given fault group detected,*
P_2: *Number of flip-flops that carry fault effects at the end of simulation,*
P_3: *Number of new states visited,*
P_4: *Number of hard-to-control flip-flops set to specific values.*

Both P_3 and P_4 contribute to the expansion of the state space. P_4 guides the search by favoring sequences that set the hard-to-control flip-flops to values that have not yet been encountered. Consequently, a new state is likely to be visited. All four parameters will be evaluated but given different weights in the fitness computation for the three stages of ALT-TEST as follows:

Stage 1: $fitness = 0.8P_1 + 0.1P_2 + 0.1(P_3 + P_4)$
Stage 2: $fitness = 0.1P_1 + 0.45P_2 + 0.45(P_3 + P_4)$
Stage 3: $fitness = 0.4P_1 + 0.2P_2 + 0.4(P_3 + P_4)$

In the first stage, the aim is to detect as many faults as possible with short sequences and minimal time. Thus, more weight is given to the number of faults detected. In the second stage, the goal is to maximize visitation of new states and fault effect propagation to flip-flops. According to a study by Wah et al. [234], which used a meta-level GA to learn the best fitness function for GA-based test generation, a heavier weighting given to fault effect propagation is preferable during the second stage of test generation. In the final stage of ALT-TEST, the focus is shifted once again to targeting the remainder of the faults that have been hard to detect by HITEC and the GA in the prior two stages. Therefore, the fitness function weights fault detections and new state identifications more heavily.

Interaction with HITEC

Once a test sequence is generated by HITEC to detect a fault, the test sequence is added to the test set and also used to seed the next GA run. The time limit for running HITEC starts at 0.4 seconds per fault; if HITEC fails to generate a test for the fault(s), the time limit is increased by 5 times. Test vector sequences derived by HITEC may have different lengths than those used for individuals in the GA. If the HITEC test sequence length is smaller, random vectors are added to the end of the HITEC test sequence. However, if the HITEC test sequence length exceeds the GA test sequence length, the sequence is split into two or more portions, as illustrated in Fig. 6.14. Then, multiple seeds are used to initialize the next GA run, in addition to the random test sequences normally used.

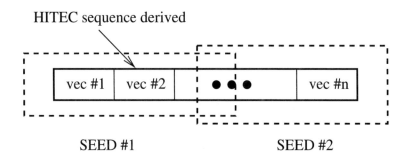

Figure 6.14. Splitting HITEC sequence into multiple seeds

Experimental Results

ALT-TEST was implemented in 6500 lines of C++ code, not including the modifications made to the existing HITEC source code. In addition to the ISCAS89 sequential benchmark circuits [23], several synthesized circuits described in Section 6.2.5 were used for evaluating the performance of ALT-TEST, as well as an unoptimized version of piir8 with 29,689 faults. All circuits were evaluated using ALT-TEST on an HP 9000 J200 with 256-MB RAM. Table 6.16 compares the results with those obtained by various other test generators. Note that these results were obtained on different machines. The number of faults detected and the test-set lengths are shown. In addition, the number of calls made to deterministic test generators are also reported for the CRIS-hybrid and ALT-TEST. For the synthesized circuits, the test set lengths and the number of detected faults were not reported for the CRIS-hybrid [196]; thus, those results are not included. However, the ATG effectiveness measure was reported for three of the circuits. ATG effectiveness is defined as the percentage of faults detected or identified as untestable. The ATG effectiveness of the CRIS-hybrid was 81.9%, 98.8%, and 99.6% for circuits div16, piir8o, and mult16, respectively, while the ATG effectiveness of ALT-TEST was 89.2%, 98.9%, and 98.0% for the same circuits.

Table 6.17 shows the performance of ALT-TEST for the large circuits at various checkpoints. Checkpoints were placed at the end of the first GA iteration before HITEC is called and at the end of the first, second, and third ALT-TEST stages. The total number of ATG calls listed in the table is cumulative. From this table, it can be observed that high fault coverages are achieved quickly for most circuits. This is due to several factors. Fault simulation of vector sequences instead of single vectors in the first stage improves the quality of test sets tremendously; fitness functions used for evolving the test sequences across generations also play a major role in convergence of high-quality test sets. Note that many of the fault coverages at the end of a few alternating phases of the GA and HITEC are already higher than the final results of the other hybrid test generators. Furthermore, test sets generated by ALT-TEST are much more compact than those generated by the previous test generators. ALT-TEST performs significantly better in terms of both fault coverage

6.4. Deterministic/Genetic Test Generator Hybrids

Table 6.16. Comparison of Results for Various Test Generators

Circuit	HITEC Det	HITEC Vec	GA-HITEC Det	GA-HITEC Vec	CRIS-Hybrid Det	CRIS-Hybrid Vec	CRIS-Hybrid ATG Calls	ALT-TEST Det	ALT-TEST Vec	ALT-TEST ATG Calls
s298	265	306	265	415	266	331	58	265	221	6
s344	328	142	328	169	329	308	15	**329**	75	4
s349	335	137	335	188	334	367	21	335	104	4
s382	363	4931	328	716	353	443	48	**364**	1823	6
s386	314	311	314	359	310	636	85	314	241	7
s400	383	4309	346	704	372	1461	63	**384**	2196	4
s444	414	2240	381	880	416	646	78	**424**	1046	3
s526	365	2232	376	873	433	1191	120	**453**	2109	4
s641	404	216	404	292	404	990	75	404	140	3
s713	476	194	476	294	475	918	110	476	185	3
s820	813	984	814	1108	-	-	-	814	1243	62
s832	817	981	818	1064	-	-	-	818	1109	59
s1196	1239	453	1239	377	1239	2467	25	1239	948	27
s1238	1283	478	1283	409	1175	2371	70	1283	880	19
s1423	776	177	928	477	1296	1773	220	**1335**	2058	5
s1488	1444	1294	1444	1369	1443	3677	44	1444	1191	50
s1494	1453	1407	1453	1224	1389	2856	130	1453	985	28
s5378	3238	941	3238	683	3503	3227	1205	**3506**	7270	7
s35932	34,898	439	34,862	425	34,899	1204	4020	**35,095**	825	4
am2910	2166	1286	2190	1214	-	-	-	**2196**	1392	29
mult16	1551	122	1633	421	-	-	-	**1652**	223	3
div16	1679	212	1741	359	-	-	-	**1765**	1120	11
pcont2	3354	7	6757	208	-	-	-	**6829**	790	2
piir8o	14,221	347	-	-	-	-	-	**15,072**	1223	2
piir8	-	-	-	-	-	-	-	**18,141**	1338	2

Det: # of faults detected   Vec: # of test vectors generated
ATG calls: # of calls to a deterministic test generator

HITEC was performed on a SPARCstation 2 with 64-MB RAM
GA-HITEC was performed on a SPARCstation 20 with 64-MB RAM
CRIS-Hybrid was performed on a SPARCstation SLC with 16-MB RAM
ALT-TEST was performed on a HP 9000 J200 with 256-MB RAM

and test set size for all of the circuits when compared with the results of either GA-HITEC [190] or the CRIS-hybrid [196]. The number of untestable faults identified contributes to the ATG effectiveness, and ALT-TEST makes significantly fewer calls to the deterministic test generator as compared to the CRIS-hybrid. Thus, ALT-TEST has fewer chances of identifying untestable faults. Despite this fact, ALT-TEST still gave better results for all circuits. In addition, execution times for the larger circuits are also smaller for ALT-TEST.

Table 6.17. ALT-TEST Results for Large Circuits

Circuit	Check Point	Det	Vec	Time	Unt	Eff	Total # ATG Calls
s1423	0	1217	919	21.5m	0	80.33	0
	1	1304	1045	25.6m	9	86.67	2
	2	1323	1255	1.7h	14	88.25	4
	3	1336	2058	2.5h	14	88.51	5
s5378	0	3200	1183	2.82h	0	69.52	0
	1	3319	1976	5.80h	117	74.65	3
	2	3503	3280	9.72h	155	75.69	6
	3	3506	7270	32.3h	155	75.69	7
s35932	0	34979	697	3.89h	0	89.47	0
	1	34,981	788	4.36h	3984	99.67	2
	2	35,095	825	8.60h	3984	99.96	3
	3	35,095	825	13.9h	3984	99.96	4
mult16	0	1636	193	2.23m	0	95.78	0
	1	1652	223	4.02m	15	97.60	1
	2	1652	223	25.0m	21	97.95	2
	3	1652	223	1.03h	21	97.95	3
div16	0	1703	538	33.5m	0	79.54	0
	1	1725	692	1.01h	136	86.92	5
	2	1762	784	2.47h	136	88.65	7
	3	1780	852	4.56h	136	89.24	11
pcont2	0	6823	540	6.9m	0	60.38	0
	1	6826	753	10.0m	2852	85.64	1
	2	6829	790	1.49h	2852	85.67	2
	3	6829	790	3.28h	2852	85.67	2
piir8o	0	15025	670	1.61h	0	75.43	0
	1	15,038	691	1.83h	4793	99.50	1
	2	15,072	1223	2.86h	4793	99.66	2
	3	15,072	1223	4.16h	4793	99.66	2
piir8	0	18104	1253	3.46h	0	60.98	0
	1	18,119	1314	6.43h	11234	98.87	1
	2	18,141	1338	9.70h	11234	98.94	2
	3	18,141	1338	12.8h	11234	98.94	2

Unt: number of untestable faults identified   Eff: ATG effectiveness
Checkpoint   0: end of first GA iteration, before first HITEC call
             1, 2, 3: end of ALT-TEST stage 1, 2, or 3, respectively

6.5 Use of Finite State Machine Sequences

The majority of the time spent by automatic test generators for sequential circuits is used to find test sequences for hard-to-test faults. Deterministic test generators often require large numbers of backtracks for the hard faults, and simulation-based test generators often fall short when targeting the hard faults because they lack information about state justification. While deterministic/genetic hybrid test generators may be effective for the hard-to-test faults, the use of deterministic algorithms necessitates backtracing values through circuit components, and handling complex component types is difficult. Simulation-based test generators, on the other hand, avoid the complexity of backtracing by processing in the forward direction only.

The deficiencies of the deterministic approach motivate the development of new techniques that use GAs in a simulation-based manner but which also incorporate problem-specific knowledge to increase the fault coverage. In the case of sequential circuit test generation, the problem-specific knowledge takes the form of finite state machine sequences and in particular, distinguishing sequences and state transfer sequences.

Previously, homing, synchronizing, and distinguishing sequences have been used to aid the test generator in improving the fault coverage [39], [76], [169], [174]. In [39], [76], [169], symbolic and state-table-based techniques were used to derive these sequences in the fault-free machine. Specifically, in [76], cube intersections of ON/OFF-set representations were used to derive distinguishing sequences. Binary Decision Diagrams (BDDs) and implicit state enumeration were used in [39] to derive synchronizing sequences. In the work by Park et al. [169], functional information was used to pregenerate sequences that simplified propagation of fault effects from the flip-flops to the POs, and state justification was done by using BDDs. Since these sequences are generated using the fault-free machine only, they may become invalid in a faulty machine. Homing sequences composed of specifying and distinguishing portions were used to aid ATG in [174], but they had to be recomputed for each target fault.

Distinguishing sequences are used to assist in fault-effect propagation in our DIGATE test generator [98], [102]. By definition, a distinguishing sequence, when applied to the finite state machine, distinguishes two states (or two sets of states) by producing two unique output sequences. This is especially helpful to test generation. A fault that is activated essentially has its fault effects propagated to at least one flip-flop, resulting in a 1/0 or 0/1 good and faulty circuit value pair at the flip-flop. Propagation of the fault-effect to a primary output can be interpreted as distinguishing the fault-free state from the faulty state. Depending on which flip-flop the fault-effect propagates to, different distinguishing sequences will be needed. Furthermore, we are distinguishing a fault-free state from a specific faulty state, so maintaining and pruning of these sequences for different faults will be needed.

Instead of targeting groups of faults as in the previous approaches, the DIGATE test generator [98], [102] targets one fault at a time. The reason for targeting single faults is that hard faults will not be neglected as before. To clarify this point, suppose we are trying to detect as many faults as possible in a group of N faults; the fitness objective will give a higher fitness to a sequence that detects $N/2$ faults in the group rather than a sequence that detects only 1 hard fault. Furthermore, hard faults are detected by unique sequences; thus, maximization of the number of faults detected in a fault group does not work very well in this case.

In the first phase of DIGATE, the target fault is activated, i.e., the fault is excited and propagated to the flip-flops, using a GA. Some faults may already be activated by previous test vectors already added to the test set. In this case, the activated fault that has fault effects at the most observable flip-flops can be selected as the target fault, and the first phase may be skipped. In the second phase, the fault effects are propagated to the POs using a GA and knowledge about distinguishing sequences.

We pregenerate a class of distinguishing sequences statically for the fault-free machine and also dynamically capture distinguishing sequences for the fault-free and faulty machines during the test generation process. These sequences are then used as seeds for the GA during fault propagation. If these seeds are valid for the present situation, no further processing is required. Otherwise, we genetically engineer valid sequences from the seeds. In addition to this, the difficulty of deriving a distinguishing sequence is also taken into account at run time in the computation of flip-flop observability. This measure of observability is much more accurate than the conventional observability metric and helps to guide the test generator much more effectively.

Fault effects may reach multiple flip-flops at the end of the fault activation phase, which will require the use of several distinguishing sequences in the fault propagation phase; therefore, the list of distinguishing sequences is pruned adaptively over time to increase the power and accuracy in distinguishing the states. When a sequence is found that successfully propagates the fault effects to the POs, a fault simulator is invoked to remove any additional faults detected by the sequence. Flip-flops that do not have distinguishing sequences are identified during the process, and propagating fault effects to these hard-to-observe flip-flops is avoided. DIGATE targets all faults until little or no more improvement is made.

Results of DIGATE on the ISCAS89 sequential benchmark circuits and several synthesized circuits show very high fault coverages. However, DIGATE requires that faults be activated in order for it to be effectively applied. The hard-to-activate faults in some circuits may require specific states and justification sequences in order for them to be activated, and the previous GA-based test generators have failed to drive the circuit to these specific states for fault excitation, resulting in low fault coverages. For instance, GA-based test generators have obtained low fault coverages for ISCAS89 circuits s820, s832, s1488, and s1494 due to frequently deep and specific sequences necessary to excite the faults, but deterministic test generators have been quite successful in generating tests for them. The differences in fault coverages were as high as 30% for such circuits. Even when a GA was specifically targeted at state justification, the simple fitness function used was inadequate for these circuits [190], [192].

Storing the state information for large circuits is impractical; similarly, keeping a list of sequences capable of reaching each reachable state is infeasible. Our STRATEGATE test generator [99] uses the *linear* list of states dynamically obtained during the derivation of test vectors to guide state justification. The storage necessary is on the order of the number of vectors generated in this case. When justifying states that have not been visited, several candidate sequences that lead to previously visited states are used to help find the target unvisited state. The candidate states are chosen such that they are similar to the target state. The sequences that reach the candidate states may be viewed as partial solutions to finding a sequence that justifies the target state. Genetic algorithms have been demonstrated to be effective in combining useful portions of several candidate solutions to a given problem. Therefore, we have chosen to use genetic algorithms in this work, both to derive and manipulate dynamic state-transfer sequences and in the overall test generation process. Utilizing the dynamic state traversal information allows us to overcome the limitations of the previous genetic approaches and close the gap of 30% fault coverage difference in some of the circuits; for the other circuits, extremely high fault coverages have been obtained compared to other test generators.

6.5.1 Algorithm Overview

Our test generation strategy uses several passes through the fault list, with faults targeted individually in two phases. The two-phase strategy is illustrated in Fig. 6.15. The first phase focuses on activating the target fault, while the second phase tries to propagate the fault effects from the flip-flops to the POs. A target fault is selected from the fault list at the beginning of the fault activation phase, and an attempt is made to derive a sequence that excites the fault and propagates the fault effects to a PO or to the flip-flops. Once the fault is activated, the fault effects are propagated from the flip-flops to the POs in the second phase with the assistance of distinguishing sequences. The target fault is detected at the POs when the faulty machine state is distinguished from the fault-free machine state.

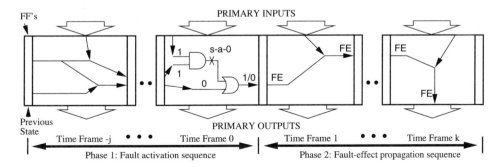

Figure 6.15. Two-phase strategy

The GA of Fig. 6.5 is used for both fault activation and fault propagation in which each individual represents a test sequence. Tournament selection without replacement and uniform crossover are used. A binary coding is used, and mutation is done by simply flipping a bit. In the first stage of test generation, the sequence length is set equal to the structural sequential depth. The sequence length is doubled in the second stage and doubled again in the third stage, since harder faults may require longer sequences for activation and/or propagation. The population size used is a function of the string length, which depends on both the number of PIs and the test sequence length. The population size is set equal to $4 \times \sqrt{sequence\ length}$ when the number of PIs is less than 16 and $16 \times \sqrt{sequence\ length}$ when the number of PIs is greater than 15. Evolution from one generation to the next is continued until a sequence is found to activate the target fault or propagate its effects to the POs or until a maximum number of generations is reached.

During the fault activation phase, single-time-frame mode is entered if no activation sequence can be found directly from the state in which the previous sequence left off. The aim of single-time-frame fault activation is to find a test vector, composed of primary input and flip-flop values, that can activate the target fault in a single time frame. Single-time-frame fault activation is illustrated in Fig. 6.16 and will be described in detail in the next section. Once a vector (PI and flip-flop values) is successfully derived, the state (FF values) is first relaxed to one that has as many don't-care values (X) as possible and still can activate the target fault; this improves the success rate of state justification which immediately follows [160], [163].

State justification is performed by using a GA with an initial population composed of random sequences and any useful state-transfer sequences. If the relaxed state S_r matches a previously visited state S_p, and a state-transfer sequence $T_{S_p}^{S_c}$ exists that drives the circuit from the current state S_c to state S_p, the state justification sequence $T_{S_p}^{S_c}$ is simply seeded into the GA. However, if the relaxed state S_r does not match any of the previously visited states, genetic engineering of several sequences is performed to try to justify the target state. Several candidate states are selected from the set of previously visited states that most closely match the relaxed state S_r. The selection is based on the number of matching flip-flop values in the states. Let the set of selected candidate states be $\{S_i\}$; the set of sequences that justify these states from current state S_c is $\{T_{S_i}^{S_c}\}$. These sequences are used as seeds in the GA to aid in evolving an effective state justification sequence. Candidate sequences in the GA population are simulated, starting from the current state. The objective is to engineer a sequence that justifies the required state by genetically combining the candidate justification sequences. Consider the situation shown in Fig. 6.17 in which an attempt is being made to justify state 1X0X10. Sequence T_1 successfully justifies all but the third flip-flop value; on the other hand, sequence T_2 justifies all but the final flip-flop value. These two sequences T_1 and T_2 may provide important information in evolving the complete solution, T_3, which justifies the complete state. They are used as seeds for the GA in an attempt to genetically engineer the sequence T_3. Presence of the fault may invalidate a sequence that was

6.5. Use of Finite State Machine Sequences

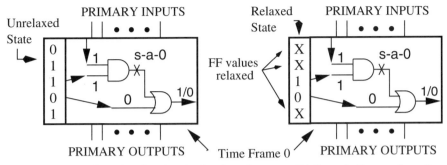

(a) Step 1: Fault activation in single time frame   (b) Step 2: State relaxation

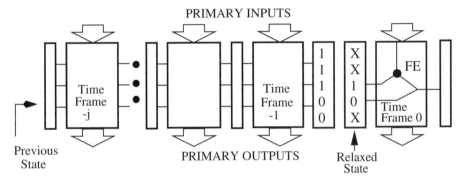

(c) Step 3: State justification using state-transfer sequences

Figure 6.16. Multistage fault activation

previously used to traverse a set of states. However, since the GA performs simulation in the presence of the fault to derive a sequence for the current situation, any sequence derived will be valid.

To facilitate the dynamic state traversal algorithm, a table of visited states is mapped to the list of vectors in the test set, as shown in Fig. 6.18. During state justification, the goal is to generate a sequence that will justify the desired state from the current state. In Fig. 6.18, the starting state (also the current state) is reached at the end of vectors i, k, and m, and the desired state (labeled *End State*) is reached at the end of vectors j and l. Therefore, either sequence T_1 (vectors $i+1$ to j) or T_2 (vectors $k+1$ to l) is sufficient to drive the circuit to the ending state. However, if the desired state has not been visited, a set of *ending states* that closely match the desired state is formed from the visited state table; the sequences corresponding to these states are seeded in the GA in an attempt to engineer a valid justification sequence for the desired state. If the current state was not visited before the ending states were reached, the n vectors that lead to the ending states are seeded into the GA, where n is the current GA sequence length.

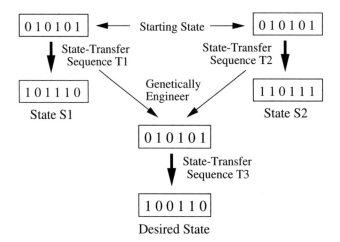

Figure 6.17. Genetic justification of desired state

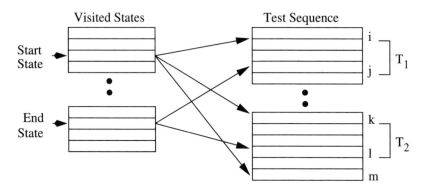

Figure 6.18. Data structure for dynamic state traversal

During state justification, a sequence that correctly justifies one portion of the required state may simultaneously set an incorrect value on the other portion(s), resulting in conflicts. Nevertheless, the justification sequences for each partial state may be viewed as partial solutions in finding the justification sequence for the complete state. Because important information about the assignment of PIs for justifying a specific part of a state is intrinsically implied by each sequence, this information may be important and useful in searching for the complete justification sequence. Stated differently, each partial solution is a chromosome in the evolutionary process; the desired solution may be evolved from the population of chromosomes with appropriate fitness functions. The GA is capable of combining several partial solutions, under arbitrary constraints, to form a complete solution to a problem via the evolutionary processes.

6.5. Use of Finite State Machine Sequences

If a sequence is found that justifies the required state, the sequence is appended to the test set, a fault simulator is invoked to remove any additional faults detected by the sequence, and the test generator proceeds to the fault propagation phase. Otherwise, the current target fault is aborted, and test generation continues for the next fault in the fault list. In the fault propagation phase, the GA is seeded with distinguishing sequences for the flip-flops to which fault effects have propagated, and an attempt is made to engineer a valid distinguishing sequence for propagating the fault effects to a PO in the fault propagation phase. If a sequence that drives the fault effects to the POs is successfully obtained, the sequence is added to the test set, and the fault simulator is used to identify other detected faults that may be dropped. Test generation continues with the next target fault.

Fig. 6.19 illustrates the different types of distinguishing sequences. A distinguishing sequence of type A for flip-flop i is defined as a sequence that produces two distinct output responses when applied to the fault-free machine for two initial states, and the initial states differ in the ith position and are independent of all other flip-flop values. A type-B distinguishing sequence for flip-flop i is a sequence that, when applied to the fault-free machine with ith flip-flop $= 0$ (or 1) and applied to the faulty machine with the same flip-flop $= 1$ (or 0), produces two distinct output responses independent of the values of all other flip-flops. A type-C distinguishing sequence is similar to type B except that a subset of flip-flops are assigned to specific logic values. For some flip-flops, neither type-A nor type-B distinguishing sequences exist, or distinguishing sequences may exist but are expensive to compute. Generating distinguishing sequences for finite state machines is already a difficult task. Since our main goal is test generation, we do not want to spend too much time on generating all possible distinguishing sequences, so a relaxed version, i.e., type-C sequences, are obtained dynamically during test generation. This speeds up our ATG process significantly.

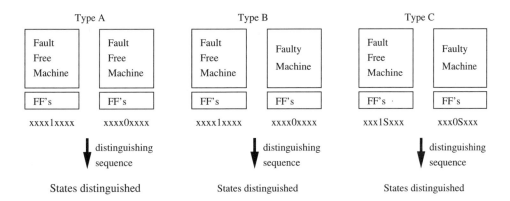

Figure 6.19. Types of distinguishing sequence

The value **x** in the state denotes an unknown or more precisely a *don't-care* value, and the character **S** in a state represents a string of known values (e.g., 1 or 0). The distinguishing sequences of type A are pregenerated statically for the fault-free machine only, while sequences of types B and C are derived dynamically for both the fault-free and faulty machines during test generation. Note that the distinguishing sequences of type C may depend on a partial state of the machine, so they cannot necessarily be applied directly. With these distinguishing sequences of various types seeded, the GA is used to evolve a valid distinguishing sequence to propagate the fault effects from the flip-flops to the POs. The sequences generated in [169] are similar to the type-A distinguishing sequences, except that they were generated using BDDs; no pruning of sequences was done, and sequences of types B and C were absent. When the sequences fail to distinguish the states for specific faulty machines, no procedure was given to modify the sequences. In contrast, we use a variety of distinguishing sequences and modify them to get valid sequences for each fault.

6.5.2 Single Time Frame Mode

When a hard-to-activate fault is targeted and the GA fails to generate an activation sequence, a second attempt is made to activate the fault in a single time frame. The aim here is to engineer a vector, composed of PI and flip-flop values, capable of activating the target fault (i.e., exciting the target fault and propagating its effects to at least one flip-flop or PO) in a single time frame. Initially, the GA is seeded with random vectors, and the evolution process continues until a vector is found or a maximum of eight generations is reached.

To improve the chances of activation for hard faults, dynamic fitness objectives are set up for each target fault. Fig. 6.20(a) illustrates the justification frontier for a stuck-at-0 fault at the output of gate k. The justification frontier consists of values necessary for justifying a desired value. During the single-time-frame fault activation, the fitness function for fault excitation tries to maximize the number of justification frontier values justified. Once a target fault is excited, its fault effects need to be propagated to at least one PO or flip-flop. Fig. 6.20(b) shows the propagation frontier for this case. The fitness function aims to dynamically advance the propagation of fault effects beyond the current propagation frontier. In the example shown in the figure, the fault effects are not yet propagated beyond gates b and d. Therefore, the dynamic fitness objectives will place emphasis on setting a **1** on line A and a **0** on line C to advance the fault effects beyond gates b and d. The dynamic objectives are updated after every generation of the GA is evolved.

Because an unjustifiable state is undesirable, the fitness function also uses the dynamic controllability values of the flip-flops to guide the search toward more easily justifiable states. The dynamic controllability measures correspond to the frequency of setting/clearing a particular flip-flop. With these measures, there is still no guarantee that the resulting state is indeed justifiable. Therefore, a further

6.5. Use of Finite State Machine Sequences

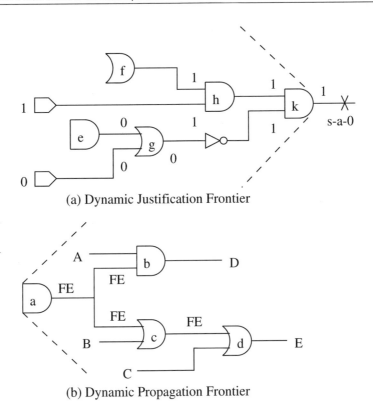

Figure 6.20. Dynamic objectives for the target fault

relaxation step is performed. Let S denote a state to be justified and s_i the ith flip-flop in state S. The state S' is obtained by inverting the value of s_i. If the target fault is still activated by S', then the ith flip-flop can be relaxed to the unknown value X. This implies that activation of the target fault does not depend on the assignment of s_i. The order in which the flip-flops are relaxed is determined in a greedy fashion: from the least controllable to the most controllable flip-flop. When state relaxation is finished, the relaxed state has to be justified. The GA is seeded with the state-transfer sequences. The remaining individuals, if any, are seeded with random sequences. The GA then attempts to combine the partial solutions to form the complete solution that justifies the relaxed state.

6.5.3 Test Generation Procedure

The STRATEGATE test generator [99] is comprised of three stages; each stage involves several passes through the fault list, and a stage is finished when little or no improvement in fault coverage is achieved. Faults are targeted individually within each stage, and GAs are used to activate a fault and propagate the fault effects to the POs. Different test sequence lengths for individuals are used in the GA population for the different stages. Since the time required for the fitness evaluation is directly proportional to the test sequence length, the shorter sequences are tried first, and faults are removed from the fault list once they are detected. Test generation for a target fault is divided into *fault activation* and *fault propagation* phases as described earlier. The GA is initialized with random sequences, and any useful state-transfer or distinguishing sequences are used as seeds in place of some of the random sequences in the fault activation and fault propagation phases.

Fitness Functions

Since the fault activation and fault propagation phases target different goals, their corresponding fitness functions differ. The parameters that affect the fitness of an individual in the GA are as follows:

P_1: *Fault detection*
P_2: *Sum of dynamic controllabilities*
P_3: *Matches of flip-flop values*
P_4: *Sum of distinguishing powers*
P_5: *Induced faulty circuit activity*
P_6: *Number of new states visited*

Parameter P_1 is self-explanatory, in particular during the fault propagation phase. It is included in the fault activation phase to cover faults that propagate directly to the POs in the time frame in which they are excited. To improve state justification and fault detection, flip-flops that are hard to control or hard to observe are identified dynamically during the process and are given lower controllability values; as a result, justification of difficult states in the first phase and propagation of fault effects to the hard-to-observe flip-flops in the first and second phases may be avoided. P_2 indicates the quality of the state to be justified. Maximizing P_2 during single-time-frame fault activation makes the state more easily justifiable and also avoids unjustifiable states. On the other hand, minimizing P_2 during state justification expands the search space by forcing hard-to-justify states to be visited. A sequence that justifies a hard-to-justify state is favored during test generation, since the GA is more likely to bring the circuit to previously unexplored state spaces as a consequence. P_3 guides the GA to match the required flip-flop values in the state to be justified during state justification, from the least controllable to the most controllable flip-flop value. P_4 measures the quality of the set of flip-flops reached by the fault effects. Maximizing P_4 increases the probability that the fault effects reach flip-flops having more powerful distinguishing sequences and thus indirectly improves the chances for detection. P_5 measures the number of events generated in the faulty circuit, with events on more observable gates weighted more heavily. Partial cones are computed and set up for the POs and flip-flops as in DIGATE [98], [102]. Events are weighted more heavily inside the partial cones of the POs or flip-flops with more powerful distinguishing sequences; events inside the partial cones of the hard-to-observe flip-flops are weighted more lightly. P_6 is used to expand the search space. It was suggested in [175] and [144] that visiting as many different states as possible helps to detect more faults. The fitness functions thus favor visiting more states when the fault detection count drops very low. Hence, P_6 is considered in the final stage only. Different weights are given to each parameter in the fitness computation during the two phases:

Fault activation phase:
  Multiple time frame:
  1. $fitness = 0.2P_1 + 0.8P_4$
  Single time frame:
  2. $fitness = 0.1P_1 + 0.5P_2 + 0.2P_4 + 0.2(P_5 + P_6^\dagger)$
  State justification:
  3. $fitness = 0.1P_1 + 0.2((k - P_2) + 0.5P_3 + 0.2(P_5 + P_6^\dagger)$,
  where k is a constant

Fault propagation phase:
  4. $fitness = 0.8P_1 + 0.2(P_4 + P_5 + P_6^\dagger)$
  $\dagger$: included in the final GA stage only

In the fault activation phase, the aim is to excite the fault and propagate the fault effects to as many good flip-flops as possible, where good flip-flops are those with more powerful distinguishing sequences; thus, the fitness function places a heavier weight on the quality of flip-flops reached by the fault effects. Any positive value on parameter P_4 implies that the target fault is excited and the fault-effects are propagated to at least one flip-flop. If no sequence is obtained to activate the current target fault, single-time-frame fault activation and state justification are used in a second attempt to activate the fault. In this case, the fitness function favors states that can be easily justified during single-time-frame fault activation, while hard-to-reach states are favored during state justification, because hard-to-reach states may be necessary in order to reach the desired unvisited target state. In the fault propagation phase, the goal is to find a sequence that will propagate the fault effects to a PO, so the emphasis is placed on fault detection.

Other Implementation Details

Because one fault is targeted at a time and the majority of time spent by the GA is in the fitness evaluation, parallelism among the individuals can be exploited. Therefore, parallel-pattern simulation [2] is used to speed up the process. During fitness evaluation, 32 candidate sequences from the population are simulated simultaneously, with values bit-packed into 32-bit words during simulation. Fault-free simulation is first performed, followed by faulty circuit evaluation, in which events start exclusively from the faulty gate.

Targeting untestable faults is a waste of time because untestable faults cannot be identified using our approach. Thus, the HITEC deterministic test generator [160] is used after the first GA stage to identify and remove many of the untestable faults.

A small time limit of 0.4 seconds per fault is used in an initial HITEC pass through
the fault list to minimize the execution time. If a large number of untestable faults
are identified or if only a small number of faults remain in the fault list, a second
HITEC pass with a time limit of 2 seconds per fault is used. Any test sequences
generated by HITEC are discarded.

6.5.4 Experimental Results

The STRATEGATE test generator was implemented in C++; ISCAS89 sequential
benchmark circuits and several synthesized circuits described in Section 6.2.5 were
used to evaluate its performance on an HP 9000 J200 with 256-MB RAM. STRATE-
GATE is compared to various other test generators in Table 6.18. Results for both
optimized (piir8o) and unoptimized (piir8) versions of circuit piir8 are shown. For
each circuit, the number of faults detected and the test set length are given for each
test generator. The number of states visited by STRATEGATE is also reported for
each circuit. The first test generator is HITEC [160], a deterministic test generator,
followed by the GA-based test generators GATEST [189], DIGATE [98], and finally
STRATEGATE.

Table 6.18. STRATEGATE Results

Circuit	HITEC		GATEST		DIGATE		STRATEGATE		
	Det	Vec	Det	Vec	Det	Vec	Det	Vec	States
s298	**265**	306	264	161	264	239	**265**	306	154
s344	328	142	**329**	95	**329**	109	**329**	86	272
s382	363	4931	347	281	363	581	**364**	1486	1159
s400	383	4309	365	280	382	3369	**384**	2424	1954
s444	414	2240	405	275	420	1393	**424**	1945	1085
s526	365	2232	417	281	446	2867	**454**	2642	1764
s641	**404**	216	**404**	139	**404**	180	**404**	166	115
s713	**476**	194	**476**	128	**476**	147	**476**	176	109
s820	813	984	517	146	621	465	**814**	590	25
s832	817	981	539	150	606	703	**818**	701	25
s1196	**1239**	453	1232	347	1236	549	**1239**	574	386
s1238	**1283**	478	1274	383	1281	504	1282	624	406
s1423	776	177	1222	663	1393	4044	**1414**	3943	3605
s1488	**1444**	1294	1392	243	1378	542	**1444**	593	48
s1494	**1453**	1407	1416	245	1354	581	**1453**	540	48
s5378	3238	941	3175	511	3447	10500	**3639**	11571	9550
s35932	34,902	240	35,009	197	**35,100**	386	**35,100**	257	195
am2910	2164	874	2163	745	2195	2206	**2198**	2509	2233
mult16	1640	273	1653	204	1664	915	**1665**	1530	943
div16	1665	189	1739	634	1802	4481	**1814**	3476	2425
pcont2	3354	7	6826	272	**6837**	3452	**6837**	194	152
piir8o	14,221	347	15,013	531	**15,072**	506	15,071	354	305
piir8	11,131	31	-	-	18140	603	**18,206**	443	326

Det: number of faults detected   Vec: test set length   States: number of states visited

6.5. Use of Finite State Machine Sequences

The best fault coverages are highlighted in bold. As shown in the table, the fault coverages achieved by STRATEGATE match or surpass those obtained by all other test generators for all circuits except two, s1238 and piir8o, where only one less fault was detected. In many cases, the fault coverages obtained by STRATEGATE are significantly higher. For the hard-to-test circuits, such as s444, s526, s1423, s5378, s35932, and the synthesized circuits, where long execution times are required by HITEC, the fault coverages achieved by STRATEGATE are much higher, and execution times are much shorter. Even in the circuits where previous GA-based test generators did not perform well, such as s820, s832, s1488, and s1494, STRATEGATE is able to detect all of the detectable faults. These four circuits contain faults that require specific and often long sequences for fault activation. None of the previous GA-based test generators could match the results of HITEC for these circuits; STRATEGATE, however, is able to reach all the required states via dynamic state traversal. STRATEGATE is able to visit more states than HITEC in the larger circuits where higher fault coverages are obtained. The test sets obtained by STRATEGATE are more compact than those obtained by HITEC, even when higher fault coverages are achieved by STRATEGATE. The test sets are more compact than those obtained by CRIS or DIGATE for most of the circuits.

STRATEGATE achieves very high fault coverages very quickly using a small number of vectors. This phenomenon is illustrated in Table 6.19 for 12 circuits at different checkpoints placed at the end of each GA stage. Recall that sequence

Table 6.19. Results at Various Checkpoints for STRATEGATE

Circuit	$\sqrt{}$	Det	Vec	Time	Circuit	$\sqrt{}$	Det	Vec	Time
s382	1	361	601	1.07m	s1423	1	**1410**	2065	13.2m
	2	362	1285	5.9m		2	**1410**	2965	40.1m
	3	**364**	1486	8.1m		3	**1414**	3943	1.27h
s444	1	408	354	38.5s	s1494	1	1393	295	5.34m
	2	**420**	753	2.3m		2	**1453**	540	7.50m
	3	**424**	1945	20.1m		3	**1453**	540	7.60m
s526	1	431	486	1.37s	s5378	1	**3562**	2175	4.60h
	2	442	1098	8.3m		2	**3607**	4461	25.1h
	3	**454**	2642	54.5m		3	**3639**	11571	37.8h
s713	1	475	157	1.1m	s35932	1	**35100**	257	10.1h
	2	**476**	176	1.30m		2	**35100**	257	10.2h
	3	**476**	176	1.31m		3	**35100**	257	10.9h
s820	1	812	572	3.07m	am2910	1	2190	953	6.25m
	2	**814**	590	3.60m		2	**2197**	1761	13.5m
	3	**814**	590	3.63m		3	**2198**	2509	29.4m
s1196	1	1235	521	1.12m	div16	1	1727	352	32.0m
	2	1237	536	1.21m		2	**1810**	1168	2.62h
	3	**1239**	574	1.49m		3	**1814**	3476	8.1h

Check Point k: end of GA stage k

lengths for the individuals in the population are doubled from one stage to the next, with longer sequences used to target the harder faults. The checkpoint, number of faults detected, test set size, and execution time are displayed in Table 6.19 for each circuit. The fault coverages at the end of the first or second GA stages are already higher than the final fault coverages of the other test generators for many circuits. For example, STRATEGATE detects 1410 faults in 13.2 minutes for circuit s1423; all other test generators spend hours of execution time and still do not reach this fault coverage. This phenomenon is consistent for many circuits shown in the table. The user may wish to stop the test generation process if the fault coverage has reached a satisfactory level at the end of the first or second stage.

6.6 Dynamic Test Sequence Compaction

Test sets generated by GATEST tend to be very compact compared to those generated by other test generators, such as HITEC. Nevertheless, GATEST is not successful in generating tests for all the hard-to-test faults, and therefore, other test generation approaches may be needed. Both HITEC and STRATEGATE have been successful in covering faults missed by GATEST, and both of these test generators target faults one at a time. However, since the tests generated are targeted at individual faults, they do not necessarily detect a significant number of additional faults. Test application time is an important consideration, since it directly impacts the testing cost. If shorter test sets can be used that still obtain a given fault coverage, more chips can be tested in a given time period, and fewer testers are needed. Thus, we would like to compact the test sets, either during or after test generation, but still maintain a given level of fault coverage.

Several techniques have recently been proposed for static compaction of test sequences [100], [101], [162], [177], and large reductions in test set sizes have been reported. Static compaction is performed after test generation is completed. Heuristics were used to merge test sequences in [162], but some faults incidentally detected by the original test sets were no longer detected by the compacted test sets. Therefore, additional passes of test generation and compaction were required to obtain the original fault coverage, increasing the overall execution time. Vector insertion, vector omission, and vector selection were used in [177] to compact a given test set, but execution times were high due to the large number of fault simulations performed. Faster approaches to static compaction were presented in [100], [101] that use only two fault simulation passes, but less compact test sets were obtained. Even if static compaction is performed, dynamic compaction during test generation may still be useful in reducing the final test set size and also the execution time of static compaction.

In contrast to static compaction, dynamic compaction is performed concurrently with the test generation process. Several different dynamic compaction approaches have been proposed for sequential circuits. In the first approach [176], test generation alternates between fault-independent and fault-oriented phases, and vectors are added to the test set one-at-a-time to minimize test set size. Very compact

6.6. Dynamic Test Sequence Compaction

test sets were obtained for the few circuits reported, but execution times were not given. In the second approach [28], [182], [183], heuristics are used for selection of *secondary* target faults after a partially specified test sequence is obtained for a *primary* target fault. Attempts are made to extend the test sequence generated for the primary fault to cover secondary faults using assignments to unspecified PIs only. In the third approach [132], three heuristics are used to obtain compact test sets. Selecting the best of 64 random fillings for unspecified bits in a partially specified sequence was found to reduce test set sizes significantly. In the fourth approach [178], the static compaction techniques proposed in [177] are used for dynamic compaction; test vectors are omitted from the existing test set or inserted into the existing test set so that the test set will be able to detect the current fault being targeted. Multiple passes are necessary, since faults previously covered by the test set may not be covered after the modifications. In the fifth approach [17], a symbolic algorithm is used to find all test sequences for each fault and to combine test sequences into a compact test set that covers all testable faults. This approach is suitable for control-dominant circuits that can be managed using binary decisions diagrams.

We take a different approach to dynamic test sequence compaction that uses fault simulation and GAs to reduce the number of test vectors in a generated test sequence and increase the number of faults detected. Three techniques are used that are simple additions to an existing fault simulator, and two of the three techniques use GAs. The first technique attempts to reduce the length of the test sequence provided by a test generator. Some of the test vectors at the beginning and end of the test sequence may not be needed to detect a target fault if the circuit state is known. The test generator may assume that the circuit starts from an unknown state, but the circuit may have already been initialized by previous vectors in the test set, and a fault simulator interfaced to the test generator can be used to remove any unnecessary vectors. The second technique attempts to increase the number of faults detected by a given sequence while ensuring that it detects the original target fault. For test sequences generated by deterministic test generators, which contain many unspecified bits, the unspecified bits are typically filled randomly with ones and zeros. Instead, we use a GA to optimally fill the unspecified bits so that more faults are detected. The specified bits are fixed at the values provided by the test generator, and thus, the approach is limited to test generators such as HITEC [160] that can provide partially specified test sequences. The third technique uses a more general approach to genetic compaction that removes this restriction on the type of test generator used. It uses a GA to evolve the entire test sequence, including the bits specified by the test generator. The test sequence provided by the test generator is simply used as a *seed* in the initial population of the GA. This method does not depend upon the presence of unspecified bits, and therefore, it can be used for various different test generation algorithms.

The genetic approach has the advantage of being a simulation-based approach in which processing occurs in the forward direction only. Thus, constraints on the test sequences generated are easily handled. The GA is able to make use of the previous good and faulty circuit states reached after all previous test vectors in the test set have been applied, often improving the fault coverage. Increasing the number of faults covered by a given test sequence reduces the number of faults that must be specifically targeted by the deterministic test generator. Therefore, reductions in the overall execution time can be expected as well. The approach described here is much simpler than many of the previous approaches, more scalable to larger circuits, and can easily be added to an existing test generator.

6.6.1 Overview

Test generators often target faults individually. Attempts are made to excite the target fault and propagate its effects to the primary outputs, possibly through several time frames via the flip-flops. Deterministic test generators use backtracing operations for fault excitation, fault propagation, and state justification. In the process, values are assigned to a subset of the bit positions in the test vectors that make up the test sequence generated. The remaining bits are unspecified. Often several additional faults are detected by the same partially specified sequence, but for other faults, particular values must be assigned to the unspecified bits if the faults are to be detected. In HITEC [160], the unspecified bits are filled randomly with ones and zeros in an attempt to cover additional faults. However, more faults can often be detected if several alternative random fillings are fault simulated, and only the best one is added to the test set. Even a greater number of faults can sometimes be detected if a GA is used to evolve the best filling of ones and zeros. In a simulation-based test generator such as STRATEGATE [99], all bits in each test sequence generated are specified, but not all values are necessarily required to detect the target fault. For both deterministic and simulation-based test generators, more faults may be detected with small changes to the given test sequences. The modified sequences may no longer detect the original faults targeted, but these faults may eventually be covered by sequences generated later in the process.

One way to perform dynamic test sequence compaction is to try to maximize the number of faults detected by each sequence. Thus, we can use a GA for this purpose. Two different approaches are possible. First, we can use the GA to fill the unassigned values in a partially specified test sequence. Second, we can simply seed the GA with the sequence provided by the test generator and evolve an optimized sequence. With either of the two approaches, we set the GA fitness function to maximize fault detection. The GA will explore several alternative sequences through a number of generations, and the best one is added to the test set.

Genetic filling of unassigned values is illustrated in Fig. 6.21. A partially specified test sequence that detects the primary target fault is generated by a deterministic test generator. The test sequence is then sent to the dynamic compactor, which is a constrained GA-based test generator. The objective of the dynamic compactor is to find a filling of ones and zeros in the unspecified bit positions that detects the greatest number of secondary faults. The GA of Fig. 6.5 is used. All individuals in the GA population are initialized with random values, and unspecified values

6.6. Dynamic Test Sequence Compaction

in the test sequence are filled using values in successive bit positions from an individual in the GA population. In the example shown in the figure, two unspecified bits in the first vector of the test sequence are filled using the first two values in the GA individual, one unspecified bit in the second vector is filled using the third value in the GA individual, and so on, until all unspecified bits are filled. The fully specified test sequence is then fault simulated to obtain its fitness value; the fitness value measures the quality of the corresponding solution, primarily in terms of fault coverage. The same fitness evaluation procedure is used for all individuals in the population, and then a new generation is evolved. Processing continues for a set number of generations, and the best test sequence found in any generation is added to the test set.

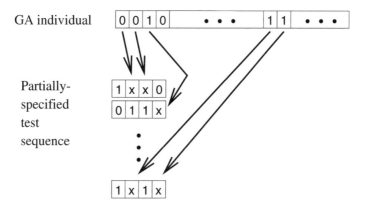

Figure 6.21. Genetic filling of unspecified bits

The second approach to genetic compaction is illustrated in Fig. 6.22. The GA is seeded with copies of the test sequence provided by the test generator. If every individual in the population is seeded with the same sequence, the specified bits will all be the same for every individual. Bits that are not specified are filled randomly. In the example shown in Fig. 6.22, the partially specified test sequence provided by the test generator is x11 xx1 x0x 01x, where **x** represents an unspecified bit; therefore, all individuals in generation 1 have the same values in positions 2, 3, 6, 8, 10, and 11. Each fully specified test sequence is then fault simulated to obtain its fitness value; the fitness value measures the quality of the corresponding solution, primarily in terms of fault coverage. The GA is evolved over several generations, and the best test sequence found in any generation is added to the test set. By the time the last generation is reached, several of the values specified by the test generator may have changed in many of the individuals, as illustrated in Fig. 6.22. Thus, the best sequence evolved may not detect the original target fault, but test generators often make several passes through the fault list, and it is likely that the fault will be detected in a later pass, if it is not covered by a sequence generated for a different target fault.

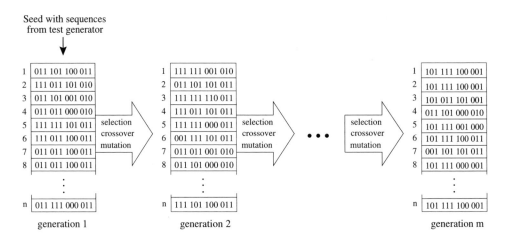

Figure 6.22. Test sequence evolution

Before the GA is used, unnecessary vectors at the beginning and end of the test sequence provided by the test generator may be identified and removed using a fault simulator. These extra vectors are generated if the test generator assumes that the initial circuit state is unknown. However, the fault simulator may have information about the circuit state if any test vectors have been previously added to the test set. Thus, for a test sequence of length N, fault simulations are performed for the target fault in which the first k vectors are removed from the test sequence, where k ranges between 1 and $N - 1$. The shortest subsequence that detects the target fault is selected. Then any vectors at the end of the sequence that do not contribute to detecting the fault are removed. Note that this technique is only useful for test generators such as HITEC [160] that assume that the circuit starts in an unknown state. Test generators such as FASTEST [113] that make use of the circuit state would not benefit from this procedure, although they would benefit from the genetic compaction.

The techniques described were implemented in the GA-COMPACT and Squeeze dynamic test sequence compactors, which can be interfaced to existing test generators. GA-COMPACT uses a GA for optimal filling of unspecified bits in each partially specified test sequence, while Squeeze uses a GA to evolve the entire test sequence, including bits specified by the test generator. Both tools were implemented as minor modifications to the GATEST test generator, and the same fitness functions, selection, crossover, and mutation schemes, and GA parameters were used. Fault samples of size 100 were used in the fitness evaluation to speed up the computation, and the original target fault was always included in the fault sample in Squeeze to ensure that at least one fault was detected by any sequence generated. The overall algorithm for test generation with dynamic compaction is illustrated in Fig. 6.23. The test generator and the dynamic compactor run independently,

6.6. Dynamic Test Sequence Compaction

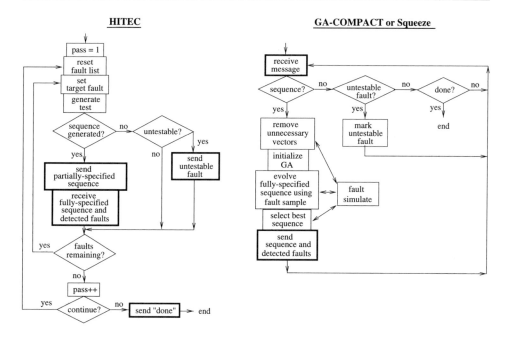

Figure 6.23. Test generation with dynamic compaction

communicating through UNIX sockets. The test generator acts as a client, and the dynamic compactor acts as a server. In our implementation, we used the HITEC deterministic test generator. HITEC makes several passes through the fault list, with increasing time limits per fault for successive passes. The next fault in the fault list is selected as the target fault, and test generation is attempted. If HITEC is successful, the partially specified test sequence is sent to the dynamic compactor, which returns a fully specified test sequence and a list of detected faults. The detected faults are removed from the fault list, and the process continues for the next fault in the fault list. If the targeted fault is identified as untestable, this information is also sent to the dynamic compactor. Otherwise, if no test is generated in the given time limit, HITEC proceeds with the next fault. After each pass through the fault list, the user is prompted about whether to continue with the next pass. Processing terminates when the user responds negatively.

The dynamic compactor receives messages from the test generator until it receives a message indicating that processing is finished. When a test sequence is received, unnecessary vectors at the beginning and end of the sequence are first removed. The GA is initialized, and a fully specified test sequence is evolved by the GA using a small fault sample. This sequence is then fault simulated using the full fault list. For Squeeze, any vectors at the end of the sequence that do not contribute to the fault coverage are removed. This step is skipped for GA-COMPACT, since the last vector in the sequence will always detect the original target fault. Finally, the test sequence is sent back to the test generator, along with the list of detected faults obtained through fault simulation using the full fault list. The PROOFS sequential circuit fault simulator [161] is used to identify the unnecessary vectors, to evaluate the fitness of each candidate test sequence, and again to update the state of the circuit after the best test sequence is selected. If the message sent by the test generator is an untestable fault identification, the corresponding fault is marked as untestable to avoid including it in any fault sample.

6.6.2 Experimental Results

The GA-COMPACT and Squeeze dynamic test sequence compactors were implemented using the existing PROOFS [161] source code and 1700 additional lines of C++ code. The HITEC deterministic test generator [160] was used to provide test sequences for compaction. Tests were generated for several of the ISCAS89 sequential benchmark circuits [23] and several synthesized circuits on an HP 9000 J200 with 256-MB RAM. Experiments were performed to evaluate the three proposed techniques for dynamic test sequence compaction: (1) genetic filling of unspecified bits in each partially specified test sequence, (2) removal of unnecessary vectors at the beginnings and ends of test sequences, and (3) evolution of test sequences.

In the first set of experiments, the effects of the genetic filling of unspecified bits in the partially specified sequences were studied. Thus, unnecessary vectors at the beginnings and ends of the test sequences were not removed before the GA was invoked. Instead, the fully specified test sequences evolved by the GA were fault simulated with the entire fault list, and noncontributing vectors at the ends of the sequences were removed before the sequences were sent back to the deterministic test generator. Results of these experiments are shown in Table 6.20 for a GA with a population size of 8 and 8 generations. Results for HITEC with a single random filling of unspecified bits and for the best of 64 random fillings are shown for comparison. GA-COMPACT, was successful in reducing the test set size for all circuits except s526, am2910, and pcont2, but for these circuits, the fault coverages increased. Fault coverages can drop if HITEC is unsuccessful in generating tests for hard-to-test faults, as is the case for s344 and mult16. However, in most cases, the fault coverage increases in the first two passes through the fault list, and a higher fault coverage is sometimes observed after the third pass, e.g., s1423, s35932, div16, and pcont2. Comparing results for the genetic filling of unspecified bits to the best of 64 random fillings, the genetic approach outperforms the random approach for a majority of the circuits, even though the same number of fillings are simulated for each partially specified test sequence. These results demonstrate the power of genetic algorithms. However, the random approach sometimes provides more compact test sets or higher fault coverages. Either approach is a significant improvement over the current practice of using a single random filling. When GA-COMPACT is used, more faults are typically covered by the test sequences generated by the computation-intensive deterministic test generator, and fewer faults have to be targeted. Consequently, the execution time for HITEC often

6.6. Dynamic Test Sequence Compaction

Table 6.20. Genetic Compaction Results

Circuit	HITEC+GA-COMPACT			HITEC			HITEC+RANDOM		
	Det	Vec	Time	Det	Vec	Time	Det	Vec	Time
s298	265	**207**	16.5m+14s	265	322	16.2m	265	225	16.5m+13s
s344	321	80	10.7m+6.7s	324	115	8.07m	324	**77**	7.56m+6.2s
s349	332	**51**	5.37m+4.6s	332	128	7.73m	331	56	6.19m+5.0s
s382	301	**1193**	1.51h+2.0m	301	1463	1.52h	301	**1193**	1.51h+1.9m
s386	314	237	7.86s+9.1s	314	286	7.25s	314	**218**	6.57s+6.8s
s400	341	**1367**	1.20h+2.1m	341	845	1.21h	341	1594	1.19h+2.0m
s444	373	**1532**	1.70h+2.2m	373	1761	1.49h	373	1733	1.49h+2.4m
s526	329	**678**	5.55h+52s	316	436	5.79h	337	**678**	5.44h+52s
s641	404	**103**	4.61s+11s	404	209	4.84s	404	116	4.70s+12s
s713	476	**86**	6.99s+10s	476	173	6.71s	476	109	7.15s+12s
s820	814	**842**	3.68m+1.3m	813	1115	3.50m	814	905	3.71m+1.4m
s832	818	860	5.33m+1.1m	817	1137	5.75m	818	**814**	5.39m+1.0m
s1196	1239	**268**	3.73s+34s	1239	435	5.50s	1239	280	3.82s+34s
s1238	1283	**293**	6.50s+37s	1283	475	8.23s	1283	475	6.73s+38s
s1423	931	140	12.1h+30s	723	150	13.9h	1023	153	11.4h+33s
s1488	1444	831	17.6m+2.8m	1444	1170	16.5m	1444	**821**	15.2m+2.5m
s1494	1453	**812**	10.2m+2.7m	1453	1245	9.59m	1453	845	9.07m+3.0m
s3271	3237	**582**	42.3m+2.7m	3227	641	39.4m	3236	542	36.5m+2.5m
s3330	2108	386	10.5h+1.7m	2097	551	10.6h	2110	**483**	10.5h+2.0m
s3384	3028	**113**	5.52h+49s	2996	161	6.06h	3026	101	5.54h+44s
s4863	4629	**293**	2.25h+2.3m	4621	477	2.38h	4629	302	2.28h+2.2m
s5378	3240	**568**	18.2h+3.5s	3231	912	18.4h	3238	589	18.3h+2.9m
s6669	6664	150	25.1m+1.8m	6655	319	33.3m	6670	**183**	19.3m+2.2m
s35932	34925	**341**	4.28h+20m	34901	496	4.73h	34910	402	4.42h+21m
am2910	2175	1006	58.8m+4.6m	2171	871	56.4m	2190	**594**	37.0m+2.7m
div16	1683	**183**	4.98h+43s	1667	228	5.35h	1679	177	5.04h+42s
mult16	1560	73	2.10h+25s	1582	**111**	1.90h	1571	85	1.91h+25s
pcont2	4343	**7**	8.22h+19s	3514	7	9.58h	4270	7	8.31h+20s

drops when GA-COMPACT is used in conjunction. Because small fault samples are used by GA-COMPACT, execution times for the GA are minimal. Therefore, execution times for GA-COMPACT are usually significantly smaller than the HITEC test generation times, and the overall time for HITEC + GA-COMPACT is smaller than the execution time for HITEC alone for many of the circuits.

Experiments were performed to evaluate the effects of GA population size on the quality of results. Population sizes of 8, 16, 32, and 64 were used. A population size of 32 gave good results for the largest number of circuits, but population sizes of 64, 16, and 8 provided good results as well [193].

Experiments were carried out to determine if removing unnecessary vectors at the beginning and end of a partially specified test sequence before the GA is invoked is effective in reducing the test set size. Results are shown in Table 6.21. Significant reductions in test set size occurred for many of the circuits, especially for the highly sequential circuits such as s382 [193]. Test sequences of 60 vectors or more were often

Table 6.21. Dynamic Compaction: Squeeze vs. GA-COMPACT

Circuit	Total Faults	GA-COMPACT			Squeeze		
		Det	Vec	Time	Det	Vec	Time
s298	308	265	145	16.3m+28.2s	265	**133**	16.2m+31.1s
s344	342	326	54	5.31m+15.2s	329	**62**	3.65m+17.1s
s349	350	332	61	5.53m+17.3s	334	**53**	3.40m+15.9s
s382	399	323	272	1.17h+1.65m	359	**485**	36.7m+2.00m
s386	384	314	182	7.25s+23.0s	314	**128**	5.35s+16.8s
s400	426	364	389	51.3m+2.35m	375	**454**	40.1m+2.41m
s444	474	374	228	1.59h+1.51m	418	**598**	39.3m+3.73m
s526	555	275	76	6.18h+35.3s	359	**84**	5.04h+28.2s
s641	467	404	83	4.18s+33.8s	404	**72**	3.92s+28.4s
s713	581	476	60	6.09s+27.5s	476	**56**	6.07s+27.4s
s820	850	814	567	2.75m+2.75m	814	**462**	2.18m+2.45m
s832	870	817	561	4.42m+2.50m	818	**438**	3.60m+2.33m
s1196	1242	1239	232	4.62s+1.85m	1239	**219**	4.31s+1.62m
s1238	1355	1283	248	7.25s+2.10m	1283	**232**	6.86s+1.73m
s1423	1515	963	*119*	11.6h+1.64m	1047	137	8.97h+1.71m
s1488	1486	1444	475	14.3m+5.76m	1444	**408**	12.4m+5.31m
s1494	1506	1453	502	8.52m+6.51m	1453	**359**	6.73m+4.51m
s3271	3270	3242	593	35.4m+8.35m	3253	**521**	26.1m+7.77m
s3330	2870	2104	246	10.5h+4.08m	2117	**258**	10.3h+5.20m
s3384	3380	3047	*85*	5.21h+2.30m	3066	118	5.23h+3.52m
s4863	4764	4626	263	2.18h+7.48m	4629	**249**	2.11h+7.09m
s5378	4603	3239	*189*	18.2h+5.82m	3245	198	18.1h+6.51m
s6669	6684	6666	**129**	21.4m+5.60m	6663	135	22.9m+6.32m
s35932	39,094	35,051	*160*	2.04h+34.3m	35,084	178	1.55h+53.6m
am2910	2391	2184	531	49.5m+9.96m	2192	**264**	28.0m+6.23m
div16	2147	1678	151	5.17h+2.61m	1695	**121**	6.50h+1.89m
mult16	1708	1594	*82*	1.77h+1.79m	1599	92	1.61h+2.00m
pcont2	11,300	2847	3	10.5h+44.0s	3011	3	9.99h+29.5s

reduced to less than 5 vectors. In addition, fault coverages for four of the circuits were highest when this procedure was used: s400, s444, s35932, and mult16. For most of the remaining circuits, test set sizes were very close to the minimal obtained by GA-COMPACT. For a few circuits, including s526, s1423, and pcont2, the results were not as good as in some of the previous experiments, but this was due to a drop in fault coverage rather than a failure in compaction. HITEC is unable to generate tests for many of the faults in these circuits when they are specifically targeted. A simulation-based test generator such as GATEST [188] is more capable of generating tests for them. Thus, to handle these circuits, we added an extra phase at the end of each pass through the fault list in which completely unspecified sequences are passed to GA-COMPACT, and sequences are generated until no more faults are detected.

6.6. Dynamic Test Sequence Compaction

Unnecessary vectors at the ends of the sequences are removed using fault simulation with the full fault list before the fully specified sequences are returned to HITEC. Test sequence lengths of x, $2x$, and $4x$ were used in the first, second, and third passes through the fault list, respectively, where x is equal to half the structural sequential depth of the circuit (or the next larger integer). Fault coverages increased to their highest values for four of the five circuits, and the fault coverage for the fifth circuit, s526, was almost as high as the best of the previous experiments. Furthermore, test generation times dropped significantly, since faults were detected much earlier in the test generation process, and test sets remained reasonably compact, since the GATEST approach tends to produce compact test sets [188].

Experiments were performed to evaluate test sequence evolution as implemented in Squeeze and compare it to the genetic filling of unspecified bits. As for GA-COMPACT, the fault simulator was used by Squeeze to remove vectors at the beginning and end of the sequence that were not necessary for detection of the target fault before genetic compaction was performed for each sequence. The resulting sequence was then used as a seed in the GA. The GA was evolved over eight generations, and the best sequence found in any generation was selected. Finally, any vectors at the end of the sequence that did not contribute to the fault coverage were removed before the sequence was sent back to the test generator. A GA population size of 32 and a mutation rate of $1/64$ were used. Squeeze is compared to GA-COMPACT in Table 6.21. Three passes through the fault list were made by HITEC for all circuits except pcont2, for which only two passes were used to reduce the execution time. Time limits for the three passes were 0.5, 5, and 50 seconds per fault. For each circuit, the total number of collapsed faults and the number of faults detected (**Det**), the number of test vectors generated (**Vec**), and the execution time using HITEC with GA-COMPACT and with Squeeze are shown. Execution times for HITEC and for GA-COMPACT or Squeeze are separated in columns 5 and 8. The test set size is shown in bold if it is the most compact of the test sets that achieved the highest fault coverage. The test set size is shown in italics if the highest fault coverage was not achieved, but the test set is more compact than the one that did achieve the highest fault coverage.

The major difference between the approaches used by GA-COMPACT and Squeeze is that GA-COMPACT requires that the bits specified by the test generator remain at fixed values, while Squeeze allows these values to change. Values not specified by the test generator are filled randomly for both dynamic compactors, and the same deterministic sequence is used as a seed for all 32 individuals in the initial GA population in Squeeze. The best results were achieved by Squeeze for all circuits except s6669, if fault coverage is considered to be the primary metric of quality, and test set size is secondary. HITEC is often unsuccessful in generating tests for hard-to-test faults, and a simulation-based compactor is particularly useful in these cases. In fact, for four of five circuits, Squeeze achieves fault coverages comparable to those obtained for GA-COMPACT with a GATEST phase. The last circuit, pcont2, is one example where HITEC is unable to generate sequences for most faults targeted, and a simulation-based approach is much more effective.

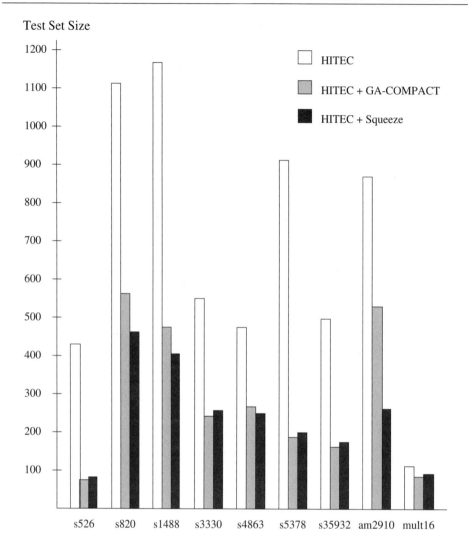

Figure 6.24. Dynamic compaction results

The improvements in compaction are illustrated graphically in Fig. 6.24 for nine circuits. Test set sizes are shown for the following three test generation approaches: (1) HITEC with a single random filling of unspecified bits, (2) HITEC with GA-COMPACT (including removal of unnecessary vectors), and (3) HITEC with Squeeze (including removal of unnecessary vectors). GA-COMPACT provides significant reductions in test set sizes, but Squeeze provides even better compaction for a majority of the circuits. In addition, Squeeze provides better fault coverages and lower overall test generation times.

6.6. Dynamic Test Sequence Compaction

The GA-COMPACT test sets are slightly more compact for five circuits: s1423, s3384, s5378, s35932, and mult16. Here, compaction was measured by performing fault simulation using the test set that achieved the higher fault coverage and noting the number of vectors required to achieve the lower fault coverage. For circuits having differences in fault coverage for the two techniques, the number of vectors needed from each of the test sets to obtain the lower fault coverage is shown in Table 6.22. Since the test sequences evolved by Squeeze detect more faults than those evolved by GA-COMPACT, fewer faults had to be targeted by the test generator, and the test generation time often dropped as a result. Compaction times for the two approaches were comparable and were dependent on the test set size. Since higher fault coverages were often achieved by Squeeze, the test sets were sometimes longer, and consequently, the compaction times were also longer. GA-COMPACT provides better results than a random approach, and since Squeeze is clearly better than GA-COMPACT, it is also better than a random approach.

Table 6.22. Comparison of Test Set Compactness

Circuit	Detected Faults	Number of Vectors	
		GA-COMPACT	Squeeze
s344	326	54	54
s349	332	61	**49**
s382	323	272	**156**
s400	364	389	**271**
s444	374	228	**124**
s526	275	76	**48**
s832	817	561	**431**
s1423	963	*119*	132
s3271	3242	593	**493**
s3330	2104	246	**228**
s3384	3047	*85*	111
s4863	4626	263	**242**
s5378	3239	*189*	191
s6669	6663	**128**	135
s35932	35,051	*160*	172
am2910	2184	531	**247**
div16	1678	151	**92**
mult16	1594	*82*	92
pcont2	2847	3	**3**

Test sets generated by HITEC/Squeeze are more compact than those generated in [175] for four of five circuits; results were not reported for the larger circuits in [175]. The HITEC/Squeeze test sets are significantly smaller than any reported in [28], [132], [182], [183] for comparable fault coverages. Finally, the HITEC/Squeeze test sets are smaller than those reported in [178] for five of the six circuits having comparable fault coverages. Further compaction may be possible by performing the static compaction techniques of [100], [177] in a postprocessing step.

6.7 Conclusions

A genetic algorithm framework was developed for use in sequential circuit test generation. Populations of candidate tests are evolved by the GA starting from a random initial population, and the best test evolved is added to the test set in a given time frame. A highly accurate fitness function is used to evaluate candidate tests in order to achieve good quality test sets. Results for the ISCAS89 sequential benchmark circuits indicate that the selection and crossover schemes used have a significant impact on fault coverage. The best results were obtained for tournament selection without replacement and uniform crossover. Variations in mutation rate had a much smaller effect on fault coverage, and binary codings tended to give higher fault coverages when small population sizes of 16 or 32 were used. Nonoverlapping populations gave the highest fault coverages, but average speedups of 1.3 were obtained by using overlapping populations, with only a 0.4% drop in fault coverage. More significant reductions in execution time were obtained by using small fault samples in the fitness evaluation.

The GA-based test generator was extended to generate tests for full scan and partial scan circuits, with the objective of reducing test application time. A genetic approach was used to generate compact test sets that limit the scan operations. The genetic approach provided significant improvements over the traditional method.

While the GA-based test generator often gives better results as compared to deterministic approaches, limitations do exist. In a simulation-based approach, untestable faults cannot be identified. Furthermore, the GA-based test generator is best suited for data-dominant circuits, and higher fault coverages are obtained for highly sequential circuits using a deterministic test generator. Combining the deterministic and genetic approaches can be beneficial. One such hybrid test generator combines deterministic algorithms for fault excitation and propagation, and GAs for state justification. Deterministic procedures for state justification are used only if the genetic approach is unsuccessful. High fault coverages were obtained using this approach. Another approach to combining the deterministic and genetic approaches is to alternate repeatedly between them. Even higher fault coverages were obtained using this approach.

Despite the good results achieved when using a simple GA, problem-specific knowledge is needed to achieve the best results for all circuits. Targeting the hard-to-test faults one at a time in a two-phase strategy was found to provide the highest fault coverages. The first phase excites a fault and propagates the fault effects to the flip-flops; the second phase drives the fault effects from the flip-flops to the POs. Finite state machine sequences are used in both phases, including state-transfer sequences in the first phase and distinguishing sequences in the second phase. The GA is able to combine various sequences, which are partial solutions to the problem, to engineer a complete solution.

6.7. Conclusions

High test application time can be a problem for sequential circuits, and therefore, compact test sets are desirable. In our approach to dynamic compaction for sequential circuits, we use fault simulation and GAs to compact the sequences generated by a test generator and increase the number of faults detected by the sequences. Significant reductions in test set sizes were observed for the benchmark circuits studied, corresponding to reductions in test application time. Fault coverages sometimes improved, and ATG execution times dropped for many of the circuits. Test sets generated using these techniques are more compact than any previously reported for most circuits, yet the execution times for dynamic compaction are considerably smaller. The proposed techniques can easily be added to existing commercial test generators with little or no detrimental effect on performance and are expected to provide a significant improvement in the quality of the test sets generated.

Genetic algorithms are particularly amenable to parallel implementations, so very good speedups are expected for a parallel GA-based test generator. In addition, the GA-based test generator is not limited to the single stuck-at fault model, and other fault models can easily be accommodated with appropriate fitness functions.

SUMMARY

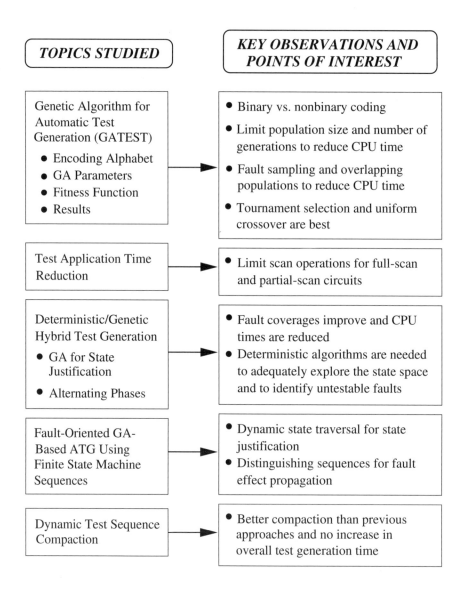

CHAPTER 7
at a glance

- Peak power estimation
  - Delay models
    - Zero delay
    - Unit delay
    - Variable delay
  - Measurement duration
    - Peak single-cycle power
    - Peak n-cycle power
    - Peak sustainable power

Chapter 7

PEAK POWER ESTIMATION

M. S. Hsiao, E. M. Rudnick, J. H. Patel

Peak power dissipation is a critical design factor and can determine the thermal and electrical limits of a design. A sequence that is capable of generating *peak power* over a given period of time may be applied during stress-testing to ensure reliability. It was pointed out in [217] that large instantaneous power dissipation can cause overheating (local hot-spots), and that the failure rate for components roughly doubles for every 10° C increase in operating temperature. In addition, the continuing decrease in feature size and increase in chip density in recent years have reinforced the concerns about excessive power dissipation in VLSI chips.

Peak power estimation involves finding a specific vector sequence that will produce maximum power dissipation in a circuit. This estimate enables circuit designers to optimize the circuit to reduce peak power in order to avoid circuit failures due to power-related hazards. In this chapter, a genetic-algorithm-based tool, K2, is described that generates such sequences. K2 was developed for the purpose of measuring peak power for both combinational and sequential circuits. Optimizations based on genetic algorithms will be explained in detail. Furthermore, various delay models can influence the estimation of peak power dissipation. Studies of these delay-related effects using genetic algorithms are also reported.

First, three measures of peak power in the context of sequential circuits will be discussed. Then, the problem formulation using genetic algorithms will be explained. K2 is used to obtain very good lower bounds on these measures as well as the actual input vectors that attain such bounds. Results of K2 are compared against the estimates made from randomly generated sequences as well as previous non-GA-based approaches. Issues concerning delay models and the corresponding effects will be discussed, and the correlation between the various delay models and power estimation will be addressed.

7.1 Problem Description

The power dissipated in CMOS logic circuits is a complex function of the gate delays, clock frequency, process parameters, circuit topology and structure, and the input vectors applied. Once the processing and structural parameters have been fixed, the measure of power dissipation is dominated by the switching activity (toggle counts) of the circuit.

Much effort has been invested in estimating average power dissipation [41], [42], [77], [154], [155], [227], [228]. Most commonly, the average power is estimated from signal switching probabilities. Average power dissipation, however, does not provide accurate boundary conditions for estimating limits of the design. On the other hand, the guidelines for boundary conditions and limits can be provided by the peak power dissipation.

Estimating peak power involves optimization of the circuit's switching function such that maximal activity results. The problem is further complicated by the fact that switching of a gate is heavily dependent on the gate delays, since multiple switching events at internal nodes can occur due to uneven circuit delay paths in the circuit. Finally, the initial state of the circuit in sequential circuits is an important factor in determining the amount of switching activity and must be taken into account.

The problem of estimating the worst-case power dissipation in CMOS *combinational* circuits has been addressed in [56]. The worst-case power is transformed into a weighted max-satisfiability problem on a set of multi-output Boolean functions, obtained from the logic description of the circuit. Either a disjoint cover enumeration or branch-and-bound algorithm is used to solve the complex NP-complete max-satisfiability problem. The largest circuit reported in [56] has only 733 gates, and several hours of CPU time are required for computation. Results for neither ISCAS85 nor ISCAS89 benchmark circuits are reported. Peak current estimation for combinational circuits is addressed in [123], [124]. The authors' approach aims to find the time window during which a gate in the circuit could switch. Partial input enumerations are performed to resolve correlations and to make the time window smaller. Maximum power cycles are computed using symbolic transition counts in [143]; the maximal average cycle in the State Transition Graph (STG) is found, where the edges in the STG indicate the power dissipation between two adjacent states. In a small circuit, s208 for instance, the dual graph necessary to compute the power dissipation contains 71 million edges [143]; the largest circuit reported has less than 500 gates, and handling large circuits with large numbers of flip-flops is impractical. Finally, peak power estimation is computed for sequential circuits using a test generation based technique in [142], [233]. Attempts are made in [233] to create toggles in the circuit for gates with the greatest numbers of fanouts. The estimates are based on a zero-delay fault-simulation model and do not extend to cover peak sustainable power. On the other hand, [142] computed power dissipation for three different delay models: zero, unit, and variable delay. It also used an ATG-based technique in which the circuit is expanded by adding multiple copies of internal gates at various propagation times for each gate. For a small combinational circuit with 6 gates, c17, the expanded circuit has a total of 31 gates, more than 500% increase in gate count. No results were reported for sequential circuits.

Estimation of maximum power was proposed in [107] using a genetic algorithm framework. The authors used a GA-based approach similar to the approach described in this chapter; however, only single-cycle measures of maximum power are reported. In addition, the underlying delay model used was not described clearly, making it difficult to compare the results and to verify if the vectors which attain maximum power will still produce near peak power when the gate delays in the circuit change slightly.

Instead of basing our approach on the previous techniques, new measures of peak power in the context of sequential circuits are proposed, and an automatic procedure is developed to obtain very good lower bounds on these measures, as well as providing the actual input vectors that attain such bounds. The peak powers that we will define are peak average powers, where the average is computed over different time periods. The three measures that we will use are: *Peak Single-Cycle Power*, *Peak n-Cycle Power*, and *Peak Sustainable Power*, covering time durations of one clock cycle, several consecutive clock cycles, and an infinite number of cycles, respectively. The unit of power used throughout this chapter is energy per clock cycle and will simply be referred to as power.

Peak Single-Cycle Power is the maximum total power consumed during one clock cycle.

Peak n-Cycle Power is the maximum average power of a contiguous sequence of n clock cycles, assuming that the initial state of the machine is fully controllable.

Peak Sustainable Power is the maximum average power that can be sustained indefinitely over many clock cycles.

The definitions are illustrated in Fig. 7.1, where a sequential circuit is shown unrolled into several clock cycles, commonly known as an Iterative Logic Array (ILA) representation of the sequential circuit. In a typical sequential circuit, the switching activity is largely controlled by the state vectors and less influenced by input vectors, because the number of flip-flops far outweighs the number of primary inputs. For this reason, it is important to understand the differences in the three measures defined above.

As illustrated in Fig. 7.1(a), the peak single-cycle power is controlled by initial state S_1 and input vectors V_1 and V_2. The state S_1 and input vector V_1 initialize all gate outputs and determine the next state S_2. Then vector V_2 and state S_2 switch some of the gates, which accounts for the power dissipation. We will obtain a three-tuple (S_1, V_1, V_2) that tries to maximize this power. Since the procedure to obtain this three-tuple is imperfect, the result will be a lower bound on this measure. What is the utility of this measure? In full-scan circuits, the state S_1 can be initialized to any arbitrary value, and therefore, this bound is attainable in practice. However, in cases where the initial state is not fully controllable, we can only speculate that during the operation of the circuit, the machine may reach state S_1, and only then can we be assured that the bound is attainable. If the computed

7.1. Problem Description

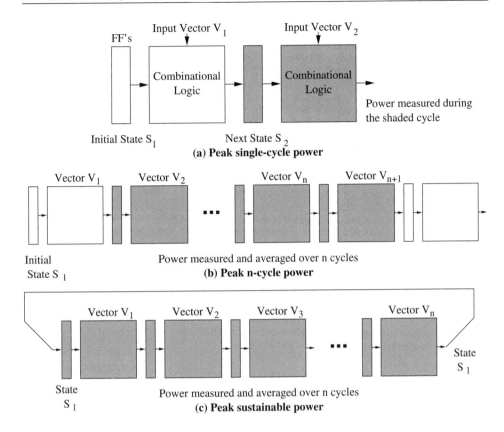

Figure 7.1. Definitions of peak power measures

state S_1 is not a valid state (i.e., not reachable in the normal operation), then we may not be able to reach that power level. If we put the restriction that the state S_1 be a valid state, the peak single cycle power obtained could be lower than the value obtained without any restriction. In the method proposed later in this chapter, we can obtain peak power with or without the restriction of reachability of the state S_1.

Peak n-Cycle power is illustrated in Fig. 7.1(b). We will search for an $(n+2)$-tuple $(S_1, V_1, ..., V_n, V_{n+1})$ that maximizes the power over n cycles. Sequential circuits place considerable constraints on the sequence of consecutive states that can be traversed. Therefore, this peak power will always be less than or equal to the peak single-cycle power. Utility of this measure is in thermal management of the package. Single-cycle power is close to the instantaneous peak power, which is mostly a transient event. However, for a reasonable size n, the peak n-cycle power could represent a practical worst case for heat dissipation. We could also restrict the initial state S_1 to a known valid state, in which case the peak could be lower than the peak obtained without any restriction on S_1.

Peak sustainable power is illustrated in Fig. 7.1(c). The state S_1 is repeated at the end of the input sequence. As a result, the power level can be maintained by applying this input sequence again and again. Clearly, this peak power measure is very important for thermal management of a chip.

In all cases, the power dissipated in the combinational portion of the sequential circuit can be computed as

$$P = \frac{V_{dd}^2}{2 \times clock\ period} \times \sum_{for\ all\ gates\ g} [toggle(g) \times C(g)],$$

where $toggle(g)$ is the number of switches (0 to 1 or vice versa) for gate g in a clock period, and $C(g)$ represents the output capacitance of gate g. Since the output capacitance can be approximated by a constant times the number of fanouts of the gate, the power expression can now be rewritten as

$$P = \frac{V_{dd}^2 \times C_{load}^1}{2 \times clock\ period} \times \sum_{for\ all\ gates\ g} [toggle(g) \times fanout(g)],$$

where C_{load}^1 is a unit-load capacitance per node. Switching rate per node is compared across different circuits, so we report switching frequency per node instead of total power; the switching frequency is computed as

$$SF = Q/(number\ of\ capacitive\ nodes),$$

where $Q = \sum [toggle(g) \times fanout(g)]$, over all gates g.

7.1.1 Adding Delay Models to the Problem

With the various peak power dissipations defined, we will now proceed to the specifics of accurately estimating peak power. It has been shown in [156] and [224] that, due to uneven circuit delay paths, multiple switching events at internal nodes can result, and power estimation can be extremely sensitive to different gate delays. Both [156] and [224] computed the *upper bound* of maximum transition (or switching) density of individual internal nodes of a combinational circuit via propagation of uncertainty waveforms across the circuit. However, these measures cannot be used to compute a tight upper bound of the overall power dissipation of the entire circuit. Power dissipation for three different delay models based on an ATG-based technique for combinational circuits were computed in [142]: zero, unit, and variable delay. However, the circuit expansion from this algorithm makes it impractical for larger circuits.

Although maximum switching density of a given internal node can be extremely sensitive to the delay model used, it is unclear whether peak power dissipation of the entire circuit is also equally sensitive to the delay model. Since glitches and hazards are not taken into account in a zero-delay framework, the power dissipation measures from the zero-delay model may be off greatly from the actual powers. Will the peak

7.1. Problem Description

power be vastly different among various nonzero delay models (where glitches and hazards are accounted for) in sequential circuits? Moreover, the delays used for internal gates in most simulators are mere estimates of the actual gate delays. Can we still have confidence in the peak power estimated using these delay measures considering that the actual delays for the gates in the circuit may be different? Since computing the exact gate delays for every gate in the circuit during simulation is expensive, does there exist a simple delay model for which the execution time for estimating power is reduced and an accurate peak power estimate can be obtained (i.e., the peak powers estimated will not vary due to a variation in the gate delays)? These questions can be summarized as follows: First, given an initial state S_i and a vector sequence Seq_i, (S_i, Seq_i)-tuple, that generates peak power under delay model DM_i, is it possible to obtain another tuple (S_j, Seq_j) that generates equal or higher dissipation under a different delay model DM_j? Second, will the (S_i, Seq_i)-tuple generated for peak power under delay model DM_i also produce near peak power under a different delay model DM_j?

Four different delay models are studied: *zero delay*, *unit delay*, *Type-1 variable delay*, and *Type-2 variable delay*. The three delay models used in [142] are the same as the first three delay models of this work, where zero-delay assumes no delays for all circuit elements, unit-delay assigns equal delay to every circuit element, and Type-1 variable delay assigns a delay for a gate that is proportional to the number of gate fanouts. This model is more accurate than the unit delay model; however, fanouts that feed bigger gates are not taken into account, and inaccuracies may result. This is where the fourth model comes in. It is a variable-delay model based on the fanouts as well as the sizes of successor gates. The gate delay data for various types and sizes of gates are obtained from a VLSI library. The difference between the Type-1 and Type-2 variable-delay models for a typical gate is illustrated in Fig. 7.2. From the figure, the output capacitance of gate G1 is estimated to be 2 (the number of fanouts) in the Type-1 variable-delay model, while the delay calculated using the

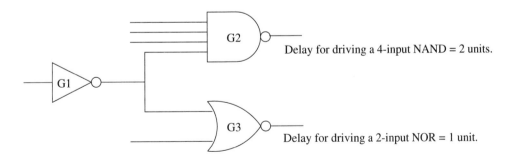

Type-1 variable:
Delay for gate G1 is 2 units.

Type-2 variable:
Delay for G1 is 2 + 1 = 3 units

Figure 7.2. Variable delay models

Type-2 variable-delay model is proportional to the delay associated with driving the successor gates G2 and G3, or simply $2 + 1 = 3$. Since the Type-1 variable-delay model does not consider the size of the succeeding gates, the delay calculations may be less accurate. The delay model used determines the switching activity of internal nodes and thus affects power estimates.

7.2 Application of Genetic Algorithms to Peak Power Estimation

Because peak power estimation involves the maximization of a switching-activity function $toggle(g) \times fanout(g)$, optimization algorithms are needed. Genetic algorithms are well suited to solving such optimization problems. The estimates obtained are compared with the estimates from randomly generated sequences and the results obtained in [233]; the GA-based estimates will be shown to achieve much tighter lower bounds on the peaks.

The GA framework used in the implementation of the algorithm K2 is similar to the simple GA described in Chapter 1. Peak n-cycle power estimation requires a search for the $(n+2)$-tuple $(S_1, V_1, ..., V_n, V_{n+1})$ that maximizes power dissipation. This $(n + 2)$-tuple is encoded as a single binary string, as illustrated in Fig. 7.3. The population size used is a function of the string length, which depends on the number of primary inputs, the number of flip-flops, and the vector sequence length n. Larger populations are needed to accommodate longer vector sequences in order to maintain diversity. The population size is set equal to $32 \times \sqrt{sequence\ length}$ when the number of primary inputs is less than 16 and $128 \times \sqrt{sequence\ length}$ when the number of primary inputs is greater than or equal to 16. The *sequence length* parameter of the equation is set to 2 for combinational circuits because only two vectors are needed to set up the state of the circuit and measure the activity induced by the second vector. The fitness of an individual measures the quality of the vector sequence in terms of switching activity. Because each fanout on a gate contributes to the output capacitance of that gate, the number of gate fanouts is taken into account in the fitness function: $fitness = \sum (toggle\ count(g) \times fanout(g))$ over all gates g.

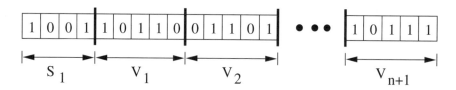

Figure 7.3. Encoding of an individual

The population is first initialized with random strings. A logic simulator (given a specific delay model) is then used to compute the fitness of each individual. The evolutionary processes of *selection*, *crossover*, and *mutation* are used to generate an entirely new population from the existing population. Evolution from one generation to the next is continued until a maximum number of generations is reached. In this work, 32 generations are allowed. Tournament selection without replacement, uniform crossover, and a small mutation probability of 0.01 are used. Since a binary coding scheme is used, mutation is done by simply flipping the bit. A crossover probability of 1 is used; i.e., the two parents are always crossed in generating the two offspring. Because the best individual may appear in any generation, we save the best individual found.

Since the majority of time spent by the GA is in the fitness evaluation, parallelism among the individuals can be exploited. Parallel-pattern simulation [2] is used to speed up the process; thus, 32 candidate sequences from the population are simulated simultaneously, with values bit-packed into 32-bit words.

7.3 Estimation of Peak Single-Cycle and n-Cycle Powers

We estimate the power dissipation in CMOS circuits by measuring the amount of gate switching activity; static power dissipation is neglected. Two components make up the switching activity: zero-delay activity and spurious activity such as glitches and hazards. The output capacitance (measured by the number of fanouts on a given gate) is accounted for by the GA in the fitness function. Here the assumption is made that gate capacitance is proportional to the number of fanouts. However, assigned output capacitances for the gate output nodes can be handled by our optimization technique as well.

Peak single-cycle switching activity occurs when the greatest number of nodes are toggled between two consecutive vectors. For combinational circuits, the task is to search for a pair of two vectors (V_1, V_2) that generates the most gate transitions. For sequential circuits, on the other hand, the activity depends on the initial state as well as the primary input vectors. For the pair of vectors to be useful, the initial state needs to be reachable. The estimate for peak power dissipation from a reachable state can be used as a lower bound for worst-case power dissipation in the circuit in any given time frame. The goal is to find such a bound for the peak power dissipation of the circuit.

Peak n-cycle switching activity is a measure of the maximum average power dissipation over a contiguous sequence of n vectors. This measure serves as an upper bound to *peak sustainable power*, and it is considered only for sequential circuits. As with peak single-cycle power, the initial state of the sequence must be reachable.

The vector sequence that produces peak single-cycle power may not simultaneously generate the peak n-cycle power, or vice versa, as shown in Fig. 7.4. Sequence T_1 generates a higher single-cycle power but has a lower n-cycle power compared to sequence T_2. Conversely, sequence T_2 produces higher n-cycle power but generates a lower single-cycle power. The reason for this phenomenon is that the (S, V_1, V_2)-tuple which generates peak single-cycle power P_1 may not be sustainable, while an n-cycle sequence $(S, V_1, V_2, ..., V_{n+1})$-tuple may generate higher peak n-cycle power even when all of the single-cycle powers are lower than P_1.

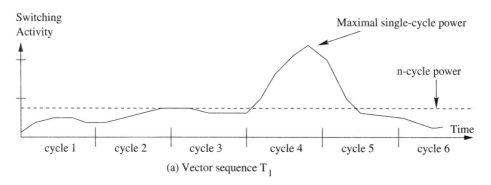

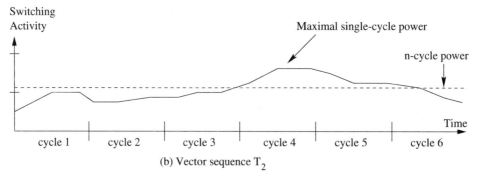

Figure 7.4. Contrasting peak and maximal power estimates

The n-cycle power dissipation varies with the sequence length n. When n is equal to 1, the power dissipation is the same as the peak single-cycle power dissipation, and as n increases, the average peak power is expected to decrease if the peak single-cycle power dissipation cannot be sustained over the n vectors in the sequence. Fig. 7.5 illustrates a typical curve for the peak power dissipation as a function of vector sequence length. The peak levels off as the sequence length approaches infinity or earlier when a loop is found that is capable of maintaining the given power.

7.3.1 Resolving the Problem of State Reachability

The initial circuit state plays an important role in determining the amount of switching activity in a sequential circuit. In order to take the initial state into consideration during power estimation, the *state* portion of the individuals in the GA may be seeded using a previously computed set of reachable states of the circuit, S_{reach} (this set may not be complete). After several generations of the evolutionary processes, the *optimized* individuals may contain states that are not in S_{reach}. At this point, attempts can be made to prove the reachability of the required state with the use of a sequential circuit test generation state justification procedure. State justification, however, is a very complex problem involving a large number of

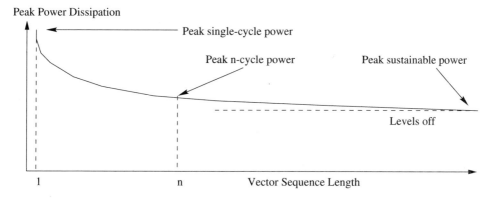

Figure 7.5. Lower bound of peak power dissipation

backtracks [160], [163]; even when a simulation-based technique is used [99], [191], state justification remains a very complex problem. In order to reduce the execution time, an alternative to state justification is taken. A set of states, S_{sim}, is formed by selecting states from S_{reach} that are similar to the required target state. A state S_i is similar to another state S_j if the *Hamming distance* between their encodings is small. Hamming distance is the number of bits having different values. For instance, the states 101101 and 011001 have a Hamming distance of 3. The best vector-sequence generated by the GA is simulated from every state in S_{sim}, and the maximum power obtained from the set of states is taken as the peak power. Again, parallelism is exploited among the different starting states during simulation. The state from which the most switching activity is obtained is selected to be the initial state for the sequence.

7.4 Peak Sustainable Power Estimation

Average power dissipation gives an estimate of the power dissipated in the chip under *normal* operation [41], [42], [77], [154], [155], [227], [228]. The peak single-cycle and n-cycle power estimates discussed in the previous section, however, only last as long as the length of the sequence. Another valuable and useful measure is the peak *sustainable* power. A sequence capable of generating and sustaining high levels of power dissipation over a long period of time can be used to test and evaluate the limits of the circuit design. An interesting question, therefore, is how to estimate the peak sustainable power, and whether a long sequence of vectors can be derived such that a very high power dissipation is generated and sustained.

Symbolic transition counts based on Binary Decision Diagrams (BDDs) have been used to compute maximum power cycles in [143]; the goal is to find the maximum average cycle in the state transition graph, i.e., the cycle with peak sustainable power. Edge weights in the STG are used to indicate the power dissipation in the combinational portion of the circuit between two adjacent states. However, the huge sizes of STGs and BDDs in large circuits make the approach infeasible and impractical. Our approach, on the other hand, avoids the STG and symbolic techniques entirely.

From the curve of the lower bound of peak power dissipation illustrated in Fig. 7.5, the lower bound for maximal power dissipation reaches a *steady state* when the sequence length goes to *infinity*. The term *peak sustainable power dissipation* denotes the steady-state value. Consider the case where state A, together with vectors V_i and V_j (i.e., the (A, V_i, V_j) 3-tuple) generate the peak single-cycle power dissipation for the chip; in addition, state A is reached again at the end of applying vector V_j, as illustrated in Fig. 7.6. If the initial vectors of a sequence T_{sus} take

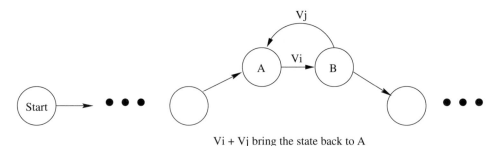

Vi + Vj bring the state back to A

Figure 7.6. Ideal case involving self-loops

the circuit to state A, and the (V_i, V_j) input pair is repeated for the remainder of T_{sus}, the peak power dissipation can be sustained for every time frame after state A has been reached, and sequence T_{sus} becomes the peak-power-sustaining sequence. Unfortunately, it is nontrivial to find such loops, especially when we are working without a state diagram. This problem is complicated further in that we also want to maximize the power dissipation in loops. We could approach this problem in the following way. First, find a peak n-cycle power sequence. Then, try to close the loop with as few additional state transitions as possible. The problem with this approach is that closing the loop is a hit-or-miss proposition. Moreover, the additional transitions will reduce the peak power, since their selection is primarily based on the objective to close the loop, not to maximize power. Another approach is to derive a peak n-cycle power sequence starting from an initial state S_{easy}, which is easy to reach from any state, then close the loop with only a few additional transitions. We took this approach to an extreme, starting with the entirely *don't care* state. Since this state is a superset of any state, it can be reached in just one transition from any state. We derive a maximum n-cycle power sequence starting from the all-unknown state; the sequence is always a loop because the final state of the sequence is covered by the initial state. Typically, the final state is a fully specified state, and therefore, the derived sequence is also a synchronizing sequence. In fact, any synchronizing sequence is a loop, as shown in Fig. 7.7, when the initial state is set equal to the final state of a synchronizing sequence. Power dissipation is

7.5. Experimental Results

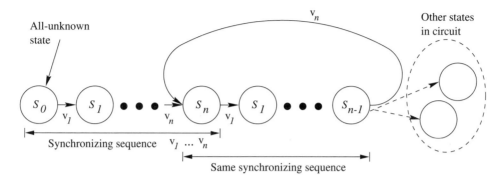

Figure 7.7. Power estimation with synchronizing loops

measured inside the loop, starting from state S_n, when the flip-flops are initialized. This approach restricts the search of peak power loops to a subset of all loops and thus may not be a very tight lower bound. However, our experiments will show that this approach still yields peaks higher than extensive random search.

7.5 Experimental Results

The power estimation algorithms presented were implemented in a tool called K2 using the C++ language; ISCAS85 combinational benchmark circuits, ISCAS89 sequential benchmark circuits [23], and several synthesized circuits were used to evaluate the speed and accuracy of the power estimator K2. All computations were performed on an HP 9000 J200 with 256-MB RAM. Table 7.1 displays the characteristics of the synthesized circuits. Structural sequential depth, number of flip-flops, number of primary inputs, number of primary outputs, and total number of gates are given in the table. Am2910 is a 12-bit microprogram sequencer [4]; mult16 is a 16-bit 2's-complement shift-and-add multiplier; and div16 is a 16-bit divider. Table 7.2 lists the total number of gates and capacitive nodes for both combinational and sequential circuits. The total number of capacitive nodes in the circuit is necessary for computing the power dissipation and is computed as the total number of gate inputs in the circuit.

Table 7.1. Characteristics of Synthesized Circuits

Circuit	Sequential Depth	FFs	PIs	POs	Gates
am2910	4	87	20	16	1053
mult16	9	55	18	34	751
div16	19	50	33	33	973

Table 7.2. Numbers of Capacitive Nodes in Benchmark Circuits

Circuit	Gates	Cap Nodes	Circuit	Gates	Cap Nodes	Circuit	Gates	Cap Nodes
c432	204	343	s298	143	264	s1196	576	1041
c499	276	440	s344	196	295	s1238	555	1073
c880	470	755	s382	189	333	s1423	754	1243
c1355	620	1096	s400	195	349	s1488	687	1412
c1908	939	1523	s444	212	379	s1494	681	1418
c2670	1567	2216	s526	224	472	s5378	3043	4440
c3540	1742	2961	s641	458	582	s35932	18149	30317
c5315	2609	4509	s713	471	633	am2910	1053	1998
c6288	2481	4832	s820	332	781	mult16	751	1323
c7552	3828	6252	s832	330	793	div16	973	1760

All power estimates from K2 are compared against the estimates obtained from randomly generated vector sequences. In addition, the peak single-cycle power estimates for sequential circuits are compared against the results obtained in [142] and [233] for combinational and sequential benchmark circuits, respectively. Note that [233] reported results only for the zero-delay model. All powers are expressed in *Peak Switching Frequency per node (PSF)*, which is the average frequency of peak switching activity of the nodes (ratio of the number of 0-to-1 and 1-to-0 transitions on all nodes to the total number of capacitive nodes) in the circuit. In evaluating this metric, we made the assumption that the output capacitance for each gate is equal to the number of fanouts for all four delay models; however, assigned gate output capacitances can be handled by our optimization technique as well. Notice that for the nonzero delay models, PSFs greater than 1.0 are possible due to repeated switches on the internal nodes within one clock cycle. For comparisons with randomly generated vectors, the K2 estimates are compared against the results obtained from the *same* number of random simulations; the total number of random vector simulations is *identical* to the number of vector simulations in K2.

Before we begin to discuss the results, we must show that the estimates made by our GA-based technique are indeed good estimates of peak power. In other words, how close are our estimates to the theoretical peaks? To partially answer this question, extensive simulations were performed using the unit-delay model for the sequential circuits. Millions of random state-vector tuples were simulated for eight small sequential circuits. For instance, if exhaustive search were performed for s400, which has 21 flip-flops and 3 primary inputs, the total number of simulations for the three-tuple (S_1, V_1, V_2) would be $2^{(21+3+3)} = 2^{27} \simeq 134$ *million*. The results for millions of random simulations are shown in Table 7.3. Two exhaustive simulation data points are chosen for comparison; they are taken at 2 and 100 million random vector-pairs for peak power estimation, shown in the first four columns of the table. The rightmost columns show the results obtained by the GA-based

7.5. Experimental Results

Table 7.3. GA-based vs. Random Estimates of Peak Single-Cycle Power

Circuit	Random				GA-based	
	2 million		100 million		(~ 64,000)	
	PSF	Time	PSF	Time	PSF	Time
s298	1.015	6.5 min	1.015	5.39 hr	1.015	5.32 sec
s344	1.468	10.9 min	1.529	10.7 hr	1.275	12.7 sec
s382	1.045	8.8 min	1.063	7.29 hr	1.081	10.3 sec
s400	1.049	9.1 min	1.072	7.58 hr	1.106	13.4 sec
s444	1.069	10.9 min	1.113	9.01 hr	1.150	12.7 sec
s526	0.903	10.5 min	0.922	8.65 hr	0.930	11.3 sec
s641	2.567	27.3 min	2.749	22.3 hr	2.869	56.2 sec
s713	2.649	30.3 min	2.749	24.8 hr	2.815	50.2 sec

PSF: peak switching frequency per node

technique. For most circuits, the GA-based peak power estimates are equal to or slightly greater than the estimates made from 100 million random vector-pairs, which took many additional hours of simulation time. For the last two circuits, s641 and s713, the results from simulation of 100 million random sequences (over seven hours of execution) still lag 4.4% behind the GA-based estimates, which took less than 60 seconds each. From this experiment, we observe that if random-based methods were to achieve similar levels of peak power estimates, several orders higher execution times would be needed over the GA-based technique.

Now that we have confidence in our peak power estimates, we will proceed to discuss the experiments on peak power using various delay models. For the ISCAS85 combinational circuits, Table 7.4 compares the GA-based results against the randomly generated sequences as well as those reported in [142]. Since the results for the previous GA-based technique [107] were reported in mW, and a different underlying delay model was used, it is very difficult to compare the results. For each circuit, we report results for all four delay models. The estimates obtained from the best of randomly generated vector-pairs, an ATG-based approach [142], and our GA-based technique are shown. The ATG-based approach used in [142] attempts to optimize the total number of nodes switching in the expanded combinational circuit; its results were compared against the best of only 10,000 random simulations. In our work, on the other hand, the number of random simulations depends on the GA population size, which is a function of the number of primary inputs in the circuit. Typically, the number of random simulations exceeds 64,000. The GA-based estimates are the highest for all of the circuits and for all delay models except for one zero-delay estimate for c1355, where a 1.1% lower power was observed. The average improvements made by the GA-based technique over the random simulations were 10.8%, 27.4%, 38.5%, and 35.5% for the four delay mod-

Table 7.4. Peak Single-Cycle Power for ISCAS85 Combinational Circuits

Circuit	Zero			Unit		
	Rnd	[142]	GA	Rnd	[142]	GA
c432	0.636	0.536	**0.805**	1.985	1.985	**2.362**
c499	0.648	0.480	**0.659**	**1.268**	0.950	**1.268**
c880	0.632	0.495	**0.809**	1.367	0.995	**1.689**
c1355	**0.539**	0.402	0.533	1.681	1.268	**3.260**
c1908	0.628	0.596	**0.638**	2.641	1.961	**3.446**
c2670	0.587	0.561	**0.623**	1.690	1.676	**2.251**
c3540	0.534	0.402	**0.600**	2.351	1.646	**2.684**
c5315	0.584	-	**0.631**	1.859	-	**2.733**
c6288	0.553	-	**0.660**	19.30	-	**23.60**
c7552	0.572	-	**0.602**	2.052	-	**2.821**
Impr			10.8%			27.4%
Circuit	Type 1 Variable			Type 2 Variable		
	Rnd	[142]	GA	Rnd	[142]	GA
c432	2.557	1.181	**3.399**	2.286	-	**2.522**
c499	**1.345**	0.575	**1.345**	1.345	-	**1.409**
c880	1.546	0.613	**1.710**	1.515	-	**1.951**
c1355	1.671	0.608	**2.707**	1.956	-	**2.676**
c1908	2.863	1.226	**4.000**	3.006	-	**4.182**
c2670	1.925	1.363	**2.750**	1.931	-	**2.825**
c3540	3.137	1.178	**3.555**	3.478	-	**4.678**
c5315	2.180	-	**3.205**	2.308	-	**3.263**
c6288	16.31	-	**24.25**	17.56	-	**25.18**
c7552	2.387	-	**2.833**	2.838	-	**3.238**
Impr			38.5%			35.5%

Impr: Average improvements over the best random estimates
Power reported in peak switching frequency per node

els, respectively. Since [142] estimated power on the expanded circuits, backtrace using zero-delay techniques was used. Consequently, only hazards are captured; zero-width pulses, on the other hand, are not taken into account in the nonzero delay models. For this reason, our GA-based estimates are higher than all the estimates obtained in [142]. Furthermore, results from [142] suggested that peak power estimated from the unit-delay model consistently gave higher values than both zero-delay and Type-1 variable-delay estimates. We do not see such consistency from our results in Table 7.4. The estimated powers from random simulations do not show such a trend either. One plausible reason for this could be that zero-width pulses are not accounted for in [142]. Nevertheless, in both approaches, peak pow-

7.5. Experimental Results

ers for some circuits are very sensitive to the underlying delay model. For instance, in circuit c3540, the peak power estimates by the GA are 0.600, 2.684, 3.555, and 4.678 for zero, unit, Type-1 variable, and Type-2 variable-delay models. It is also observed that estimates made from nonzero delay models always give higher values than from the zero-delay model, since extra events due to glitches and hazards are accounted for in the nonzero delay models.

The GA-optimized peak power estimates using various delay models for sequential circuits are shown in Tables 7.5 and 7.6 for single-cycle and multicycle periods. For each circuit, the peak power dissipation obtained for each of the four delay models is given in terms of peak switching frequency per node (PSF). In Table 7.5, results for the best random estimates and [233] are also reported, and the average improvements achieved by the GA-based estimates are shown at the bottom of the table. For 10-cycle and sustainable power estimates, only the GA-based estimates

Table 7.5. Peak Single-Cycle Power Estimates for ISCAS89 Sequential Circuits

Cir-	Random				[233]	GA-based			
cuit	Z	U	V1	V2	Z	Z	U	V1	V2
s298	0.829	0.989	1.023	**1.057**	0.761	0.848	1.015	1.064	**1.095**
s344	0.708	**1.275**	1.027	1.078	0.627	0.797	**1.275**	1.017	1.057
s382	0.802	0.970	1.021	**1.111**	0.799	0.826	1.081	**1.129**	**1.129**
s400	0.779	0.957	1.032	**1.057**	-	0.822	1.106	**1.158**	**1.158**
s444	0.707	0.968	1.053	**1.150**	-	0.752	1.150	**1.198**	**1.198**
s526	0.754	**0.858**	0.824	0.847	0.631	0.790	**0.930**	0.905	0.905
s641	0.680	**2.462**	1.263	1.113	-	0.844	**2.869**	1.354	1.206
s713	0.654	**2.667**	1.207	1.215	-	0.823	**2.815**	1.932	1.308
s820	0.778	0.936	**0.963**	0.953	0.784	0.822	0.974	**1.047**	1.013
s832	0.457	0.923	**0.971**	0.957	0.781	0.823	0.984	**1.038**	1.016
s1196	0.591	0.899	0.904	**0.963**	0.571	0.605	1.027	1.003	**1.104**
s1238	0.587	0.851	**0.971**	0.913	0.568	0.622	**1.058**	1.042	1.042
s1423	0.714	**1.275**	1.202	1.130	0.551	0.854	**1.736**	1.455	1.438
s1488	0.674	**1.228**	1.053	1.070	0.664	0.674	**1.258**	1.069	1.074
s1494	0.669	**1.230**	1.063	1.065	0.661	0.673	**1.253**	1.072	1.078
s5378	0.560	0.866	0.910	**0.972**	0.675	0.753	1.114	1.351	**1.377**
s35932	0.635	**1.545**	1.534	1.521	0.655	0.641	1.543	**1.579**	1.521
am2910	0.474	2.961	4.124	**5.280**	-	0.509	3.128	4.945	**6.329**
mult16	0.797	**1.751**	1.725	1.512	-	0.881	2.543	2.688	**2.748**
div16	0.601	3.414	3.028	**4.430**	-	0.701	3.803	4.439	**8.103**
Avg	0.673	1.451	1.335	1.470	0.671	0.753	1.634	1.624	1.845
Impr						11.9%	12.6%	21.6%	25.5%

Z: Zero   **U**: Unit   **V1**: Type-1 Variable   **V2**: Type-2 Variable
Impr: Average improvements over the best random estimates
Power reported in peak switching frequency per node

Table 7.6. Ten-cycle and Sustainable Peak Power Estimates for ISCAS89 Sequential Circuits (Peak Switching Frequency per Node)

Circuit	10-Cycle				Sustainable			
	Z	U	V1	V2	Z	U	V1	V2
s298	0.434	0.482	**0.505**	**0.505**	0.363	0.413	**0.416**	0.415
s344	0.537	0.698	0.723	**0.732**	0.502	**0.683**	0.637	0.677
s382	0.326	0.389	0.402	**0.405**	0.246	0.262	0.277	**0.280**
s400	0.320	0.390	**0.403**	0.399	0.220	0.252	0.271	**0.279**
s444	0.313	0.395	0.430	**0.439**	0.217	0.281	0.306	**0.309**
s526	0.305	0.335	0.335	**0.339**	0.217	0.241	0.235	**0.246**
s641	0.477	**1.252**	0.815	0.798	0.453	**0.902**	0.738	0.749
s713	0.494	**1.421**	0.881	0.840	0.443	**0.964**	0.779	0.824
s820	0.579	0.682	**0.695**	0.670	0.511	0.628	**0.645**	0.637
s832	0.565	**0.696**	0.675	0.687	0.553	0.647	**0.653**	0.645
s1196	0.558	0.775	**0.837**	0.818	0.536	0.775	**0.837**	0.796
s1238	0.529	0.796	**0.840**	0.820	0.507	0.796	**0.840**	0.811
s1423	0.557	**0.934**	0.898	0.852	0.411	0.604	0.640	**0.675**
s1488	0.593	0.778	**0.816**	0.776	0.490	0.663	**0.670**	0.646
s1494	0.586	0.790	**0.806**	0.804	0.486	**0.657**	0.649	0.648
s5378	0.550	0.727	0.760	**0.814**	0.337	**0.465**	0.432	0.464
s35932	0.522	**1.107**	1.097	1.100	0.522	**1.107**	1.093	1.097
am2910	0.366	1.662	2.231	**3.068**	0.381	1.295	2.012	**2.417**
mult16	0.441	0.872	**0.902**	0.857	0.347	**0.787**	0.753	0.759
div16	0.521	1.941	2.301	**3.249**	0.446	1.781	2.131	**2.829**
Avg	0.479	0.852	0.854	0.943	0.409	0.710	0.750	0.810
Impr	17.4%	32.5%	26.7%	32.0%	12.3%	23.8%	24.4%	30.2%

Z: Zero   U: Unit   V1: Type-1 Variable   V2: Type-2 Variable
Impr: Average improvements over the best random estimates

are shown since [233] did not report results for multicycle power, and the best random estimates are significantly lower. However, the average improvements achieved by the GA-based estimates are still shown at the bottom of the table. The length of the sequence in these two cases is set to 10. The number of random simulations is the same as the number of simulations required in the GA-based technique. For the small circuits, the number of simulations is about 64,000. The GA-based estimates surpass the best random estimates for all circuits for all three peak power measures under all four delay models. Up to 32.5% improvement is obtained on the average for peak 10-cycle powers.

Across the four delay models, estimates made from the zero-delay model consistently gave significantly lower power, since glitches and hazards are not accounted for. However, there is not a clear trend as to which of the other three delay models would consistently give peak or near peak power estimates. In circuits such as s641,

7.5. Experimental Results

s713, and am2910, for example, the peak powers are very sensitive to the delay model. Over 100% increase in power dissipation can result when a different delay model is used. For some other circuits, such as s400, s444, s526, and s1238, the peak powers estimated are insensitive to the underlying delay model; less than 5% difference in the estimates are observed from the three nonzero delay models.

The execution times needed for the zero-delay estimates are typically smaller than for the other delay models, since fewer events need to be evaluated; no significant differences in the execution times among the three nonzero delay models were observed. Table 7.7 shows the execution times for the GA optimization technique for the unit-delay model. The execution times are directly proportional to the number of events generated in the circuit during the course of estimation. For this reason, peak 10-cycle power estimates do not take 10 times as much computation as peak single-cycle power, since the amount of activity across the 10 cycles is not 10 times that of the peak single cycle. For circuits in which peak power estimates differ significantly among various delay models, the computation costs will also differ significantly according to the number of activated events. Since we do not see any delay model for which highest peak powers are consistently generated, the overall execution times are comparable across the three nonzero delay models. The execution times required for the random simulations are very close to the GA-based technique since identical numbers of events are evaluated.

Table 7.7. Execution Times for the GA-based Technique (seconds)

Circuit	Single cycle	10-cycle	Sustain-able	Circuit	Single cycle	10-cycle	Sustain-able
s298	5.32	19.56	20.93	s1196	44.14	286.8	408.9
s344	12.72	61.10	70.50	s1238	53.99	301.8	421.7
s382	10.33	36.45	25.07	s1423	164.5	514.8	508.8
s400	13.36	38.09	26.38	s1488	78.37	405.8	472.6
s444	12.67	40.79	27.15	s1494	78.96	355.2	470.5
s526	11.34	40.98	24.35	s5378	563.4	2065	1920
s614	56.18	213.8	259.4	s35932	23200	84449	66411
s713	50.17	237.0	293.8	am2910	343.9	1695	1216
s820	28.11	145.0	217.2	mult16	155.8	365.2	338.2
s832	29.34	144.4	212.9	div16	542.1	2117	3450

When examining the relationships between peak single-cycle, 10-cycle, and sustainable powers, it is intuitive that neither peak 10-cycle nor sustainable power estimates would be as high as the peak single-cycle estimate, unless the peak single-cycle power can be sustained in a 3-tuple-loop fashion as described in the previous section. Furthermore, the peak sustainable power estimates are expected to be lower than the peak 10-cycle power estimates because the sequences that produce peak 10-cycle power are not restricted to loops. Fig. 7.8 illustrates graphically

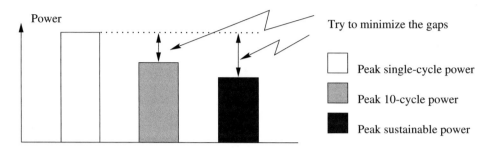

Figure 7.8. Relationship among peak single-cycle, 10-cycle, and sustainable powers

the relationship among the estimates of peak single-cycle, 10-cycle, and sustainable power. Peak 10-cycle power is obtained using a sequence of 11 vectors, starting from some initial state, that produces maximal total toggle counts throughout the entire sequence. Peak sustainable power, on the other hand, is obtained by repeating a synchronizing sequence starting from the all unknown state that maximizes switching activity.

Now with all the peak single-cycle, 10-cycle, and sustainable power estimates computed, Fig. 7.9 illustrates the relationship among these estimates for eight circuits under the unit-delay model. In many circuits, significant gaps between the peak single-cycle and 10-cycle power estimates exist, indicating that the peak single-cycle power dissipations are difficult to sustain for these circuits. However, among these eight circuits, five of them, namely s344, s526, s832, s1238, and s35932, all have peak sustainable power estimates very near their corresponding peak 10-cycle power lower bounds; this suggests that the power estimated for the peak 10-cycle power can be sustained. In other words, the lower-bound peak sustainable power dissipation curves (i.e., Fig. 7.5) are likely to level off near the sequence length $n = 10$ for these eight circuits. For the remaining three circuits, the lower-bound curves level off at longer sequence lengths.

The peak single-cycle and 10-cycle power estimates discussed so far are valid only if the initial states for these sequences are reachable. How much difference in the estimate would result if reachability of the state were accounted for? The results for state reachability analysis are shown in Table 7.8 for sequential circuits using the unit-delay model. The reachability analysis was performed for the initial state, and the respective power dissipations were reported for estimates computed before and after the reachability analysis. The time required to perform reachability analysis was minimal, so it is not reported. The set of reachable states was computed prior to power estimation, and the execution times for reachable state computation are not included in the table. The set of reachable states may be derived by several different methods: from the state table, random vector simulation, an automatic test generator, or a GA-based approach. Reachability analysis is not needed for peak sustainable power because sustainable power is estimated using synchronizing

7.5. Experimental Results

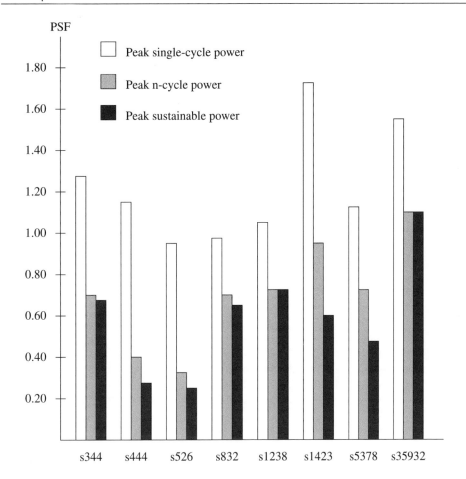

Figure 7.9. Results from K2 on unit-delay peak single-cycle, 10-cycle, and sustainable powers

sequences which take the circuit from the *don't-care* state to a known state. The initial don't-care state is reachable from any state.

In many practical circuits, the portion of states reachable out of the 2^N possible states becomes smaller when the number of flip-flops, N, increases. Thus, the optimized initial state is frequently unreachable for circuits with many flip-flops, resulting in a lower peak power measure after the reachability analysis for most circuits. However, there are cases where the power dissipations increase slightly when sequences are applied starting from reachable states that had not been considered previously in the huge search space. Nevertheless, the estimates computed by K2 after reachability analysis are still higher than the best estimates from randomly generated vector-pairs after reachability. Average drops of 17.6% and 20.1% in

Table 7.8. Effect of State Reachability on Peak Power for Sequential Circuits

Circuit	Single-Cycle				10-Cycle			
	Random		K2		Random		K2	
	Before	After	Before	After	Before	After	Before	After
s298	0.989	0.856	1.015	0.792	0.440	0.450	0.482	0.483
s344	1.275	0.810	1.275	0.810	0.581	0.586	0.698	0.668
s382	0.970	0.838	1.081	0.931	0.340	0.347	0.389	0.368
s400	0.957	0.877	1.106	0.908	0.332	0.344	0.390	0.375
s444	0.968	0.913	1.150	0.953	0.360	0.349	0.395	0.372
s526	0.858	0.625	0.930	0.642	0.309	0.314	0.335	0.321
s641	2.462	1.503	2.869	1.701	0.587	0.677	1.252	0.727
s713	2.667	1.534	2.815	1.599	0.594	0.617	1.421	0.855
s820	0.936	0.834	0.974	0.855	0.527	0.673	0.682	0.682
s832	0.923	0.810	0.984	0.874	0.559	0.544	0.696	0.685
s1196	0.899	0.915	1.027	1.022	0.573	0.576	0.743	0.745
s1238	0.851	0.865	1.058	1.043	0.585	0.588	0.742	0.745
s1423	1.275	1.263	1.736	1.459	0.628	0.811	0.934	0.867
s1488	1.228	0.847	1.258	1.176	0.728	0.728	0.778	0.778
s1494	1.230	1.157	1.253	1.171	0.745	0.745	0.790	0.790
s5378	0.866	0.798	1.114	1.011	0.470	0.399	0.727	0.648
s35932	1.545	0.792	1.543	0.893	1.070	1.074	1.107	1.100
am2910	2.961	2.998	3.128	3.224	1.163	1.200	1.662	1.715
mult16	1.751	1.184	2.543	1.238	0.523	0.563	0.872	0.941
div16	3.414	3.519	3.803	3.774	1.708	1.767	1.941	1.996
Avg	1.451	1.196	1.634	1.304	0.641	0.667	0.852	0.793
Red		17.6%		20.2%		-4.1%		6.9%

Red: Reductions in peak power after reachability analysis
Power reported in peak switching frequency per node

peak power are observed after the reachability analysis for the random and GA-based peak single-cycle powers, respectively. For the peak 10-cycle powers, on the other hand, an interesting phenomenon for the random simulation is seen in that the average peak random power after reachability analysis is greater than before the analysis is performed; a 4.1% improvement is made on average. For the GA-based estimates, a drop in peak power is still observed; however, this time, the drop is only 7.2% instead of 20.1% seen in the single-cycle power estimates.

Further experiments were performed to determine whether the vector sequences optimized for a given delay model produce peak or near-peak switching activities for other delay models. For the combinational circuits, the vectors optimized for the zero-delay model produced significantly lower power when they were simulated in the nonzero delay environments; over 50% drops were observed from the GA-optimized powers. The vectors that were optimized on the remaining three delay models, on the other hand, deviated less significantly when simulated using other *nonzero* delay models, all less than 10% deviations. These results suggest that the zero-delay model is insufficient for estimating peak power. On the other hand, estimates made from nonzero delay assumptions produce near-peak powers when simulated using other nonzero delay models.

For small sequential circuits, the results show that the zero-delay-optimized vectors produced peak or near-peak powers for the unit and Type-1 variable delay models as well. One plausible explanation for this phenomenon is that the Peak Switching Frequencies (PSFs) for these circuits are small, typically less than 1.2, indicating that most nodes in the circuit do not toggle multiple times in a single clock cycle. For the other circuits, especially for the circuits where high PSFs are obtained, the vectors obtained which generated high peak powers under zero-delay assumptions do *not* provide peak powers under nonzero delay assumptions. Similarly, the vectors optimized under the nonzero delay models were simulated using the other delay models. For most of the sequences optimized under the Type-1 variable-delay model, near-peak powers were also produced under the Type-2 variable-delay model. When the optimized powers are great, i.e., PSFs are greater than 2, a greater deviation is observed. Less significant deviation is observed in circuits for which smaller PSFs are obtained. The overall average deviations are small between vectors derived using the Type-1 and Type-2 variable-delay models, indicating that these two delay models are more closely correlated.

7.6 Summary

A GA-based power estimation framework, K2, was developed. The lower bounds for peak power dissipation obtained by this tool can aid the circuit designer in defining the limits and boundary conditions of the design with respect to three levels of peak power estimates, and in turn optimizing the circuit accordingly. Excessive overheating of the chip, as well as other power-related risks, can be avoided as a consequence.

Estimates for peak single-cycle, n-cycle, and sustainable power dissipation were computed for various delay models; the role of the initial state in sequential circuits has also been taken into account for a more accurate measure. K2 provides much tighter bounds when compared to the estimates made from randomly generated sequences and previous approaches. In addition, the execution times of K2 were orders of magnitude lower than those for random-based estimates, if they were to achieve similar tightness of lower bounds. Furthermore, when estimating peak power under one delay model, it is crucial to have confidence that the estimate will not vary significantly when the actual delays in the circuit differ from the delay model assumed. Peak power estimation under four different delay models was evaluated. For most circuits, vector sequences optimized under the zero-delay assumption do not produce peak or near-peak powers when they are simulated under a nonzero delay model. Similarly, the vectors optimized under nonzero delay models do not produce peak or near-peak zero-delay powers. However, the vector sequences optimized under nonzero delay models provide good measures for other nonzero delay models with only a slight deviation for most combinational circuits. For sequential circuits, small deviations are observed when the optimized peak powers are relatively small; on the contrary, when the optimized peak powers are large, i.e., nodes switch multiple times in a single cycle on average, the estimated peak powers will be sensitive to the underlying delay model. Finally, when the delay models are more closely correlated, the deviations among the peak powers are also smaller.

SUMMARY

TOPICS STUDIED

KEY OBSERVATIONS AND POINTS OF INTEREST

Genetic Algorithm for Peak Power Estimation (K2)
- Zero, Unit, & Variable Delay Models
- Peak Single-Cycle, n-Cycle, and Sustainable Powers
- Results

- Binary encoding of vectors in sequence
- Parallel-pattern simulation in fitness evaluation to reduce CPU time
- Better power estimates obtained than for previous approaches
- Orders of magnitude reduction in CPU times compared to a random approach
- Sequences optimized under nonzero delay models provide good measures for other nonzero delay models
- Sequences optimized under zero delay model do not produce peak powers for nonzero delay models
- Sequences optimized under nonzero delay models do not produce peak powers for zero delay model

CHAPTER 8
at a glance

- WOLVERINES: Distributed tool for standard cell placement
  - Implemented on a network of workstations
  - Effect of communication pattern on placement quality
  - Load balancing

- Parallel GA for test generation
  - Parallelization using data decomposition
  - Parallel migration-based genetic search
  - Subpopulation-based GA with migration

Chapter 8

PARALLEL IMPLEMENTATIONS

S. Mohan, P. Mazumder
D. Krishnaswamy, P. Banerjee, E. M. Rudnick

The typical Computer-Aided Design (CAD) environment consists of a number of workstations connected together by a high-speed local area network which allows a team of designers to access the design database. Individual CAD problems are typically solved on a single workstation. Compute-intensive jobs such as circuit simulation or maze routing are sometimes executed on a remote machine which is more powerful than the regular workstations; this remote machine may be a supercomputer or a special-purpose computer designed to efficiently execute a single algorithm such as routing or simulation. However, the loosely coupled parallel computing environment provided by the network is often not effectively used by CAD algorithms. The effective use of this distributed environment is an active area of research in software engineering and algorithm design; various paradigms such as asynchronous or synchronous message passing, remote procedure calls, and generative communication have been defined [10].

This chapter explores the use of genetic algorithms which can make use of this *freely available* resource to achieve speedup by spreading the computational effort over all the available processors in a network. The particular CAD problems addressed here are standard cell placement and automatic test generation. Parallel solutions have been developed for both of these problems, but they require extensive communication between the processors or produce lower quality results than a serial implementation. The key to the success of the parallel implementations of GA-based CAD tools presented here is the inherently parallel nature of GAs.

We begin by describing the Wolverines tool for standard cell placement on a network of workstations. The serial algorithm is discussed, and the distributed genetic placement algorithm is presented, followed by experimental results. Next, parallelization of GA-based test generation is addressed. Three parallel implementations are described, and experimental results are presented.

8.1 Wolverines: Standard Cell Placement on a Network of Workstations

As discussed in Chapter 3, standard cell placement is a crucial step in the layout of VLSI circuits which has a major impact on the final speed and cost of the chip. This problem has been studied for several years, and the best solutions have been obtained by iterative improvement algorithms, such as simulated annealing [206], simulated evolution [117], and stochastic evolution [198], which have some mechanism for escaping from local minima. While these algorithms produce good solutions, they are typically characterized by long run times, prompting many researchers to search for parallel implementations on many different types of parallel machines.

Parallel simulated annealing has been implemented with good speedup on many different kinds of parallel machines, including shared-memory machines, message passing machines with local memory [108], and massively parallel machines such as the connection machine [62]. A study by Kravitz and Rutenbar [122] showed that simulated annealing on a shared-memory multiprocessor could be accelerated in two ways: by performing several moves in parallel and by performing the subtasks for each move in parallel. The amount of parallelism within a move is limited, and good speedup can be achieved only by performing several moves in parallel. However, if moves are performed in parallel, the system can get into an inconsistent state unless the processors are synchronized after every set of noninteracting moves is evaluated in parallel. Parallel moves are effective when the number of accepted moves is low, in the low temperature regime of annealing; this fact has been utilized by Rose et al. [187] to achieve excellent speedup by using a fast min-cut approach to avoid the slow high-temperature annealing part. In the low-temperature phase, the chip area is partitioned and assigned to different processors such that each processor moves cells in a particular area and, whenever a move is accepted, broadcasts the result to all processors. There is a trade-off between the communication costs and the need to broadcast information after each accepted move; typically, several moves are accepted before a broadcast, and this leads to some errors. While error estimates for standard cell placement algorithms are not available, error control for the related problem of macro cell placement has been studied by various authors [27], [106], and the efficiency of the multiprocessor is reduced to less than half when the error is held within 1%. Since communication costs for simulated annealing are high even on a tightly coupled system, it is not desirable to speed up simulated annealing on a distributed system where communication time is of the order of several milliseconds.

While special-purpose parallel machines required for accelerating simulated annealing or some other CAD algorithm may be available in a few CAD environments, the typical CAD environment consists of a few workstations connected together by a local area network such as Ethernet. The workstations typically can support several user tasks, one of which can run in the foreground while the others run in the background. When the workstation is used for a text editing or graphics editing program, it may use its CPU only a fraction of the time and spend most of its cycles

8.1. Wolverines: Standard Cell Placement on a Network of Workstations

waiting for user input. The placement program described in this chapter makes use of the idle cycles on all the networked machines to speed up the placement process. The communication cost is low because the communication is handled by the operating system through the network interface, and the CPU does not have to wait for the communication to complete; i.e., the communication is asynchronous, with a nonblocking *send*. The cumulative error due to this mode of operation has been found to be negligible in all test cases. The program is robust and dynamically adapts to run only on the lightly loaded machines.

The genetic algorithm for standard cell placement which forms the basis for the distributed algorithm is discussed in the next two subsections. The distributed algorithm is then described, followed by a discussion of the placement and speedup results obtained, the parameter settings used in running the algorithm, and the performance of the algorithm in dynamically changing, heterogeneous computing environments.

8.1.1 Standard Cell Placement Using the Genetic Algorithm

The basic serial genetic algorithm of Section 3.1 for standard cell placement forms the basis of the parallel implementation. The serial genetic algorithm for standard cell placement is shown in Fig. 8.1. An initial population of randomly generated placement configurations is the starting point of the algorithm. Once the population size p is known (generally specified by the user, since there is no *best choice* for every case), p random permutation strings are generated over the set of integers from 1 to n, where n is the number of cells in the circuit. These permutation strings are used to create strings of cell records with the cell ID corresponding to the particular element in the permutation string. The cell-position records are then filled in based on the ordering of cells in the string. The net length corresponding to each placement configuration is computed, and the *fitness* value for each individual is recorded as $\frac{1}{net\ length}$. This completes the process of creating the initial population.

The first step in the iterative loop of the serial algorithm is the selection of individuals for reproduction. This process is carried out as follows. First, the mean fitness value for the entire population is calculated as f_{mean}. Then the crossover rate parameter is used to determine the number of offspring or new individuals to be generated. If the crossover probability is P_C and the population size is p, the number of offspring is $p_{off} = p \times P_C$, where P_C is a fraction in the interval (0, 1). Next, individuals are chosen for reproduction in a probabilistic manner, so that the expected number of offspring of an individual I with fitness f_I is $\frac{f_I}{f_{mean}} \times \frac{p_{off}}{p}$. Hence, the fitness of the individual relative to the rest of the population determines the individual's chances of reproduction; the higher the relative fitness, the better the chances, and vice versa. Once the number of offspring for each individual is determined, the offspring population is created by making copies of the string representations of the parent individuals. The individuals in this offspring population are then subjected to the genetic operations of inversion, crossover, and

Read netlist and cell library files;
Read parameter values, CrossoverRate, MutationRate,
InversionRate, PopulationSize, NumberOfGenerations;
NumberOfOffspring = PopulationSize × CrossoverRate;

Generate initial Population randomly;
For j = 1 **To** PopulationSize **Do**
    evaluate (Population, j);

For i = 1 **To** NumberOfGenerations **Do**
    Select NumberOfOffspring individuals from Population;
    Copy selected individuals into NewPopulation;
    For j = 1 **To** NumberOfOffspring **Do**
        invert (NewPopulation, j, InversionRate);

    Perform crossover on individuals in NewPopulation;
    Perform mutation on individuals in NewPopulation;
    For j = 1 **To** NumberOfOffspring **Do**
        evaluate (NewPopulation, j);

Population = reduce (Population, NewPopulation);
Solution = individual with highest fitness in final Population;

Figure 8.1. Serial placement algorithm

mutation, in that order. The inversion operation is performed on each individual with a probability specified by the inversion rate parameter. Then, the individuals in the offspring population are paired off at random and crossed over. Crossover is performed according to the procedure for cycle crossover described in Section 3.1.2. It may be noted here that since the crossover operation generates two offspring from two parents, the population size does not change due to crossover. Each individual in the offspring population is then mutated with a probability that is specified by the mutation rate parameter. The cell positions are updated at this stage, and the fitness values of all new individuals are computed. Then p individuals are chosen from this set of $p + p_{off}$ individuals for the next generation. This reduction is done in the same manner as the selection operation, with the probability of choosing an individual being dependent on its relative fitness.

The individual with the highest fitness in the final population is the solution. The fitness of an individual is a function of the net length associated with the placement: the higher the net length, the lower the fitness. The initial population is randomly chosen. Hence, repeated runs of the algorithm could produce different solutions. Selection of individuals for the next generation is based on the fitness criterion: the higher the fitness of an individual relative to the population average fitness, the higher its probability of survival into the next generation. The number of offspring to be generated and the fraction of the population to be mutated are decided by the values assigned to the genetic parameters *crossover rate* and *mutation rate*, respectively. Optimal values for these parameters were obtained by using meta-genetic optimization, as described in Sections 3.1.3 and 3.1.5. Typical values of various parameters for a population size $p = 48$ are $P_C = 0.33$ (crossover probability), $P_M = 0.05$ (mutation probability), $P_I = 0.15$ (inversion probability), and $g = 30,000$ (number of generations).

The termination condition for the genetic algorithm is typically the identification of convergence. The population is said to have converged when all the individuals have the same genetic representation, i.e., the entire population consists of multiple copies of the same individual. This condition was rarely observed in our experiments, except when the population size was reduced to the barest minimum for effective crossover (3). Hence, the primary termination conditions were determined empirically, based on the rate of improvement in fitness.

A placement configuration is represented by a string of records representing cells. Each record contains the following information: cell ID, cell position x, and cell position y. The position of the cell record in the string does not always determine the physical location of the cell in the layout. However, at certain points in the algorithm, the cell position is recalculated based on the ordering of the cell records in the string, as follows. Starting from the first row, the cells are listed in order from left to right; at the end of the row, we proceed to the next higher numbered row and list cells in reverse order. This process continues, with the direction changing after each row, until all cells are listed. Given a string of records, cells can be assigned positions in rows by reversing the above process. The first cell in the string is assigned to the leftmost position in the first row, $start_x$, the next cell is assigned to a location $start_x + sizeof(first\_cell)$, and so on, until the row is filled. Then the row number is incremented, and the process goes on until all cell coordinates are assigned. There is no cell overlap; row length variation is kept within $\pm sizeof(largest\_cell)$ for all rows except the last, which could be somewhat shorter. The area does not change with different placement configurations since the number of rows is fixed, and the spacing between rows is assumed constant. Hence, the only variation is in the net length, and the cost function is based purely on net length.

The inversion operator changes the positions of cell records in the string representing a placement but does not change the actual physical locations of the cells in the layout; this operator is always used in conjunction with the crossover operator to introduce some variation in the kinds of schemata transferred to the offspring by the crossover operator and is always applied before the crossover operator. The inversion operator takes a string and two randomly chosen points in the string and reverses the substring between the inversion points. For example, given a string AB.CDE.F, with the randomly chosen inversion points shown as dots, the result of the inversion operation is the string ABEDCF. However, the cell position in the records is not updated at this stage, so the new string still represents the same placement. The inversion operator arbitrarily increases the length of the schema transferred from one parent or the other. One rationale for the use of the inversion operator is as follows: The simple one-point crossover operator which is commonly used in genetic algorithms [79] cuts two strings of equal length l_s at a randomly chosen point l_c characters away from the start of the string and then swaps the latter halves of the strings. The number of potentially different offspring that can be produced by crossing over two given strings in this manner is $2(l_s - 1)$, as l_c ranges from 1 to $l_s - 1$, where each crossover operation produces exactly two offspring. However, the cycle crossover operator always produces exactly two offspring, since the same two offspring are produced irrespective of the starting point (in the string) of the cycle crossover operation. If the two parents are P_1 and P_2, cycle crossover would always produce two offspring C_1 and C_2 regardless of whether the crossover operation started with the first position in P_1 or the l_nth position in P_2, where l_n is some arbitrary number in the range $[1, l_s]$. The inversion operator, when applied before the cycle crossover operator, allows many different offspring to be produced.

8.1.2 Analysis of Serial Algorithm

Let the run time of the serial genetic algorithm be represented by τ_{serial}. This time is a function of the population size p, the number of generations g, the problem size n (number of cells and nets), and the values assigned to the genetic parameters. Since the genetic parameters are the same for both the serial algorithm and the parallel versions to be discussed later, they are not shown explicitly in the following equations. The time to execute the serial algorithm includes the time to initialize the data structures, the time to perform the computations for all generations of the GA, and the time to output results.

$$\tau_{serial}(n, p, g) = t_{init}(n, p) + g \times (t_{rep}(n, p) + t_{eval}(n, p)) + t_{out}(n) \quad (8.1.1)$$
$$t_{init}(n, p) = t_{read}(n) + t_{setup}(n, p) \quad (8.1.2)$$

The initialization time includes the time to read in the circuit, t_{read}, which is a function of the circuit size, and t_{setup}, which is the time required to create the initial population and is a function of both the circuit size and the population size. The computation time for each generation includes t_{rep} and t_{eval}, where t_{rep} is the time required to perform all the genetic operations and create a set of new individuals, and t_{eval} is the time required to evaluate the cost function for each of the new individuals and to select the survivors into the next generation. It is known that the largest single time factor is t_{eval}, and $g \times t_{eval}$ is about 75% of τ_{serial}. The time required to write the best results out on to the disk, t_{out}, is a function of circuit size. The typical breakup of the time spent by the serial algorithm on various tasks is evaluation 77%, reproduction/crossover 22%, and initialization and input/output about 1%.

Hence, the benefits of parallelizing the reproduction and evaluation functions are obvious. Evaluation can easily be parallelized by evaluating each new individual on a separate processor or by evaluating different aspects of the cost functions on different processors. On a distributed memory system, this would entail the overhead of passing $\Theta(p \times n)$ data over the communication medium in each generation.

The reproduction/crossover phase begins with the selection of individuals, which is an inherently sequential process that reduces the achievable speedup. The actual crossover operation can be parallelized in an obvious way, but the communication cost remains.

The next subsection is devoted to presenting a simple, yet effective parallelization technique that avoids the communication costs associated with trivial parallelization schemes and achieves maximum speedup while producing solutions of comparable quality to the serial algorithm.

8.1.3 Distributed Placement Algorithm

The distributed placement procedure runs a basic genetic algorithm on each processor in the network and introduces a new genetic operator, *migration*, which transfers placement information from one processor to another across the network. Migration transfers genetic material from one environment to another, thereby introducing new genetic information and modifying the new environment. If the migrants are fitter than the existing individuals in the new environment, they get a higher probability of reproduction, and hence their genetic material is incorporated into the local population. When the population is very small, it tends to converge after a few generations, in the sense that all individuals come to resemble one another. Migration prevents this premature convergence or inbreeding by introducing new genetic material [43]. Hence, the genetic algorithm may be modified by splitting the large population over different processors and using the migration mechanism to prevent premature convergence.

While the idea of a distributed floorplanning algorithm using the migration mechanism was suggested in [44], that work did not study the speedup, since the floorplanner was presented as a distinct algorithm in itself, rather than as a parallelization of a given algorithm. Further, the floorplanner had a synchronization requirement, whereby a processor had to wait for migrants from some other processor. This requirement increases the worst case effective communication time to the actual physical time required to transmit data from one machine to another across a shared network. This time can be large and can reduce the speedup by a significant factor. Our work avoids the synchronization problem by delegating the communication work to the operating system/communication processor and by further not waiting on a *read* operation. Results presented here show that the result quality of the parallel algorithm is not degraded significantly by not waiting on *read* operations, while the speedup achieved is almost linear in the number of processors used because the communication time is now vastly reduced. Further, a detailed experimental study of the migration mechanism is done. The statistical variations inherent in the algorithm are thoroughly studied. It is shown that the distributed network environment can actually be used to speed up a successful placement algorithm to a significant extent.

The basic algorithm which runs on each processor in the network is shown in Fig. 8.2. The genetic operators – crossover, inversion, and mutation – are the same

Read netlist and cell library files;
Read parameter values, CrossoverRate, MutationRate,
InversionRate, PopulationSize, EpochLength,
NumberOfEpochs, NumberOfMigrants;
NumberOfOffspring = PopulationSize × CrossoverRate;

Generate initial Population randomly;
For j = 1 **To** PopulationSize **Do**
    evaluate (Population, j);

For i = 1 **To** NumberOfEpochs **Do**
    For j = 1 **To** EpochLength **Do**
        Select NumberOfOffspring individuals from Population;
        Copy selected individuals into NewPopulation;
        For k = 1 **To** NumberOfOffspring **Do**
            invert (NewPopulation, k, InversionRate);

        Perform crossover on individuals in NewPopulation;
        Perform mutation on individuals in NewPopulation;
        For k = 1 **To** NumberOfOffspring **Do**
            evaluate (NewPopulation, k);

        Population = reduce (Population, NewPopulation);

    Select and write NumberOfMigrants Individuals to network;
    Read Migrating Individuals from network;

    Population = reduce (Population, NewPopulation);

Solution = individual with highest fitness in the final Population;

Figure 8.2. Parallel algorithm

as those used in the serial algorithm. The new feature here is the migration mechanism, which combines the algorithms running on different processors into a single distributed genetic algorithm. The migration mechanism transfers some individuals (placement configurations) to other processors, once every few generations. The *epoch length* is defined as the number of generations between two successive migrations. The effectiveness of the migration mechanisms depends on several factors, such as the number of individuals transferred in each migration, selection of individuals for migration, and epoch length. Too much migration can force the populations on different processors to become identical, making the parallel algorithm inefficient, while too little migration may not effectively combine the subpopulations. This has been experimentally observed, as detailed in the next subsection.

The processing time τ_K with K processors on the network may be written as

$$\begin{aligned}\tau_K &= t_i(K) + t_{init}(n, \frac{p}{K}) + n_e \times (e \times (t_{rep}(n, \frac{p}{K}) + t_{eval}(n, \frac{p}{K})) \\ &+ n_{migr} \times t_{comm}(n, K)) + t_{out}(n, K)\end{aligned} \quad (8.1.3)$$

where
$$\begin{aligned}K &= \text{number of processors/workstations} \\ p &= \text{total population size} \\ n_e &= \text{number of epochs} \\ e &= \text{epoch length} \\ t_{rep} &= \text{time to perform reproduction/crossover operations} \\ t_{eval} &= \text{time to evaluate population fitness} \\ n_{migr} &= \text{number of individuals transferred at each migration} \\ t_{comm} &= \text{time to transfer one individual to communication subsystem} \\ t_i &= \text{time to startup the program on } K \text{ processors} \\ t_{init} &= \text{time required to read netlist, create \& evaluate population} \\ t_{eval} &\gg t_{comm} \\ n_e \times e &= g = \text{number of generations}\end{aligned}$$

It may be seen that there is an extra time factor, t_{comm}, not seen for the serial algorithm, and the reproduction and evaluation times are now functions of a smaller population assigned to each processor. As already observed in the case of the serial algorithm, the initialization and output times are negligible and may be safely ignored. The reproduction and evaluation times are both linear functions of the population size p and hence scale linearly when the population is divided among the available processors. Therefore, speedup of the parallel algorithm depends on the total communication cost, which in turn depends on the migration rate, the epoch length e, and the number of migrants n_{migr}. The minimum amount of information that has to be transferred in the migration process for each cell is the placement configuration, including its location (x, y) and an identification number used by the inversion operator. In addition to this, the cost of each configuration is also transmitted, since the extra communication time required to transfer one number is small, compared to the time required for cost function evaluation. Hence, the communication time is a function of the problem size n, the number of migrants n_{migr}, and the epoch length e. In a typical workstation, the actual communication is handled by a separate Ethernet subsystem, and the CPU continues processing after downloading the data to the communication subsystem for a *write*. The *read* process takes place when the CPU finds data waiting at the communication subsystem and reads it; if there are no data waiting to be read, the CPU does not wait for the data to become available and continues with the next step in the algorithm. Hence, the actual communication time for the CPU is reduced to the time required to transfer data to and from the communication subsystem, and measurements have shown this to be negligible. However, the communication pattern does load the network, and the trade-off is between network congestion and closer coupling between subpopulations on different processors. It has been empirically observed that there is an optimum level of coupling between subpopulations, and the network congestion factor at this communication rate is low.

If the total time for communication and assimilation of the migrants, $n_{migr} \times t_{comm}$, is small compared to the total time for performing all other actions in one epoch, the speedup is close to ideal. With increasing K, the speedup increases, but the efficiency goes down, and the network traffic increases. The total network traffic T due to K processors is $C_2 \times K \times n/\tau_e$, where C_2 is some constant, n is the number of cells in the circuit to be placed, and τ_e is the time for one epoch, which is given by $\tau_e = e \times (\frac{C_1}{K} \times (t_{rep}(n,p) + t_{eval}(n,p)))$. In other words, for a given circuit and population size, $\tau_e = C_3/K$, and $T = C_4 \times K^2$. Here C_1 is the linear scaling factor for the reproduction and evaluation times considered as functions of the population size. Observations indicate that C_1 is close to 1. C_3 is derived from C_1 for a fixed population size and circuit. C_2 is related to the amount of information transmitted during each migration. Hence, the traffic increases as K^2, and using empirical values for the constants in the above equations and assuming a network capacity of 1.25×10^6 bytes/second (10 Mbit/second Ethernet), a speedup close to 25 can be achieved with up to 25 processors with a 50% load on the network. A speedup of 8 can be achieved with 8 processors and less than 10% loading of the network. With a large number of processors, the initial loading on the network is high, but after a few epochs, the loading tends to stabilize to its average value. The major results of this analysis are the following:

- If the communication time is small, the speedup is close to ideal.

- The implementation of the communication mechanism ensures that the communication time is always small, and close to ideal speedup is guaranteed.

- The speedup can be increased by increasing the number of processors, but this results in increased network loading, and up to 25 processors may be used without unduly loading the network.

- The network loading factor depends on the genetic parameters as well as on the problem size.

It may be noted that the speedup here is specified in terms of the time required to perform a fixed number of genetic operations and fitness evaluations on a constant overall population size. The migration mechanism is assumed to ensure the result quality, and empirical observation of this is presented in the next subsection. An alternative measure of speedup that considers the time taken to achieve a given fitness value is also discussed.

Implementation Issues

The distributed placement algorithm has been implemented on a network of workstations (Sun systems). In order to facilitate the evaluation of various topologies, the program has been organized as follows: There is one master program that provides the user interface and also manages all other processes (processors). The master program uses the *rexec* facility to create processes on different systems. Processors communicate by using the socket mechanism. The master processor determines the communication pattern. In order to explore different communication patterns, the master program also acts as the routing controller and routes all messages between any two processors. Hence, actual physical communication is between the master processor and the other processors only. However, with a fixed routing pattern, there need be no communication bottleneck, and the steps executed by the master and slave processors are shown in Fig. 8.3.

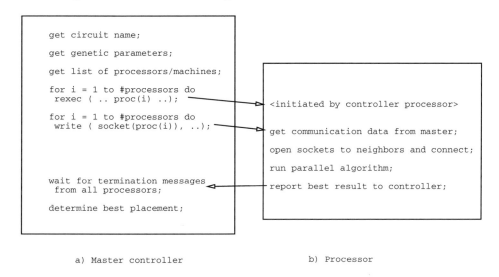

Figure 8.3. Implementation of the distributed algorithm

The master controller obtains the circuit name and the processors or machines to use. It then uses the *rexec* function to start the slave programs on all the specified machines. Once this is done, the controller sends each processor some information about the other processors with which it should communicate and then goes to sleep. Each processor executes the parallel algorithm of Fig. 8.2 and sends a message to the master controller before terminating. When all the processors have terminated, the master controller identifies the best overall solution. The data structures used in the genetic algorithm running on each processor are the same as those used in the serial algorithm.

Both the master controller and the slave programs are written in C and make use of the *rexec* and *socket* functions provided by the *UNIX* system. Typically, the master controller runs on the same machine as one of the slave programs since there is not much overlap between the two. All experimental observations have been made with the machines running their normal daily loads in addition to the distributed placement algorithm.

8.1.4 Experimental Results

The following major issues in the behavior of the distributed algorithm were studied experimentally:

- *Result Quality*: The placement results of the serial algorithm were compared with the multiprocessor placements to determine if the implementation of the migration mechanism was effective. Multiple runs with different circuits showed that the result quality was the same.

- *Statistical Variations in Result Quality*: The serial and parallel versions of the placement algorithm were run many times, and the mean and standard deviation of the final results were computed. It was found that the migration mechanism managed to preserve the essential characteristics of the serial algorithm results while providing good speedup.

- *Speedup*: Speedup was measured in two ways. The first speedup was based on the time required to examine a fixed number of configurations. By keeping the communication costs low, this speedup was measured to be close to ideal. At the same time, the result quality was also monitored to ensure that the speedup was not at the cost of result quality. The second speedup calculation was based on the statistical observations showing that for any desired final result achievable using the serial algorithm, the parallel version could provide linear speedup (close to K, for K processors).

- *Migration*: The effect of migration rate variations and the interplay between migration and the crossover mechanism were studied. Good parameter settings for migration and crossover were identified.

- *Communication Patterns*: While the bus-based network provides full connectivity, allowing any machine to communicate with any other machine, the effects of the actual communication pattern on the result quality and speedup were studied. It was seen that as the number of processors was increased, an efficient communication pattern was required to make the migration mechanism effective.

- *Coping with Network Heterogeneity*: The effects of wide differences in machine capabilities and loads on the performance of the placement algorithm were studied. It was seen that the algorithm was robust. Static and dynamic load balancing schemes were implemented.

8.1. Wolverines: Standard Cell Placement on a Network of Workstations 265

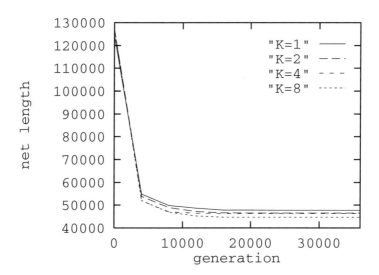

Figure 8.4. Net length as a function of generation number for uni- and multi-processor runs with same total population for *cktA*

All experiments were performed on the set of standard cell circuits containing from 100 to 5814 cells, including the following eight circuits used in benchmarking the serial genetic algorithm [208]: *cktA*, 100 cells; *cktB*, 183 cells; *cktC*, 469 cells; *cktD*, 752 cells; *cktE*, 800 cells; *cktF*, 2357 cells; *cktG*, 2907 cells; and *cktH*, 5814 cells.

Convergence and Result Quality

Fig. 8.4 plots the best net length obtained against the generation number for the genetic algorithm running on 1, 2, 4, and 8 processors. These placement results are for circuit *cktA*. Similar results were obtained for the other circuits. It was seen that in all the plots, there was a region of rapid improvement followed by a region of more gradual improvement. The total population size and the total number of configurations examined were kept constant in all these experiments. Thus, for K = 1, the uniprocessor case, the population size was 48, while for K = 8, the 8-processor case, the subpopulation on each processor was 6, making the total population 48. The other genetic parameters, such as crossover rate, mutation rate, inversion rate, and epoch length were the same on all processors. The curves for the uniprocessor and multiprocessor experiments all lie close together for all the circuits. In order to get a more accurate comparison of the results, the difference in the best net length obtained in the uniprocessor and multiprocessor experiments in each generation has been plotted.

Fig. 8.5 plots $\Delta netlength = (netlength_1 - netlength_K) \times 100/netlength_1$ as a function of generation number for K = 2, 4, and 8. It can be seen that $\Delta netlength$

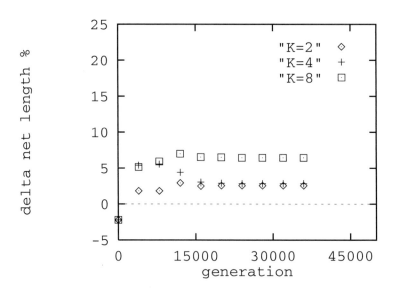

Figure 8.5. Result quality vs. generation showing percentage difference in net length between uniprocessor and multiprocessor runs for *cktA*

is initially negative. This means that, in the initial stages, the uniprocessor algorithm does slightly better than the multiprocessor algorithm. However, $\Delta netlength$ becomes positive later on for all three cases in this experiment. It was observed that in some other experiments, $\Delta netlength$ varied somewhat from $+5\%$ to -5%. In order to test the hypothesis that this variation was due to the randomness inherent in the algorithm and not due to the migration mechanism itself, the serial algorithm was run repeatedly (1000 times), and the results obtained were compared with the results from repeated runs of the multiprocessor algorithm running on 8 processors. Fig. 8.6 shows the final net length on the X-axis and the fraction of runs in which that net length was the result on the Y-axis. It can be seen that the 1-processor and 8-processor results are similar. The means and standard deviations of the distributions were 49.708 and 2.712 for the 8-processor case and 48.107 and 2.956 for the 1-processor case. When this factor is taken into account, it can be said that the parallel algorithm produces results of the same quality as the serial algorithm (to within 5%). Fig. 8.7 shows the final net length difference in another set of runs on 1 to 16 processors. Table 8.1 shows the consolidated result quality and speedup results for three representative circuits. It was observed that the runs of the multiprocessor algorithm produced final results within a range of $\pm 5\%$ of the uniprocessor results.

The effect of population size on result quality was observed to determine optimum population sizes if any. Fig. 8.8 shows the final net length as a function of population size for the sequential algorithm and for the parallel version for circuit

8.1. Wolverines: Standard Cell Placement on a Network of Workstations

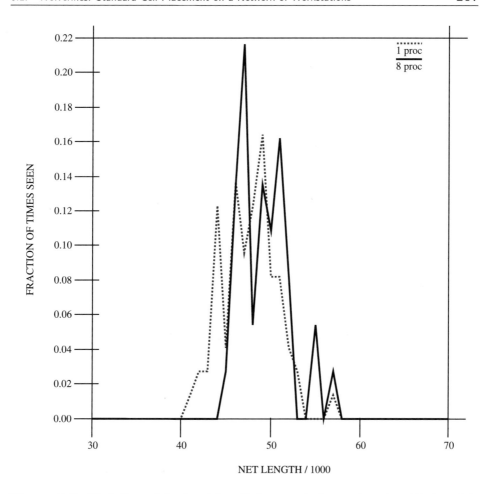

Figure 8.6. Variation of final net length in repeated experiments with 1 and 8 processors for *cktA*

cktA. All data points shown in the figure were obtained with the same set of genetic parameters except the population size p and the number of generations g. In order to keep the total number of configurations examined by the algorithm the same in all cases, the product $p \times g$ was kept constant as the population size was varied. The average final net length obtained in multiple runs with the same population size is plotted against the population size in Fig. 8.8. The best result with the sequential algorithm was obtained with a population size of 96. This population was then split across 2, 4, and 8 processors in repeated runs of the parallel algorithm. It may be observed that for $K = 2$, the parallel algorithm performs somewhat better than the sequential algorithm in terms of result quality, while in other cases, it is somewhat worse. In all cases, the parallel algorithm produces results within

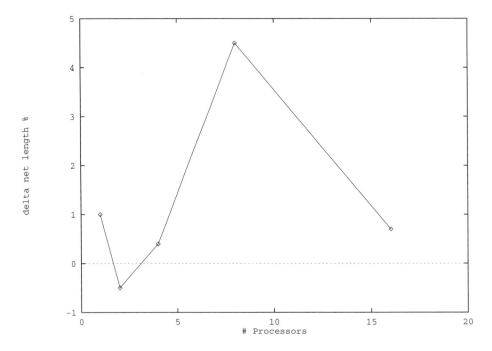

Figure 8.7. Result quality for *cktA* vs. number of processors

±5% of the best sequential result, as observed earlier; i.e., the population size does not significantly affect the solution quality. The variations seen are all within the statistical variations shown in Fig. 8.7.

Speedup

The first set of speedup measurements was based on measurements of the run time of the parallel algorithm for a fixed number of generations, τ_K. The speedup for a K-processor experiment is computed as τ_K/τ_1. Fig. 8.9 shows the speedup obtained for *cktC*. Two curves are plotted; the curve labeled *total* is the speedup computed by dividing the total run time of the serial algorithm by the total run time of the parallel algorithm, while the curve labeled *region1* is the speedup obtained in the initial stages of the algorithm where the improvement in net length is fastest. While the curve labeled *total* is smooth and shows almost linear speedup, the curve labeled *region1* is less smooth, due to the nonuniform convergence of the algorithm in the initial stages. This nonuniform convergence is due to the effect of the population size and the choice of the initial random population. Similar results were obtained for other circuits. Fig. 8.10 shows the total speedup for four representative circuits, *cktA*, *cktD*, *cktG*, and *cktH* in another form. The consolidated speedup and result quality figures for three circuits are shown in Table 8.1. It can be seen that the

Table 8.1. Speedup and Result Quality

	Num. Proc	2	4	8
cktB	net length	116,492	119,832	105,880
	delta	0.76	-2.08	9.8
	time (s)	1417	736	360
	speedup	1.9	3.67	7.51
cktD	net length	156,471	153,528	165,088
	delta	2.04	3.88	-3.34
	time (s)	6224	3248	1744
	speedup	1.88	3.61	6.72
cktF	net length	112,233	116,566	118,505
	delta	-.37	-4.24	-5.97
	time (s)	18,835	9813	5303
	speedup	1.94	3.73	6.91

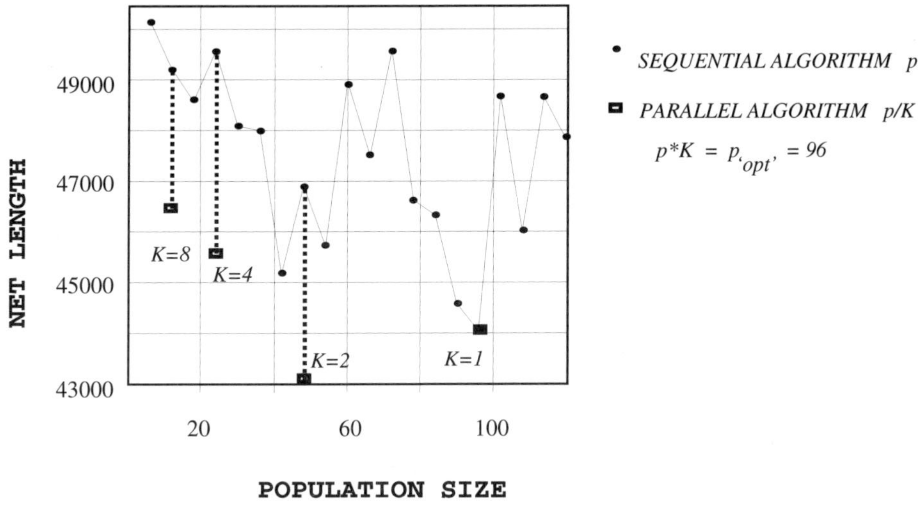

Figure 8.8. Effect of population size on result quality (dots indicate sequential algorithm averages, while boxes correspond to parallel algorithm with corresponding subpopulation size)

final speedup in all cases was close to ideal, and the final result quality was not significantly different either. The speedup was close to ideal because, among the various time factors, the communication time t_{comm} was extremely low, the initialization and setup times were negligible, and the reproduction and evaluation times $t_{rep}(n,p)$ and $t_{eval}(n,p)$ scaled properly with the population size, as expected.

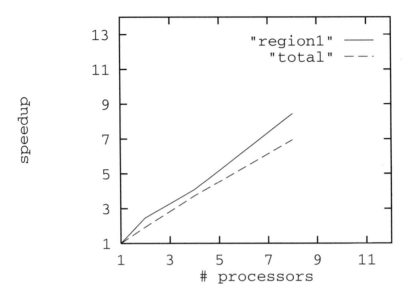

Figure 8.9. Speedup for *cktC*

Further studies on the speedup of the parallel algorithm were done to rule out scattering effects (due to the random number generator) that might lead one to erroneous conclusions about the speedup. The algorithm was run repeatedly on 1, 2, 4, and 8 processors, and the best net length obtained in each epoch was noted. Then a single best-fit curve was obtained for each of the four cases (Fig. 8.11). It was found that in all cases, the curves had the form $a + b/t^{0.5}$. The a values for all the curves were almost the same, indicating that the result quality would be the same, given sufficient run time. The b values decreased as the number of processors increased, reflecting the faster convergence and the speedup obtained by using multiple processors. The observed b values show that the time required to reach a certain given value of the cost function can be reduced by a linear factor by adding more processors. Let the required value of the cost function be y_0. Then the equations obtained by curve fitting predict the time required to obtain this value to be $(b/(y_0 - a))^2$. Let b_1 be the b value for the uniprocessor case and b_K be the b value for the K-processor case. Assuming that the a values are equal, the speedup obtained using K processors is $(b_1/b_K)^2$. It is observed that this factor is a linear function of K. This second speedup result shows that if an adaptive termination condition were to be used, stopping the algorithm once the algorithm achieved some preset solution quality, the parallelization scheme presented here would still achieve excellent speedup. For example, if the placement program were modified for gate array placement, with a cost function that emphasizes routability rather than minimum wire length, the algorithm may be modified to terminate once

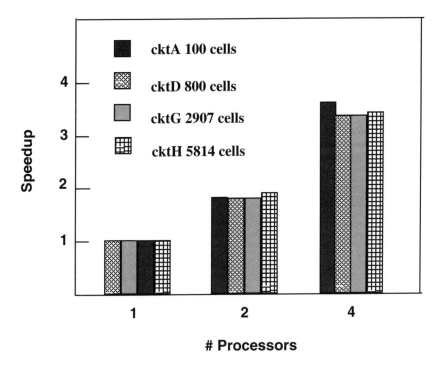

Figure 8.10. Speedup on multiple processors for three circuits

the solution is *routable* instead of examining a predetermined number of placement configurations.

Hence, the parallel algorithm attains linear speedup in two different senses. In the first case, the parallel algorithm examines the same number of configurations as the serial algorithm, and the speedup factor is close to K, since the communication, synchronization, and initialization costs are kept extremely small. At the same time, the result quality is not significantly different from the uniprocessor case. In the second case, the same result is demanded in both the uniprocessor and the K-processor runs. The time required to do this decreases linearly with the number of processors used, showing a linear speedup.

Migration and Crossover

Fig. 8.12 plots the best net length obtained on each of the 16 processors as a function of the generation number. It can be seen that the curves intersect quite often, indicating that the improvement in net length occurs in bursts, with intervening periods of stagnation. Whenever a particular processor is stagnant for a long time, a new individual introduced into the subpopulation by migration provides the impetus for a new burst of improvement.

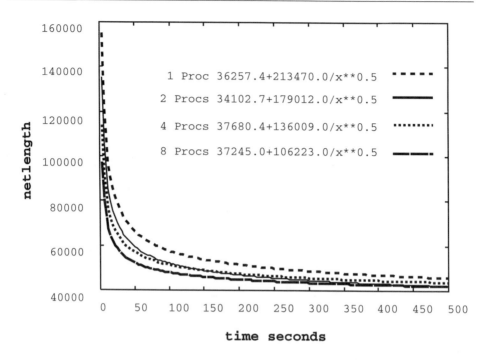

Figure 8.11. Empirical convergence curves for *cktA*

The efficacy of the migration mechanism was explored in the next set of experiments. Fig. 8.13 shows the convergence curves for different values of epoch length. The number of migrants was kept constant, while the epoch length was varied. The best net length obtained, taking into account all the subpopulations, was plotted against the generation number for different epoch lengths ranging from 10 to infinity.

It can be seen that there is a big difference in the convergence rates and the final solution for different epoch lengths. Very short epoch lengths (10 and 20) and very long epoch lengths (90 and infinity) led to poor results. Long epoch lengths imply less transfer of information between subpopulations. In the extreme case of infinite epoch length, there is no migration at all, and the parallel genetic algorithm running on K processors with a total population of p is equivalent to a sequential genetic algorithm run K times with a population of p/K. Hence, the solution obtained in the case of long epoch lengths corresponds to the best solution obtained with multiple runs of the serial algorithm using a much smaller population. On the other hand, when the epoch length is very small, the subpopulations tend to converge. This means that the effective population size is still p/K, but the final solution is worse than the case of no migration at all because all the subpopulations are identical, making this run of the K-processor algorithm equivalent to a single run of the uniprocessor algorithm with a population of p/K rather than K runs

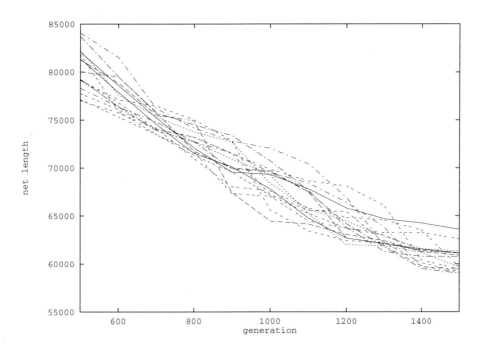

Figure 8.12. Convergence curves for 16 processors ($cktA$)

as in the case of no migration. It was observed that the K-processor algorithm performed best for some intermediate value of epoch length, as seen in Fig. 8.13.

The two main features of genetic search are exploration, the generation of new schemata in an efficient manner, and exploitation, the propagation of good schemata to subsequent generations. The crossover mechanism in the serial algorithm and the crossover and migration mechanisms in the parallel genetic algorithm are chiefly responsible for both exploration and exploitation. The migration rate is a function of the epoch length and the number of migrants transferred in each epoch. Increasing the number of migrants while keeping the epoch length and crossover rate constant corresponds to the transfer of larger fractions of the subpopulation from one processor to another. This causes the subpopulations to become almost identical after a few generations, just as in the case of extremely short epoch lengths. On the other hand, decreasing the number of migrants while keeping the epoch length and crossover rate fixed, or increasing the crossover rate while keeping the number of migrants and the epoch length fixed, reduces the beneficial influence of migration. If the epoch is too short, the subpopulations do not have enough time to generate new, highly fit individuals through crossover before the next migration; i.e., the exploration mechanism is hindered. Hence, a single highly fit individual quickly gets replicated in every subpopulation, and the net effect is of K almost identical subpopulations evolving in parallel. This represents a huge waste of search effort.

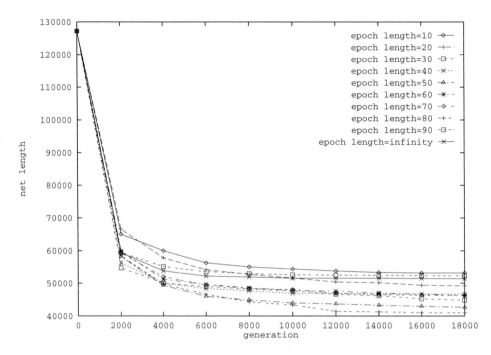

Figure 8.13. Migration rate variation (*cktA*)

Similar effects would be seen if the number of migrants increased to a large fraction of the population size. On the other hand, if the crossover rate is high, good solutions (schema) are not fully exploited because the population size is fixed; new, good solutions are generated at a high rate (due to the high crossover rate), and the survival probability of any one of the good solutions is decreased. Thus, a higher crossover rate increases the exploration part of the genetic algorithm while decreasing the exploitation part, which is actually the chief strength of the algorithm as compared to stochastic hill-climbing techniques such as simulated annealing. This means a higher crossover rate reduces the effectiveness of the exploitation part of the algorithm, and the migration mechanism, which is supposed to aid in the exploitation of good schemata, becomes irrelevant.

The crossover rate was doubled, keeping the migration rate constant, and the convergence curves for different epoch lengths were plotted. The effect of epoch length variation was seen to be minimal, and the best final solution was not as good as that obtained with a lower crossover rate. With the observation that the crossover rate could influence the effect of the migration rate (epoch length), the next experiment was an attempt to obtain the best crossover rate for a given migration rate. The epoch length and the number of migrants per epoch were kept constant while the crossover rate was varied. The results are shown in Table 8.2,

Table 8.2. Crossover Rate Variation

Offspring Per Generation	Final Net Length
10	52,611
8	46,437
6	47,166
4	46,301
2	46,225

Table 8.3. Communication and Run Times

Epoch Length	Total Run Time (seconds)	Communication Time (seconds)
30	68	10
40	68	8
50	66	6
60	67	4
70	63	3

which shows the final net length for various crossover rates indicated by the number of offspring per generation; the higher the crossover rate, the more the offspring per generation. Large crossover rates nullify the effect of migration, leading to poor solutions. Hence, the best results are obtained by keeping the crossover rate and the number of migrants low and finding an optimal epoch length.

Communication Time

The total communication time is a function of the epoch length, as stated earlier. Fig. 8.14 plots the percentage of time spent on communication as a function of epoch length. The data were obtained from several runs of the distributed algorithm on circuit *cktA*. The actual communication and run times for several runs of the parallel algorithm on 8 processors are shown in Table 8.3. It can be seen that the communication time is less than 5% of the total run time when the epoch length is greater than 50. This observation, along with the results of the previous section, shows that the communication time is not a major factor for typical runs of the distributed genetic algorithm. It may be noted that the communication time here is just the time required to communicate with the network subsystem in the workstation; the actual communication time across the network does not directly affect the speedup or the solution quality.

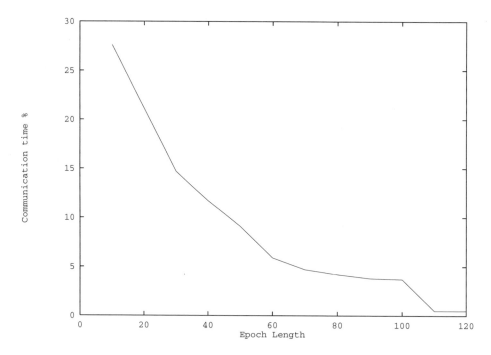

Figure 8.14. Communication time

Effect of Communication Pattern on Placement Quality

This subsection discusses the effects of different communication patterns on the distributed genetic algorithm. The communication pattern establishes a virtual network topology on top of the Ethernet bus structure. The algorithm succeeds in generating a good solution if the following conditions are met:

- the subpopulations do not converge prematurely, and
- the different subpopulations converge in the sense that the distance between two populations is constantly reduced.

The first condition is satisfied by the migration mechanism. The second condition can be satisfied if there is *sufficient* communication or migration. The value of n_{migr} should be carefully chosen so as to avoid premature convergence and ensure proper transfer of genetic material. Hence, the communication pattern and virtual network diameter are important factors in ensuring the quality of placement. For a given population size, the time to converge depends on the time to propagate an individual across the network to all the subpopulations. This time is clearly related to the diameter D of the network. The minimum number of epochs needed to propagate information to all the subpopulations is D, where D is the diameter of the network.

Five different networks of 8 processors were simulated using the simulation mechanism built into the program. The migration mechanism of the parallel algorithm has two components: *sending* individuals to other processors and *getting* individuals from other processors. *Send* always succeeds, but a *get* operation can fail if the processor which has to supply the data is not ready. When a *get* operation fails in one epoch and succeeds in the next, the data obtained in the next *get* operation correspond to the best solutions of the previous epoch. This is the same as increasing the time to propagate solutions across the network by one epoch length and is hence most critical in a ring network. In practice, at least a few, but not all, *get* operations can be expected to fail, unless the processors are either perfectly synchronized or have totally unbalanced loads. Hence, two kinds of ring networks have been simulated: One in which every *get* operation succeeds, called *ring+*, and one in which every *get* operation is delayed by 1 epoch, called *ring-*. Hypercube and square mesh topologies were simulated, as well as a totally connected network in which each processor *sends* and *gets* data to and from randomly chosen processors in each epoch. In the hypercube connection, processors communicate along one dimension in each epoch. For example, in a three-dimensional hypercube, processor (100) (100 represents the coordinates of the processor in 3-space, where each coordinate value can be either 0 or 1) communicates with processors (000), (110), and (101), respectively, in successive epochs. In a square mesh, on the other hand, a processor at location (x,y) communicates with processors at locations $(x-1,y)$, $(x,y-1)$, $(x+1,y)$, and $(x,y+1)$ in successive epochs (processors at the boundary of the mesh may thus not have any communication in certain epochs). Hence, the largest number of epochs required to transmit information from any one processor in the network to another processor in the network is just the diameter D of the network, $\lg(K)$ for an N-dimensional hypercube ($K = 2^N$) and $K^{0.5}$ for a square mesh with K processors.

With the epoch length kept small, *ring+* was better than *ring-* as expected, and all five solutions were comparable, with the mesh connection producing the best results (see Fig. 8.15). With a longer epoch length, convergence was delayed on all networks, but the results were all still roughly equal, with the hypercube producing the best results (see Table 8.4). This suggests that for small problem sizes, when network loading is not an important factor, the actual communication pattern is not important. But for large problem sizes, with significant network loading at short epoch lengths, the epochs have to be long to avoid network congestion problems, and an efficient communication pattern is essential to avoid delayed convergence.

Load Balancing

The next experiment tested the robustness of the algorithm in a distributed environment where processors of varying capabilities could be present. The population was divided unequally across six processors in such a way that the expected run time of each processor was the same (to ensure maximum efficiency). The result obtained was compared to the uniprocessor result and the result obtained by split-

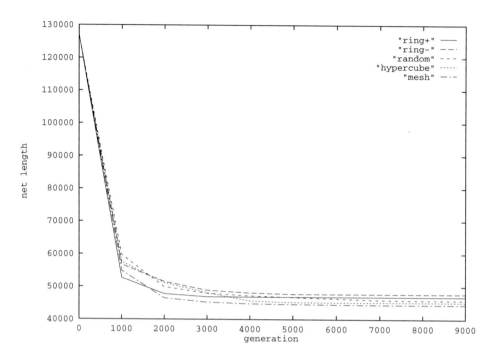

Figure 8.15. Effect of network topology on convergence

Table 8.4. Effect of Network Topology

Topology	Final Net Length
random	49,536
ring+	47,999
ring-	47,768
mesh	46,524
hypercube	46,127

ting the population evenly across six similar processors. These results are shown in Fig. 8.16, where $K = 1$ corresponds to the uniprocessor case, $K = 6$ corresponds to six unequal processors, and $K = 6eq$ corresponds to six equal processors. The total population size in all three cases was 108, and the subpopulation sizes in the case of the unequal processors were 48, 24, 12, 12, 6, and 6. It can be seen that the curves for $K = 1$ and $K = 6eq$ are almost coincident, showing that the parallel algorithm produces acceptably good results even if the population is unevenly divided across the processors. Hence, static load balancing is feasible.

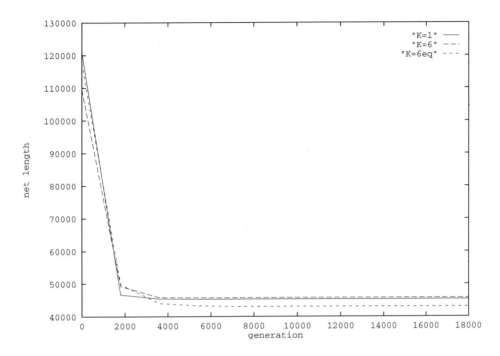

Figure 8.16. Equal and unequal division of load

With the knowledge that static load balancing is feasible, the next step was to implement dynamic load balancing. Two different schemes for dynamic load balancing were considered. In the first scheme, the number of processors used remains constant, but the population on each processor varies with time. In the second scheme, the master controller has a list of n processors from which it selects the K least-loaded machines. The load on each machine is checked periodically. When the load on any machine exceeds a specified level, the controller terminates the process running on that machine and transfers the population to the least loaded machine on its list.

The first scheme is not practicable because of the granularity of the genetic operators with respect to the population sizes used. Small changes in the population size do not change the resources consumed by the program by the same factor; for example, with a population of size 6 and a crossover rate of 0.33, the number of offspring per generation is 2, while with a population of 8 and the same crossover rate, the number of offspring is still 2 (truncated to nearest even number). Hence, the second scheme was adopted and found to work well in practice.

8.1.5 Conclusions

We have described a new distributed placement algorithm for standard cells. The algorithm runs concurrently on a network of workstations to achieve speedup linear in the number of processors used. The low communication overhead for this algorithm, as compared to the synchronization and communication overheads for conventional placement algorithms, makes this speedup possible. The parallel algorithm preserves the result quality of the serial version while achieving this speedup. Analysis of several hundred runs of the parallel algorithm with different values of K (the number of processors) has shown that, if a particular solution can be obtained in time τ_1 using a single processor, the same solution can be obtained in time $\frac{\tau_1}{K} \times C$ by running the algorithm on K processors, where C is a constant whose value is close to 1.

The parallel algorithm has been observed to be robust in the sense that it can run in a heterogeneous computing environment where different processors on the network have different speed ratings and load factors. The speedup and result quality are maintained even in this uneven environment, which is a true representative of practical distributed computing networks, by means of static and dynamic load balancing.

While the typical workstation network is just a bus with many different machines connected to the bus, the communication pattern between the machines may be random, with a machine sending migrants to a randomly selected machine in each epoch, or deterministic, with a virtual ring or hypercube network topology. The migration mechanism is not very sensitive to this virtual topology when the number of processors is low, but when the number of processors is increased, a regular virtual network topology with low diameter, such as the hypercube topology, helps to improve the performance of the parallel algorithm. An important feature of the current implementation of the placement algorithm is that it allows the communication pattern to be chosen to obtain the best result quality.

Finally, this work can be readily applied to a large class of other CAD problems and also to a host of large optimization problems in other domains, where adaptive search capabilities of the genetic algorithm have been found to yield high-quality solutions.

8.2 Parallel Genetic Algorithms for Automatic Test Generation

Genetic algorithms have been used effectively for simulation-based test generation, as discussed in Chapter 6. In this section, we present three parallel GAs for simulation-based sequential circuit automatic test generation (ATG) [126]. These parallel algorithms have been implemented using the ProperCAD II library [171], which provides a portable, parallel object-oriented platform for the development of parallel algorithms for VLSI CAD applications. The ProperCAD II library is a C++ library built around the Actor paradigm of concurrent object-oriented computing [5]. It is portable across a variety of architectures. Supported architectures include distributed memory multicomputers, such as the Intel Paragon, Thinking Machines CM-5, and the IBM SP-2; shared-memory multiprocessors, such as the SUN 4/690/MP, the SUN-SparcServer 1000E, and the SGI Challenge; and networks of workstations.

The first algorithm, PGATEST1, is a parallel version of the sequential test generation algorithm which produces the same result as the sequential algorithm. The sequential algorithm is parallelized globally, with a central processor controlling the execution. This algorithm is similar to the one used in GATTO* [45], but it is different in a few respects, as will be described later. The second algorithm, PGATEST2, uses a parallel search strategy where each processor executes the sequential genetic algorithm with a different seed and uses migration to share information between processors. The third algorithm, PGATEST3, is a subpopulation based version of PGATEST2, where subpopulations are distributed across processors, and information is migrated from one processor to another. Significant speedups have been observed for all three algorithms.

We begin with a discussion of the implementation of the sequential GA-based test generator. The three parallel genetic algorithms for simulation-based sequential circuit test generation are then presented and discussed in detail. Finally, results are presented, and the various algorithms are compared.

8.2.1 Sequential GA-Based ATG

The sequential implementation of the test generator is similar to the GA-based test generation procedure used in ALT-TEST, as described in Section 6.4.2. Processing is divided into three stages. The first stage attempts to detect as many faults as possible from the fault list. The second stage tries to maximize the number of visited states and propagate fault effects to flip-flops. The third and final stage attempts to detect the remaining faults and to visit new states. Only the fitness function changes between stages. In each of the three stages, the GA targets all faults in the fault list in groups of 31 faults, traversing the fault list several times until little or no more improvement is made. Between consecutive GA runs, useful vectors from the best sequence are added to the test set, and the two best sequences from the previous GA run are used to seed the population in the next GA run. The remaining individuals in the population are initialized with random test sequences. The test generation algorithm is given in Fig. 8.17.

The GA population evolves over several generations until no more improvements in fitness values are obtained or the maximum of eight generations is reached. The population size is a function of the string length, and the string length is equal to the number of primary inputs in the circuit multiplied by the test sequence length. A multiple of the structural sequential depth of the circuit is used as the test sequence length. The test sequence length is set equal to the sequential depth in the first stage, two times the sequential depth in the second stage, and four times the sequential depth in the third stage. Also, the population size is doubled in the third stage to further expand the search space.

```
For stage = 1 To 3 Do
    set fitness function;

    While there is improvement in this stage Do
        While there is improvement in the GA Do
            For all undetected faults, in groups of 31 Do
                target-faults = next 31 undetected faults;
                best-sequence = GA-test-generate (target-faults);
                fault-simulate (best-sequence);
                update test set;
                seed the next GA population (best 2 sequences);
            compute improvement;
        compute overall improvement;
```

Figure 8.17. Sequential ATG algorithm

Each individual in the population is simulated using a group of 31 faults, for the purpose of evaluating the fitness of the individual. The reason 31 faults are chosen is to maximize the execution speed. Fault simulation on the entire fault list is very costly, and previous work has shown that small fault samples give good quality results as well [189]. The good circuit evaluation requires only one bit position out of the 32 available positions in a computer word, and the remaining positions can be used for faulty circuit simulations.

Fig. 8.18 shows how faults are grouped for test generation. Initially the first 31 undetected faults are used as the fault group targeted for test generation. All individuals in the GA population target the same group of 31 faults, and the individuals evolve in successive generations until no more improvements in fitness values are obtained. Then a full-scale fault simulation is performed, using the entire list of undetected faults and the best test sequence found, and all detected faults are dropped from the fault list. An improved version of PROOFS [161] was used as the fault simulator. For the next GA run, the next group of 31 undetected faults is obtained, starting at the position in the fault list where the previous fault group left off. Only the good portion of the best test sequence is added to the test set to keep the test set compact.

The three stages of the test generator target different goals, and their corresponding fitness functions are different. In recent work by Wah et al. [234] and Pomeranz and Reddy [175], the results have indicated that fault effect propagation should outweigh fault detection when the easier faults have been detected, and maximization of state visitation increases fault coverage. Thus, these parameters are included for consideration when computing fitness values. The parameters that affect the fitness are as follows:

8.2. Parallel Genetic Algorithms for Automatic Test Generation

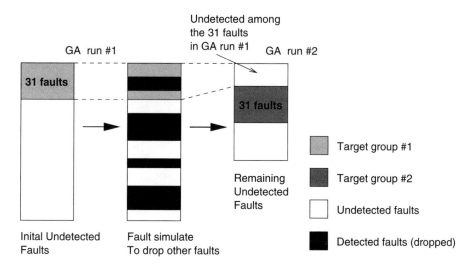

Figure 8.18. Grouping of 31 faults

P_1: number of faults in the given fault group detected,
P_2: number of new states visited, and
P_3: number of flip-flops that carry fault effects at the end of simulation.

The three parameters are given different weights in the fitness computations for the three stages:

Stage 1: $fitness = 0.9P_1 + 0.1P_3$
Stage 2: $fitness = 0.1P_1 + 0.45P_2 + 0.45P_3$
Stage 3: $fitness = 0.5P_1 + 0.4P_2 + 0.1P_3$

In the first stage, the aim is to detect as many faults as possible with short sequences and minimal time. Thus, more weight is given to the number of faults detected. In the second stage, the goal is to maximize visitation of new states and fault effect propagation to state flip-flops. In the final stage, the focus is shifted once again to target the remainder of the faults that have been hard to detect in the prior two stages. Therefore, the fitness evaluation weights fault detections and new state identifications more heavily.

A limitation of the simulation-based approach used is that it cannot identify untestable faults. Hence, although a high fault coverage may be achieved, the ATG efficiency is no better than the fault coverage. To improve the ATG efficiency, untestable fault identification is performed using a parallel version [170] of the HITEC deterministic test generator [160] after the GA-based test generation is completed. Often many untestable faults can be identified in a very short amount of time compared to the actual test generation time. By assigning a short time of 1 second per fault, one can identify many of the untestable faults.

8.2.2 Parallel GA-Based ATG

Genetic algorithms can be parallelized in various ways. A good overview for parallel GAs can be found in [3]. We shall now discuss some of the algorithms below in the context of our application.

One can implement a parallel GA using *parallelization by data decomposition*. In this model, a single population exists, and evaluation of the individuals (and sometimes application of the genetic operators) is completed in parallel. The evaluation can be parallelized by assigning a subset of individuals to each of the processors. A second approach is to use a *migration-based GA*. This model of parallelization introduces a new migration operator that is used to send individuals from one subpopulation to another. The migration operator can be defined in various ways. Classically, one tends to migrate individuals to *neighboring* processors. This approach is not very pertinent in multicomputers today, since the diameter of the network interconnecting the processors is reasonably small, and the distance between processors can be assumed to be constant, to a reasonable approximation. We have therefore introduced a randomized migration operator in two of our algorithms so that a processor may communicate with any arbitrary processor, to disperse information more quickly among the processors. This migration operator introduces a new dimension of parallel search into the algorithm and opens up exciting possibilities. It is possible that the parallel GA will terminate faster with the exchange of information between processors. It is also possible that one might be able to outperform the sequential GA in obtaining a better quality solution. A third approach to parallelization is to use a *subpopulation-based GA*. Here, the population is divided into a few subpopulations, which are relatively isolated from each other. A number of studies have been made on approaches which use subpopulation-based GAs, both with and without migration [151], [223]. Using subpopulations without migration is equivalent to performing simultaneous runs of the sequential algorithm on the subpopulations. Since there is no exchange of information, this approach is not very interesting. The execution time and results are the same as those for the execution of the sequential GA on a subpopulation. Hence, we shall only consider subpopulation-based GAs with migration. This introduces two dimensions of parallelism at once. The first dimension is that of migration of information between processors. The second dimension is that now each processor only works on a subpopulation, and hence, it has to evaluate the fitness of a smaller number of individuals. One could get superlinear speedups with this approach due to the two dimensions of parallelism in this approach, but the quality of the solution may be degraded if population sizes become too small.

Previous parallel approaches to test generation have focused on parallelizing deterministic algorithms. We have implemented parallelization with data decomposition, a migration-based parallel GA, and a subpopulation-based GA with migration. Fig. 8.19 shows a graphical outline of the approaches used in each of the algorithms for 4 processors.

For purposes of migration, a randomized migration operator was used; i.e., each

8.2. Parallel Genetic Algorithms for Automatic Test Generation

Each processor executes the same GA Algorithm
but performs N/4 fitness evaluations
P0 collects the fitness info from all processors
and broadcasts this information to all processors

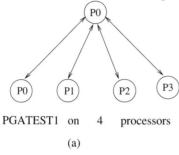

PGATEST1 on 4 processors
(a)

N individuals in population per processor
Each Processor executes GA algorithm
starting with its own seed

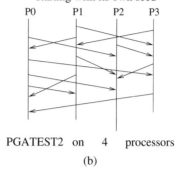

PGATEST2 on 4 processors
(b)

N/4 individuals in population per processor
Each Processor executes GA algorithm
starting with its own seed

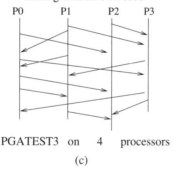

PGATEST3 on 4 processors
(c)

Figure 8.19. Comparison of three proposed parallel algorithms

processor sends an individual with a high fitness value to a single random processor. The rationale for this approach is that useful information (highly-fit individuals) will eventually migrate to all processors.

PGATEST1: Parallelization Using Data Decomposition

In this approach, we parallelize the loop which evaluates the fitness of all individuals in a given population, as illustrated in Fig. 8.19(a). Each processor maintains its own copy of the entire population. The work of evaluating the fitness is equally and statically distributed over the processors. The fault list is the same on all processors. Hence, if we have N individuals and P processors, we assign $\frac{N}{P}$ individuals to each processor. The assumption here is that the granularity of the task of evaluating the fitness of an individual is approximately the same for all individuals. The fitness values computed by the respective processors are communicated to a single processor, which collects this information and broadcasts it to all processors. Each processor now has the information regarding the fitness values of all individuals and can evolve the next generation and compute the best individual in the population for the current generation. After processing for the last generation is completed, the best individual is added to the test set if it detects one or more faults out of the current set of 31 faults targeted. Each processor generates its own test set, and the test sets for all processors are identical, since the random number generators for all processors are initialized with the same seed. An exhaustive fault simulation is then done by all processors using this best-fit individual.

The final results in terms of the number of faults detected and the test set generated are identical to those obtained on a uniprocessor when this algorithm is used. Therefore, there is no degradation in the quality of the solution obtained through parallelization. This approach is similar to the one used in GATTO* [45]. One major difference is that, while we target 31 faults at a time using a fault parallel approach, only one fault is targeted at a time in GATTO*. Secondly, the same fitness function is used at all times in GATTO*, which may slow down execution. Thirdly, GATTO* required an additional central processor, which was not required in our thread-based implementation.

Analysis of PGATEST1

Let us assume that we have a population size of N in any generation and P processors. If $S(i)$ is the sequence length of an individual i, and LS is the cost of logic simulation per time frame, then the actual cost of evaluating the fitness of an individual i is given approximately by $S(i) \times LS$, and the cost of exhaustive fault simulation for individual i is given approximately by $Y(i) = S(i) \times \lceil \frac{RF(m)}{31} \rceil \times LS$, where $RF(m)$ is the number of faults remaining in the fault list at step m of the simulation process. Assume that we take g generations on average to find an individual which can be used for the exhaustive fault simulation. Also assume that we have k individuals eventually forming the entire test sequence (i.e., the overall simulation process has k steps). Let us assume for the rest of the analysis that the cost of evaluating the fitness of an individual is X units independent of i. It must be noted that this assumption is only approximate, since the fitness evaluation of an individual depends on the activity of the gates in the circuit and on the sequential depth. Let us also assume that the exhaustive fault simulation using the entire test sequence with k individuals takes Y time units. Then the sequential execution time on a uniprocessor is

$$T_1(N) = kgNX + Y.$$

Ignoring the cost of communication involved, the parallel execution time on P processors for algorithm PGATEST1 would be approximately given by

$$T_p(N) = kg\frac{N}{P}X + Y.$$

8.2. Parallel Genetic Algorithms for Automatic Test Generation

The above model is a bit simplistic in that it has not taken into account the increase in the population size that can occur during the GA evolution and assumes that any issues related to the sequential depth of the circuit are buried in the terms X and Y. The messages being communicated are small, and hence, the cost of message passing is approximately the cost of sending zero-byte messages over the communication network. If α is the cost of sending a zero-byte message between two processors, the communication cost during each generation of the GA is approximately $O(P)\alpha$. We will choose to ignore this cost in comparison to X (for tree-based reductions/broadcasts, the cost would be $O(log(P))\alpha$). Effectively, the parallelism available depends on how large the term $kgNX$ is, compared to the term Y and to the communication costs involved.

PGATEST2: Parallel-Migration-Based Genetic Search

In this algorithm, processors interact with each other by exchanging individuals, possibly the fittest individuals, among each other, at an interval determined by the epoch for the parallel GA. The hope is that with exchange of information between processors which are exploring different parts of the search space, there will be an improvement in the overall solution (the number of faults detected will be greater, the overall test set size will be smaller, etc.). This algorithm is illustrated in Fig. 8.19(b). We always assign the same number of individuals to each processor that are assigned in the sequential algorithm, regardless of the number of processors. Hence, each processor is assigned the same amount of workload as in the sequential case. Each processor is now following a different search path. There is a possibility that the processors will finish faster due to the migration of information between processors. This is indeed the case in our implementation as processors converge to their results faster with the added advantage that the quality of the result is improved in certain cases.

Each processor maintains an independent fault list and proceeds completely independently, generating its own test set. Periodically, each processor transmits the individual with the best fitness to a random processor. A queue of messages is maintained on each processor to receive migrating individuals from other processors. These individuals are absorbed into the local subpopulation as it evolves. This gives rise to the possibility of obtaining P different solutions to the problem using a randomized parallel genetic search.

There is a small caveat to maintaining independent fault lists though. If processors have independent fault lists, the relevance of migration of individuals to the improvement of the quality of the solution is questionable, since the fitness of each individual is determined relative to the current sample of 31 faults, which is dependent on the current list of undetected faults for a given processor. Hence, the fitness function is now local, and the fitness function on one processor may not have any meaning as far as the other processors are concerned. However, it turns out that this is not necessarily the case, and individuals which have good fitness values on one processor appear to have good fitness values on other processors too.

Analysis of PGATEST2

We will introduce a function $\kappa(P)$ to capture the gain obtained through parallel search on P processors using migration. $\kappa(P)$ effectively reduces the average number of generations required from g to $\frac{g}{\kappa(P)}$. Then the execution time on P processors for algorithm PGATEST2 is given by

$$T_p(N) = k \frac{g}{\kappa(P)} NX + Y.$$

We have ignored the asynchronous communication costs for migration of individuals between processors in the above formula. It should also be noted that there will be a different solution available on each processor, and hence, the values of k and Y will differ from one processor to another. In our implementation, we take the maximum of the execution times over all processors to be the total parallel execution time.

PGATEST3: Subpopulation-Based GA with Migration

This approach is similar to the previous approach, except that each processor starts with a population of $M = \frac{N}{P}$ individuals. This algorithm is illustrated in Fig. 8.19(c). One can expect speedups for two reasons. Each processor works on a subpopulation, and therefore, the population size being used is smaller, and each processor individually has less work to do. Also, due to migration of fit individuals from one processor to another, each processor can detect faults faster than if they were to run independent GAs with a reduced population size.

One possible disadvantage of this approach is that, for a circuit with a relatively small population size (corresponding to a circuit with a small number of primary inputs and small sequential depth), the algorithm is not very scalable. If the population size for the GA becomes too low, one can expect a degradation in the results obtained. However, if the population size is large, as is the case for large circuits, then this should not be problem.

Analysis of PGATEST3

Since the population assigned to each processor is now $\frac{N}{P}$, and since we have migration, the parallel execution time on P processors for algorithm PGATEST3 is given by

$$T_p(N) = k \frac{g}{\kappa(P)} \frac{N}{P} X + Y.$$

We can see that the parallel execution time is reduced due to migration and due to the smaller population size on each processor.

8.2.3 Experimental Results

The algorithms PGATEST1, PGATEST2, and PGATEST3 were implemented in the ProperCAD II environment. All implementations are portable to a wide variety of parallel platforms. All algorithms were implemented on a SUN-SparcServer

8.2. Parallel Genetic Algorithms for Automatic Test Generation

1000E, a shared memory multiprocessor with 8 processors and 512 MB of memory. In addition, the algorithm PGATEST1 was ported to a network of SUN-Sparc5 workstations and to the Thinking Machines CM-5, a distributed memory multicomputer with a SUN-Sparc1 processor on each node and 32 MB of memory per node, to demonstrate the portability of the implementations.

Each of the parallel genetic algorithms on completion was followed by a phase of ProperHITEC [170], executed in parallel, to drop any untestable faults among the faults left undetected by the genetic algorithm. A short time of one second was assigned per fault for this purpose. All three parallel genetic algorithms execute in the same fashion on a single processor. Table 8.5 shows the performance of the genetic algorithm running on a single processor. The circuits taken are part of the ISCAS89 benchmark suite [23]. A collapsed fault list for each circuit was taken as the initial fault list. All execution times are in seconds. **Det** refers to faults that were detected by the genetic algorithm, **Unt** refers to the untestable faults identified by ProperHITEC, and **Vec** refers to the test set size. It can be seen that the amount of time spent in the ProperHITEC phase is a small fraction of the overall execution time, and it helps in improving the efficiency of the overall ATG process.

Table 8.5. PGATEST1: Uniprocessor Run on a SUN-SPARCServer1000E Shared-Memory Multiprocessor

Circuit	Total Faults	PGATEST1 Det	PGATEST1 Time	ProperHITEC Unt	ProperHITEC Time	Total Time	Vec	Fault Coverage	ATG Efficiency
s298	308	265	67.70	22	25.01	92.70	166	0.86	0.93
s344	342	329	60.08	8	5.45	65.53	55	0.96	0.995
s349	350	335	63.53	10	5.50	69.03	77	0.96	0.995
s1196	1242	1236	457.2	3	0.1	457.3	516	0.995	0.997
s1238	1355	1281	776.9	72	1.46	778.4	576	0.945	0.999
s1494	1506	1297	1378	20	107	1485	286	0.86	0.87
s5378	4603	3243	23,864	114	1037	24,901	468	0.70	0.73
s35932	39,094	34,456	43,578	3984	1155	44,733	158	0.88	0.983

Table 8.6 shows the performance of PGATEST1 on eight processors. It can be seen that the results remain unchanged from the uniprocessor run and that execution times have been reduced significantly.

As a comparison, Table 8.7 shows the results for the same circuits when a parallel deterministic algorithm such as ProperHITEC is used in isolation [170]. One must look at the total execution time, the overall test set size in terms of the number of vectors, the fault coverage, and the fault efficiency when comparing different algorithms. One can observe that PGATEST1 performs better for some circuits such as s298, s344, s349, and s5378, while ProperHITEC performs better for circuits such as s1494 and s35932. We therefore believe that a hybrid strategy needs to be adopted in practice.

Table 8.6. PGATEST1: Eight-Processor Run on a SUN-SPARCServer1000E Shared-Memory Multiprocessor

Circuit	Total Faults	PGATEST1 Det	Time	ProperHITEC Unt	Time	Total Time	Vec	Fault Coverage	ATG Efficiency
s298	308	265	15.9	22	8.4	24.3	166	0.86	0.93
s344	342	329	13.9	8	3.5	17.4	55	0.96	0.995
s349	350	335	13.5	10	1.3	14.8	77	0.96	0.995
s1196	1242	1236	86.8	3	0.1	86.9	516	0.995	0.997
s1238	1355	1281	117.2	72	0.4	117.6	576	0.945	0.999
s1494	1506	1297	241.4	20	24	265.4	286	0.86	0.87
s5378	4603	3243	3804	114	198	4002	468	0.70	0.73
s35932	39,094	34,456	6328	3984	261	6589	158	0.88	0.983

Table 8.7. ProperHITEC: One- and Eight-Processor Runs on a SUN-SPARCServer1000E Shared-Memory Multiprocessor

Circuit	Total Faults	One Processor ProperHITEC				Eight Processor ProperHITEC			
		Det	Unt	Vec	Time	Det	Unt	Vec	Time
s298	308	265	22	306	974	262	22	322	268
s344	342	324	11	121	331	318	11	103	79
s349	350	335	10	137	182	335	10	124	46
s1196	1242	1239	3	453	24	1238	3	471	18
s1238	1355	1283	72	375	7.5	1281	70	373	21
s1494	1506	1453	50	1249	345	1451	50	1374	89
s5378	4603	3155	189	844	26,141	3130	188	1002	7492
s35932	39,094	35,090	3984	207	1354	35,043	3984	303	381

Table 8.8 shows the performance of PGATEST2 on eight processors. The speedups obtained with this algorithm were purely due to the migration of fit individuals from one processor to another. Nevertheless, good speedups were obtained. This algorithm takes longer to execute than PGATEST1 but provides better quality results in terms of the fault coverage obtained when compared to PGATEST1.

The algorithm PGATEST3, whose performance can be seen in Table 8.9, provides excellent execution times, but the test set sizes are larger than for the previous algorithms. The quality of the results in terms of the fault coverage and efficiency is usually worse than that given by PGATEST2 and better than for PGATEST1 in most cases. For PGATEST3, the GA population size is inversely proportional to the number of processors. Hence, for a large number of processors, this algorithm may not always perform well.

We will now report results of portability of our parallel algorithms on various parallel platforms. For this part of the study, we did not use the untestable fault

8.2. Parallel Genetic Algorithms for Automatic Test Generation 291

Table 8.8. PGATEST2: Eight-Processor Run on a SUN-SPARCServer1000E Shared-Memory Multiprocessor

Circuit	Total Faults	PGATEST2 Det	Time	ProperHITEC Unt	Time	Total Time	Vec	Fault Coverage	ATG Efficiency
s298	308	265	21.9	22	8.3	30.2	156	0.86	0.93
s344	342	329	19.7	8	3.6	23.3	55	0.96	0.995
s349	350	335	23.5	10	1.3	24.8	70	0.96	0.995
s1196	1242	1235	154.9	3	0.1	155	548	0.994	0.997
s1238	1355	1282	362.1	72	0.4	362.5	592	0.946	0.999
s1494	1506	1365	382.5	20	26	408.5	278	0.91	0.92
s5378	4603	3261	4619	112	194	4813	548	0.71	0.73
s35932	39,094	35,006	8137	3984	214	8351	226	0.896	0.997

Table 8.9. PGATEST3: Eight-Processor Run on a SUN-SPARCServer1000E Shared-Memory Multiprocessor

Circuit	Total Faults	PGATEST3 Det	Time	ProperHITEC Unt	Time	Total Time	Vec	Fault Coverage	ATG Efficiency
s298	308	264	10.2	22	8.4	19.6	184	0.857	0.928
s344	342	329	4.7	8	3.6	8.0	75	0.96	0.995
s349	350	335	4.8	10	1.3	6.1	90	0.96	0.995
s1196	1242	1235	96.2	3	0.1	96.3	612	0.994	0.997
s1238	1355	1271	113.9	72	0.4	114.3	586	0.94	0.991
s1494	1506	1348	71.2	20	26	77.2	312	0.895	0.91
s5378	4603	3214	3227	114	233	3460	542	0.70	0.72
s35932	39,094	34,832	5924	3984	238	6162	284	0.890	0.992

identification phase of ProperHITEC; hence, the fault coverage and efficiency numbers are not reported; we report only execution times and speedups on multiple processors. Results of one of the parallel algorithms, PGATEST1, will be reported, and results of the other parallel algorithms are similar. Results for PGATEST1 running on the SUN-SparcServer 1000E are shown in Table 8.10. The numbers of faults detected in the parallel and sequential cases are the same. Hence, the quality of the result was not affected, and significant speedups were obtained on 8 processors.

Table 8.11 shows the results obtained for the algorithm PGATEST1 on a network of 6 SUN-SPARC5 workstations. Table 8.12 shows results for the same algorithm on the Connection Machine-5 (CM5). These results demonstrate the performance and portability of the code on various parallel platforms. The quality of the results was unaffected, and good speedups were obtained.

Table 8.10. PGATEST1: Run Time (in seconds) and Speedup on SUN-SPARC-Server 1000E (Shared Memory Multiprocessor)

Circuit	Faults Total	Det	1 processor Time	Speedup 2 proc	4 proc	8 proc
s298	308	265	67.7	1.91	3.09	4.24
s344	342	329	60.1	1.89	3.31	4.30
s349	350	335	63.5	1.87	3.8	4.72
s1196	1242	1236	457.2	1.98	3.89	5.26
s1238	1355	1281	776.9	1.71	3.21	6.63
s1494	1506	1297	1378.1	1.92	3.46	5.71

Table 8.11. PGATEST1: Run Time (in seconds) and Speedup on a Network of Six Workstations (SUN-SPARC5)

Circuit	Faults Total	Det	1 processor Time	Speedup 2 proc	4 proc	6 proc
s298	308	265	119.8	1.98	3.33	3.61
s344	342	329	78.6	1.84	2.17	2.42
s349	350	335	76.4	1.82	2.63	3.48
s1196	1242	1226	680.1	1.26	1.80	2.28
s1238	1355	1281	2267.8	1.56	3.03	4.50
s1494	1506	1304	1887.5	1.92	3.07	4.36

Table 8.12. PGATEST1: Run Time (in seconds) and Speedup on the CM-5 (Distributed Memory Multicomputer)

Circuit	Faults Total	Det	1 processor Time	Speedup 2 proc	4 proc	8 proc	16 proc
s298	308	265	368.5	1.90	3.67	5.94	7.12
s344	342	329	232.8	1.88	3.35	5.65	6.80
s349	350	335	271.7	1.85	3.54	5.58	7.33
s1196	1242	1226	820.7	1.26	1.87	3.31	3.47
s1238	1355	1281	2860.7	1.55	3.87	5.51	5.72
s1494	1506	1304	3307.5	1.94	4.01	6.21	8.27

8.2.4 Conclusions

We have presented three parallel genetic algorithms for simulation-based test generation: PGATEST1, PGATEST2, and PGATEST3. PGATEST1 is a good algorithm in general, as it provides significant speedups without degradation in the quality of the results. It exploits the parallelism available corresponding to fitness evaluation of the individuals in a population. PGATEST2 exploits the search parallelism available through parallel randomized genetic search with migration of information. It attempts to improve the quality of the results and is a highly scalable implementation, as it can run over any number of processors irrespective of the population size. However, the speedups for this algorithm grow at a slower rate compared to the other algorithms. PGATEST3 provides a dual degree of parallelism by using a subpopulation-based GA approach to reduce the workload among the processors and by exploiting the benefits of the randomized migration strategy used. However, there is a likelihood of degradation in the quality of the results for this algorithm for very large numbers of processors due to a corresponding decrease in the population size.

SUMMARY

TOPICS STUDIED	KEY OBSERVATIONS AND POINTS OF INTEREST
Standard Cell Placement on a Network of Workstations (Wolverines) • Serial Algorithm • Parallel Algorithm • Results	• Migration of genetic material • Epoch length • Number of epochs
	• Master/slave organization • Minimal communication time
	• Speedup is close to ideal • Placement quality of serial algorithm is preserved in parallel algorithm • Speedup and placement quality preserved in a heterogeneous computing environment
Parallel GAs for Automatic Test Generation • Serial Algorithm • Parallel Algorithms • Results	• PGATEST1: Master/slave organization with synchronization • PGATEST2: Parallel-migration-based genetic search • PGATEST3: Subpopulation-based genetic algorithm with migration
	• Portable to a wide variety of parallel platforms, including shared-memory multiprocessors and networks of workstations • Good speedups obtained • Three algorithms trade off execution time and result quality

CHAPTER 9
at a glance

- Devising a GA for a new problem
  - Problem encoding
  - Fitness function
  - Type of GA
  - GA parameters
- GAs vs. conventional algorithms
- Concluding remarks

Chapter 9

CONCLUSION

Genetic algorithms have been used successfully to solve numerous different problems in VLSI design, layout, and test automation, as detailed in earlier chapters of this book. In the final chapter, we turn our attention to the task of devising a GA for a new problem that the reader might encounter in a related area.

Various different approaches may be used in solving problems in the area of electronic design automation. In many cases, a deterministic approach may be suitable, particularly if a fast heuristic approach is known. However, for many new problems, no simple deterministic techniques are available. In these cases, GAs may provide a good solution. Even for problems where deterministic methods are available, GAs may provide better results. For a new tool, development of a deterministic solution is likely to be much more difficult and time consuming. For this reason, the software developer may consider using a genetic approach for a quick initial solution and then add deterministic procedures as time permits.

It is our hope that solutions described in the preceding chapters will provide guidance to the reader about how GAs can be used to solve new problems. Decisions must be made about problem encoding, fitness function, type of GA, and GA parameters. We will discuss each of these aspects in turn before discussing how GAs differ from conventional algorithms and offering some concluding remarks.

9.1 Problem Encoding

Each individual in the GA population encodes a solution to the problem at hand, with some solutions being better than others. The way that a problem is encoded depends on the application, and this problem encoding is the first attribute that has to be decided when applying a GA to a new problem. Along with the problem encoding, a way to evaluate each individual must be devised. However, before a problem can be encoded, it must be clearly defined, and several approaches may be possible. For example, in automatic test generation, one approach involves generating individual vectors or sequences of vectors that detect as many faults as possible from a given fault sample. Another approach involves generating a vector to activate a specific target fault and then generating sequences to justify the required state and propagate the fault effects to the primary outputs. In either case, each individual in the GA population represents a vector or sequence of vectors, and fault simulation can be used to evaluate the individuals. The specific mapping of a problem solution to a GA individual also depends on the application. Strings of bits (1's and 0's) are commonly used, but characters from a larger alphabet may also be used. For automatic generation of test vectors, the encoding is very straightforward. The value applied to each pin must be logic 1 or logic 0, so there is a one-to-one mapping of bits from the GA string to bits in the test vector. For generation of test sequences, individual test vectors can be placed at adjacent positions along the GA string. For other problems, such as automatic placement, various different encodings may be devised.

In the case of standard cell placement, cells are permuted so that the total amount of wiring needed to interconnect them according to the given netlist is minimized. In the genetic encoding of the problem, the chromosome may be considered to be a continuous string similar to the Traveling Salesman Problem (TSP). Various rows in a standard cell layout may be concatenated in a row-major fashion so that in the chromosome representing the solution, both the cell identifiers and cell positions are indicated by the alleles in the chromosome. In order to ensure that each offspring generated on a permutation class of a chromosome is a feasible solution, the crossover operator must delete all duplicate alleles and retain all unique alleles pertaining to the standard cells in a netlist. For standard cell placement, three crossover operators (PMX, order, and cycle) have been described which always generate conflict-free feasible offspring. However, the timing overhead in these crossover operations (between linear and quadratic in the number of cells) significantly impairs the speed of operation of the genetic-based placement algorithm. Thus, the main advantage of the proposed chromosome in the standard cell placement algorithm (GASP) is that the chromosome (genotype) very closely ties with the actual physical layout (phenotype), and the fitness function accurately measures the quality of the layout generated. However, the accompanying overhead in generating feasible offspring by any of these crossover operators becomes prohibitively large for standard cell layouts with over 10,000 cells. On the contrary, a superior encoding scheme is used for the macro cell placement problem that requires relatively less time to generate offspring using the proposed crossover operator. Readers may carefully experiment on various types of encoding schemes discussed in this book and develop an intuition on how to encode the solution of a problem and how to develop efficient crossover operators that are fast, yet that can retain good schemata, e.g., in transferring clusters of cells pertaining to suboptimal solutions from a parent to the offspring without mutilating the cell clusters.

9.2 Fitness Function

A good fitness function is critical in obtaining good results from a GA. In terms of solution quality, more information is better. For example, in automatic test generation, when the fitness function targets the specific tasks of fault activation, state justification, and fault effect propagation and includes information about controllabilities, observabilities, and circuit activity, as well as fault detection, much higher fault coverages are obtained than when a simpler fitness function is used that mainly targets fault detection. However, execution times can be longer since more computations are involved in evaluating each individual, and indeed this is the case for test generation. To speed up the fitness computation, approximate fitness functions can be used. An example for automatic test generation is the use of logic simulation instead of fault simulation for evaluation of candidate tests. Alternatively, a fault simulator may be used with a small sample of faults rather than the entire fault list.

9.3 Type of GA

One decision that must be made in applying a GA to a new problem is what type of GA to use: a simple GA or a steady-state GA. The simple GA was used for problems in FPGA mapping, test generation, test sequence compaction, and power estimation, whereas steady-state GAs were used for problems in placement and routing. The simple GA allows for better exploration of the search space, while the steady-state GA enables better exploitation of good solutions that already exist in a GA population. For placement and routing, the GA is run once to find a good placement or net routing, and it is important to ensure that the best solutions found are not lost. Minor improvements can often be made to a good solution using a hill-climbing algorithm, and these improvements can be made with the GA if a steady-state algorithm is used. However, fewer new individuals are explored with this approach. For automatic test generation, the GA is run repeatedly to find vectors or sequences to add to the test set, and the optimal solution does not have to be found. Thus, exploration is more important, and for this reason, a simple GA was used. For any given problem, it may be possible to determine in advance whether exploration or exploitation is most important and then choose the appropriate type of GA. If the answer is not clear cut, the software may easily be implemented to allow for either option, and then experiments may be performed to determine the best approach.

9.4 GA Parameters

An important problem in genetic algorithms is how to optimize various tuning parameters, such as crossover rate, inversion rate, mutation rate, population size, number of generations needed to get good results, etc. Unlike in simulated annealing or graph-based algorithms, GAs have many bells and whistles, and one has to carefully tune these parameters so that the collective parameter settings can yield good quality solutions by the GA. This is a nontrivial task and frequently requires a separate optimization algorithm to determine the optimal parameter settings. In the case of the GA-based standard cell placement algorithm GASP, a meta-level genetic optimization process has been used to show how one can optimize these parameters by running the GA on a smaller input set. The first-level GA tends to characterize the search space of the problem domain and tries to locate areas where good solutions lie, while the second, or meta-level, GA selects new parameter sets carefully such that sufficient information about the optimal settings of these parameters can be obtained quickly. In the meta-level GA discussed in Chapter 3, it was found that order crossover with deterministic reduction gives superior performance to other crossover operators and reduction criteria. The meta-level GA selects the crossover rate to be 33% and the mutation rate to be 0.5%.

Solution quality depends heavily upon the GA population size and number of generations. Increasing these parameters allows for better exploration of the search space, while decreasing them reduces the execution time. It may be the case that, beyond a particular population size or number of generations, no improvements or only minor improvements in solution quality are obtained, and that any improvements do not warrant the extra execution time required. A low mutation probability has been recommended in the past [79]. Experiments reported in Chapter 6 for automatic test generation confirm that low mutation rates approximately equal to the inverse of the population size work well, and that doubling the mutation rate or reducing it by one-half does not affect solution quality in most cases.

Another important consideration in devising a new GA is what GA operators (e.g., roulette wheel selection vs. tournament selection) will be used. Experiments described in Chapter 6 on automatic test generation demonstrated that selection and crossover are the most important operators, and mutation is secondary. A wider variation of fault coverages was obtained with different selection and crossover schemes than was observed for different mutation rates.

Some applications require problem-specific crossover and mutation operators to be developed to ensure that valid solutions are generated. This was the case with order crossover, PMX crossover, and cycle crossover developed for automatic placement and described in Chapter 3. Even when the traditional crossover operators are adequate, heuristic crossover operators may provide better solutions. Heuristic crossover was used in the CRIS automatic test generator for sequential circuits, and very good results were obtained [195]. Another heuristic crossover approach was used for automatic test generation in combinational circuits, and higher fault coverages were obtained than when the traditional crossover schemes were used [179].

All of the operations performed in a GA depend upon a random number generator. Therefore, different results can often be obtained by seeding the random number generator differently. Thus, if time permits, the GA may be run several times with different random seeds, and the best solution may then be selected.

While good solutions may be obtained by using the same GA parameters and operators that were used for another application, better results can often be obtained by tuning them to a specific application. This can only be done through experimentation, i.e., by performing several trial runs on representative benchmarks using various different parameters and operators.

9.5 Genetic Algorithms vs. Conventional Algorithms

Genetic algorithms are unique in the sense that they do not optimize a problem by directly applying transformation operations on the physical representation of a problem. For other algorithms, such as simulated annealing, the transformation is applied on the current solution using some well-defined criteria. Consider the problem of automatic placement, in which the transformation involves moving around cells in a layout. Conventional algorithms use iterative refinement techniques in which transformation operators, such as pair-wise cell position exchange, arbitrary movement of a cell, and mirroring of a cell, are applied on the current solution to generate a sequence of incrementally improved solutions until no further improvements are possible. Thus, these algorithms may require the application of millions of transformation operations, and at each step, a very small amount of time is required to recompute the objective function pertaining to the quality of the chip layout generated.

In contrast to these algorithms, a GA applies its transformations to its chromosome or genotype only. Thus, a user must understand the problem characteristics and select a suitable encoding scheme that will relate the transformations on the genotype to the actual improvement in solution (phenotype) quality. Thus, it is of paramount importance that the encoding scheme and the crossover, mutation, and inversion operators are highly efficient so that the algorithm does not waste too much time in eliminating any inconsistencies between the genotype and the phenotype of a problem. For example, in GASP, the crossover operators take time of $O(n)$ to remove these inconsistencies, where n is the number of cells, and significantly slow it down for large problem sizes having over ten thousand standard cells.

Another major shortcoming of GAs is that the crossover operator modifies several genes every time it is applied. Thus, the timing overhead of the evaluation process in each generation is very high. At a later phase in a GA, when the solution quality is close to optimal, many offspring are generated and evaluated before a better offspring is found. This intrinsic limitation of the GA renders it incapable of solving large standard cell placement problems in near-linear time, as is performed by the advanced version of the simulated annealing algorithm TimberWolf.

This timing overhead issue can be tackled in two ways in a GA. In a steady-state GA, the simplest way is to carefully reduce the population size and also the crossover rate, while simultaneously increasing the mutation rate after the GA reaches a point when very few offspring are entered in the new population. Since the mutation process mimics the pairwise exchange of cells in simulated annealing, and very few nets are disturbed due to mutation, the evaluation time is very small. In Chapter 3, the SAGA placement algorithm demonstrates how to combine a GA with simulated annealing so that the overall run time of a pure GA can be reduced. Further research is required in this direction, including development of superior data structures that can use pointers to those cells that are perturbed due to a crossover application. This approach may significantly reduce the wasted efforts in scanning through all cells after an offspring is generated.

The second way to accelerate the speed of a GA is to exploit its intrinsic parallelism. Since a GA works with a population of solutions and it tends to improve the quality of these solutions through iterative creation of better offspring over several generations, one can judiciously distribute the population of solutions over a number of workstations. In Chapter 8, distributed GAs for standard cell placement (Wolverines) and for automatic test generation (PGATEST) have been discussed. It was shown that a linear acceleration can be achieved by running the GA on a distributed network of workstations. Several communication patterns have been studied, and also experimentation was done on changing the epoch rate, i.e., how often data are communicated between the independently running processes. The approach used allows the overall search process over different workstations to be highly coordinated, and the communication overhead in migrating good solutions between workstations is very low. Intrinsic parallelism is a strong feature of GAs, and future research is needed to verify how the distributed GA performs over a large number of processors.

GAs also differ from other algorithms in another respect. Since a GA works on a population of solutions, it has the intrinsic ability to explore various areas of the search space concurrently. It exploits its current search information to intensively explore those regions of the search space that may quickly give good solutions and abandons other regions that are unlikely to yield good solutions. This adaptive feature of a GA enables it to be more intelligent than conventional algorithms. Considerable research is necessary to fully exploit the adaptivity of a GA to the search space and to develop suitable heuristics that may work in conjunction with the traditional GA to give fast and high-quality solutions.

9.6 Concluding Remarks

Genetic algorithms are little understood because, unlike simulated annealing, which can be rigorously analyzed using a Markovian model, the genetic transformations and iterations through generations are difficult to model by known analytical techniques. Although some probabilistic analyses were made by John Holland and others and schema theorems were proposed to explain the genetic optimization process, no clear and insightful analytical techniques yet exist that can explain rigorously the convergence of genetic algorithms or how many iterations are required to obtain the globally optimal solution or a solution within some predefined infinitesimally close range of the globally optimal solution.

Simulated annealing has enjoyed a tremendous success in solving VLSI/CAD problems after rigorous mathematical models were formulated to explain the behavior of the algorithm and how to select the tuning parameters, such as initial temperature, cooling schedule, inner-loop stopping criteria, etc., in such a way that run time of the algorithm can be reduced significantly without compromising the quality of the solution. An early version of Timberwolf (v3.2) has been found to be comparable, both in terms of run time and placement quality, to the GA-based standard cell placement algorithm GASP. However, subsequent versions of Timberwolf have applied mathematically derived parameter settings, and these versions are found to run in linear time, yielding the best quality placement solutions for problems involving several hundred thousand cells. Currently, no GA-based placement algorithm can handle such gargantuan cell-placement problems. Nevertheless, GAs have wider applicability than simulated annealing because encoding of very complex optimization problems is relatively easy with GAs. As an example, no successful attempt at applying simulated annealing to automatic test generation has been reported in the literature, while GAs have been quite effective for this application.

Since a GA applies very complex optimization techniques, it is unlikely that a comprehensive mathematical model can be built that will enable us to optimally select the GA tuning parameters or that will give a clear relationship between the timing overhead and the parameter settings. In spite of this lack of a comprehensive mathematical model, there is a large body of empirical data in various different disciplines ranging from chemical database searches to music composition showing that GAs work. Based on results from various GA applications, a combined analytical and empirical approach can be developed to explain the behavior of a GA and also how to estimate its run time from the various parameter settings. This book introduces the basic techniques that will assist readers in applying GAs to various problem domains. However, in order to exploit the full potential of a GA process, readers are encouraged to develop analytical models in blending with empirical studies.

Theoretical study can also be made with a view toward developing accurate mathematical models for characterizing the distributed GA running on a network of workstations. Wolverines and PGATEST, described in Chapter 8, merely demonstrate how GAs can run concurrently on various processors, and an almost linear speed-up can be achieved. Appropriate mathematical models should be built to explain parallel GAs, their convergence time, and also their solution quality vis-a-vis those of a GA running on a uniprocessor.

We strongly believe that GAs are not merely a voodoo art where experimentalists painstakingly personalize the GA for each type of problem separately. GAs are powerful optimization techniques that are, at this time, lacking in full mathematical understanding. In the absence of appropriate mathematical models, a meta-genetic-based approach was developed for accurate parameter setting, as discussed in Chapter 3. This type of empirical approach may not work well for many problems. It is true that a GA is not a deterministic algorithm, and its complexity analysis is almost intractable. We started examining GAs to ascertain their capability in solving constrained combinatorial optimization problems in VLSI/CAD. We have successfully applied GAs to various problems as discussed in this book. We have shown how to encode a physical problem, how to build efficient GA operators (crossover, mutation, and inversion), and then how to tune various GA parameters so that we can obtain high-quality solutions. Yet our work falls short in giving mathematically insightful explanations about the convergence model of a GA, about its complexity, and about the question: For a given VLSI/CAD problem, can a GA guarantee convergence to the globally optimal solution? We have merely developed

9.6. Concluding Remarks

engineering tools and have compared them with other successful commercial and academic tools. We have encountered many roadblocks in applying GAs to these problems, and our tools, mostly developed on an ad hoc and exploratory basis, have several shortcomings. Nevertheless, on the whole, our experimentation with GAs has shown that this ill-understood heuristic method can definitely make significant contributions in solving extremely large and complex combinatorial optimization problems that plague the fields of VLSI design, layout automation, and chip testing. Readers are encouraged to investigate in greater depth some of these problems and apply these ideas in solving problems in various other domains. Readers with strong mathematical backgrounds are especially encouraged to develop theoretical frameworks of GAs so that many of the existing shortcomings can be surmounted by employing these models. We sincerely hope that in the future, GAs will prove to be a general-purpose powerful heuristic method for solving a wider class of engineering and scientific problems.

SUMMARY

TOPICS STUDIED

KEY OBSERVATIONS AND POINTS OF INTEREST

Devising a GA for a New Problem

- Problem Encoding
- Fitness Function
- Type of GA
- GA Parameters

- Clearly define the problem
- Devise a mapping from GA individual to problem solution
- Devise a way to evaluate each individual

- Solution quality is improved by more information
- Use approximations to speed up fitness computations

- Simple GA allows for better exploration of search space
- Steady-state GA allows for better exploitation of good individuals

- Meta-level GAs are useful in optimizing GA parameters
- Selection and crossover operators are most important
- Problem-specific crossover and mutation operators may be useful
- Parameters can be tuned using representative benchmarks

GLOSSARY

allele
: the value of a gene in a GA chromosome

automatic test generation
: the production of a set of test patterns to evaluate a circuit design or fabricated integrated circuit chip; e.g., test patterns are applied to fabricated chips in order to identify or diagnose defective chips

chip fabrication
: the physical process of constructing a VLSI circuit on silicon

chromosome
: also called a string, a sequence of characters or bits that represents an individual in the GA population

Classifier System (CS)
: a class of evolutionary algorithms that take a set of inputs and produce a set of outputs which indicate some classification of the inputs

constructive algorithm
: an optimization algorithm that starts with empty sets and builds a solution by adding elements

crossover
: a key operator used in the GA to create new individuals by combining portions of two parent strings

crossover probability
: probability of performing a crossover operation, i.e., the ratio of the number of offspring produced in each generation to the population size

cut set
: the set of edges cut by a partition

cycle crossover
: a crossover scheme defined for standard cell placement

distributed algorithm
   a parallel algorithm implemented on a group of distributed processors

DNA
   a commonly-used acronym that refers to DeoxyriboNucleic Acid, a double-stranded macromolecule of helical structure. There are four nucleotide bases (A, T, C, and G) in DNA, and their sequences in DNA molecules determine the characteristics of an organism.

duplicate check
   a procedure often used in steady-state GAs to check whether a new offspring is the same as an existing individual in the GA population

dynamic test sequence compaction
   a process that is performed during test generation which results in the final test set having fewer vectors for a given level of fault coverage

epoch length
   the number of generations between two successive migrations in a parallel GA

evaluation
   the computation performed to determine the quality of an individual in the GA population

evaluation function
   also called fitness function, the function used to compute a fitness value to assign to each member of the GA population

Evolution Strategy (ES)
   a class of evolutionary algorithms that rely on mutation as the search operator and a population of size one

Evolutionary Algorithms (EAs)
   a broad class of computational models that employ evolution as a key element in the design and implementation of computer-based problem solving. The major classes of evolutionary algorithms are: Genetic Algorithms (GAs), Classifier Systems (CS), Evolution Strategies (ES), Evolution Programming (EP), and Genetic Programming (GP). They all share a common conceptual base of simulating the evolution of individual structures via processes of selection, mutation, and reproduction. The processes depend on the fitness (performance) of the individual structures which aggregate to form a population.

Evolutionary Programming (EP)
   a class of evolutionary algorithms that differ from genetic algorithms by dispensing with both genomic or chromosomal representations and with crossover as a reproduction operator

exploitation
: the process of using information gathered from previously visited points in the search space to determine which places might be profitable to visit next

exploration
: the process of visiting entirely new regions of a search space to determine if anything promising may be found; while exploitation concentrates on previously visited points to maximize the gain, exploration attempts to find new regions in the search space where the gain can be high

fault
: an abstract logical representation of a physical defect

fault excitation
: application of test vectors to the primary inputs of a circuit so that the logic value at the fault site is different for the fault-free circuit and the faulty circuit

fault propagation
: application of test vectors to the primary inputs of a circuit so that the effects of a fault can be observed at the primary outputs

fitness
: number assigned to an individual in the GA population to represent the quality of the corresponding solution

fitness function
: see evaluation function

fitness scaling
: adjusting of fitness values so that the differences between individuals are reasonable

FPGA
: field-programmable gate array

freezing point
: the stopping point in a simulated annealing algorithm

function optimization
: for an n-input, single-output function, function optimization is a technique to find the set of input parameters that maximize the output function

gene
: a character or symbol in a GA chromosome

generation
> the process of creating a new population from an existing population through reproduction

generation gap
> the fraction of individuals in the population that are replaced from one generation to the next

Genetic Algorithm (GA)
> a class of evolutionary algorithms that typically use fixed-length character strings to represent their genetic information, together with a population of individuals that undergo crossover and mutation in order to find interesting regions of the search space

Genetic Programming (GP)
> a class of evolutionary algorithms that typically use nonfixed length character strings that encode possible solutions to the problem. GP usually does not use mutation as a genetic operator.

genotype
> the encoding of a GA individual

global routing
> the assignment of each interconnection line in a circuit to specific routing regions

Hamming distance
> the number of bits from two binary strings that have different values

hybrid GA
> a GA that performs local optimization in every generation

inheritance
> passing of features from one generation to the next

inversion
> operator that changes the order of bits or characters in a GA string without changing the solution represented by that string

inversion probability
> probability of performing an inversion operation on an individual

iterative improvement algorithm
> an optimization algorithm that starts with an initial solution and repeatedly modifies it in an attempt to obtain a better solution

layout synthesis
> the transformation of a logic-level circuit description into a layout description containing a placement of circuit components, which are instances of the library primitives, and an associated routing of the component interconnections that meet the area and delay constraints of the design

logic synthesis
> the transformation of a register transfer level circuit description into an optimized gate-level circuit that uses elements from a given component library

macro cell placement
> construction of a layout of logic modules that are typically rectilinear in shape, but not of a fixed height or width; information about location and orientation of each module is included, and an acceptable placement will meet constraints on wire length and layout area

Manhattan geometry
> layout in which only horizontal and vertical interconnection lines are used

meta-level GA
> a GA used to optimize the parameters for another GA

migration
> the transfer of the genes of an individual from one subpopulation to another

min-cut placement algorithm
> a procedure for automatic placement of circuit components that uses repeated partitioning in an attempt to obtain a layout that minimizes routing

mutation
> an incremental change made to a member of the GA population

mutation probability
> the probability of mutating each gene in a GA chromosome

netlist
> a circuit description that contains the interconnections between terminals of all modules

offspring
> an individual resulting from crossover of two parent individuals and possible mutation

one-point crossover
: a crossover scheme in which one crossing point is randomly chosen, and the two parents are crossed at that point in creating the offspring

order crossover
: a crossover scheme defined for standard cell placement

parallel pattern simulation
: simulation of a number of test vectors in parallel using the bitwise parallelism of the computer word

parallelization by data decomposition
: a parallel GA in which a single population is maintained but individuals are evaluated in parallel on separate processors

parent
: an individual selected from the GA population for reproduction in the next generation

partitioning
: the process of separating the objects in a set into two or more subsets

peak n-cycle power
: the maximum average power of a contiguous sequence of n clock cycles, assuming that the initial state of the machine is fully controllable

peak single-cycle power
: the maximum total power consumed during one clock cycle

peak sustainable power
: the maximum average power that can be sustained indefinitely over many clock cycles

peak switching frequency per node
: average frequency of peak switching activity of circuit nodes, i.e., ratio of the number of 0-to-1 and 1-to-0 transitions on all nodes to the total number of capacitive nodes

phenotype
: the decoded meaning of a GA chromosome

photolithography
: the process of applying a pattern to the surface of a silicon wafer during chip fabrication, e.g., to define areas for ion implantation or metalization

placement
: the assignment of circuit components to specific locations on a substrate

PMX
 a crossover scheme defined for standard cell placement

population
 a collection of character or bit strings (chromosomes) that are candidate solutions to a given problem

population size
 the number of individuals in the GA population

power estimation
 a procedure to determine the approximate power dissipation of a circuit

ranking
 sorting individuals in the GA population from best to worst

roulette wheel selection
 a selection scheme in which each individual is selected with a probability proportional to its fitness

routing
 the process of connecting the pins of placed component modules in order to implement a logic design

scaling
 adjusting the fitness values of the individuals in a GA population to modify the differences in selection probabilities

schema
 a partial solution having a specific set of values assigned to a subset of genes in a chromosome

segmentation
 partitioning of a circuit into a number of subcircuits that are not necessarily disjoint

selection
 a genetic operator used to choose individuals for reproduction

simple GA
 also called total replacement GA, a GA characterized by nonoverlapping populations, in which every individual in the population is replaced from one generation to the next

simulated annealing algorithm
 a stochastic optimization procedure that starts with a random initial solution and makes iterative improvements; all changes that reduce the cost are accepted, and changes that increase the cost are accepted according to a probabilistic decision function that mimics the metal annealing process

single stuck-at fault
> a fault that causes one node in the circuit to be tied to logic one or logic zero

speedup
> the ratio of the time required by one procedure to the time required by an alternative approach, e.g., for a parallel implementation

standard cell placement
> construction of a layout of modules taken from a standard cell library that includes information about location and orientation of each module; an acceptable placement will meet constraints on wire length and layout area

state justification
> a search for a sequence of test vectors to bring the circuit from the current state to a desired state

steady-state GA
> a GA characterized by overlapping populations, in which only a small number of individuals are replaced from one generation to the next

Steiner problem in a graph
> given a graph and a designated subset of vertices, the problem of finding a minimum cost subgraph that spans the designated vertices

stochastic universal selection
> a selection scheme in which each individual is selected with a probability proportional to its fitness, and in which N individuals are selected at the same time, where N is the population size

string
> also called a chromosome, a sequence of characters or bits that represents an individual in the GA population

subpopulation-based GA
> a parallel GA in which the population is divided into subpopulations, which are evolved on separate processors

survival of the fittest
> competition that results in the best features surviving into the next generation and the bad features being weeded out

technology mapping
> a phase in the logic synthesis procedure in which the equations and memory elements of a circuit description are implemented by choosing gates from a fixed library of primitive components

test sequence compaction
> the process of reducing the number of vectors in a test sequence while maintaining a comparable fault coverage

TimberWolf
> a layout synthesis tool developed at the University of California, Berkeley that uses a simulated annealing algorithm

total replacement GA
> see simple GA

tournament selection
> a selection scheme in which two individuals are removed from the population, and the best one of the two is selected; can be performed with or without replacement of the two individuals back into the original population

two-point crossover
> a crossover scheme in which two crossing points are randomly chosen, and the two parents are crossed at those points in creating the offspring

uniform crossover
> a crossover scheme in which each gene position of the two parents is crossed with some probability, typically one-half, in creating the offspring

BIBLIOGRAPHY

[1] E. Aarts and J. Korst, *Simulated Annealing and Boltzmann Machines: A Stochastic Approach to Combinatorial Optimization*, Chichester, UK: John Wiley & Sons, 1989.

[2] M. Abramovici, M. A. Breuer, and A. D. Friedman, *Digital Systems Testing and Testable Design*. New York: Computer Science Press, 1990.

[3] P. Adamidis, "Review of parallel genetic algorithms bibliography," Tech. Report, Aristotle University of Thessaloniki, Thessaloniki, Greece, 1994.

[4] Advanced Micro Devices, "The AM2910, a complete 12-bit microprogram sequence controller," in *AMD Data Book*, Sunnyvale, CA: AMD Inc., 1978.

[5] G. A. Agha, *Actors: A Model of Concurrent Computation in Distributed Systems*, Cambridge, MA: The MIT Press, 1986.

[6] V. D. Agrawal, K. T. Cheng, and P. Agrawal, "A directed search method for test generation using a concurrent simulator," *IEEE Trans. Computer-Aided Design*, vol. 8, no. 2, pp. 131–138, Feb. 1989.

[7] A. V. Aho, J. E. Hopcroft, and J. D. Ullman, *Data Structures and Algorithms*, Reading, MA: Addison-Wesley, 1983.

[8] C. J. Alpert and A. B. Kahng, "Multiway partitioning via geometric embeddings, orderings, and dynamic programming," *IEEE Trans. Computer-Aided Design*, vol. 14, no. 11, Nov. 1995.

[9] C. J. Alpert and S.-Z. Yao, "Spectral partitioning: The more eigenvectors, the better," *Proc. Design Automation Conf.*, pp. 195–200, 1995.

[10] G. R. Andrews, "Paradigms for process interaction in distributed programs," *ACM Computing Surveys*, vol. 23, no. 1, pp. 49–90, March 1991.

[11] Y. P. Aneja, "An integer linear programming approach to the Steiner problem in graphs," *Networks*, vol. 10, pp. 167–178, 1980.

[12] B. Babba, M. Crastes, and G. Saucier, "Input driven synthesis on PLDs and PGAs," *Proc. European Design Automation Conf.*, pp. 48–52, 1992.

[13] J. E. Baker, "Reducing bias and inefficiency in the selection algorithm," *Proc. Second Int. Conf. Genetic Algorithms*, pp. 14–21, 1987.

[14] J. E. Beasley, "An SST-based algorithm for the Steiner problem in graphs," *Networks*, vol. 19, pp. 1–16, 1989.

[15] J. E. Beasley, "OR-Library: Distributing test problems by electronic mail," *Journal of the Operational Research Society*, vol. 41, pp. 1069–1072, 1990.

[16] M. Beardslee, J. Burns, A. Casotto, M. Igusa, F. Romeo, and A. Sangiovanni-Vincentelli, "MOSAICO: User's manual," Department of Electrical Engineering and Computer Sciences, University of California, Berkeley, 1990.

[17] R. Bevacqua, L. Guerrazzi, F. Ferrandi, and F. Fummi, "Implicit test sequences compaction for decreasing test application cost," *Proc. Int. Conf. Computer Design*, 1996.

[18] S. N. Bhatt, F. R. K. Chung, and A. L. Rosenberg, "Partitioning circuits for improved testability," *Advanced Research in VLSI: Proc. 4-th MIT Conf.*, pp. 91–106, 1986.

[19] T. Boseniuk and W. Ebeling, "Boltzmann-, Darwin-, Haeckel-strategies in optimization problems," *Proc. Workshop on Parallel Problem Solving from Nature*, pp. 430–444, Springer-Verlag, 1991.

[20] M. A. Breuer, "A random and an algorithmic technique for fault detection test generation for sequential circuits," *IEEE Trans. Computers*, vol. 20, no. 11, pp. 1364–1370, Nov. 1971.

[21] M. A. Breuer, "Min-cut placement," *Journal of Design Automation and Fault-Tolerant Computing*, vol. 4, pp. 343–382, Oct. 1977.

[22] F. Brglez and H. Fujiwara, "A neutral netlist of 10 combinational benchmark designs and a special translator in Fortran," *Int. Symp. Circuits & Systems*, June 1985.

[23] F. Brglez, D. Bryan, and K. Kozminski, "Combinational profiles of sequential benchmark circuits," *Int. Symposium on Circuits & Systems*, pp. 1929–1934, May 1989.

[24] S. D. Brown, R. J. Francis, J. Rose, and Z. G. Vranesic, *Field-Programmable Gate Arrays*, Boston, MA: Kluwer Academic Publishers, 1992.

[25] T. N. Bui, and B. R. Moon, "A fast and stable hybrid genetic algorithm for the ratio-cut partitioning problem on hypergraphs," *Proc. ACM/IEEE Design Automation Conf.*, pp. 664–669, June 1994.

[26] W. S. Carter, K. Duong, R. H. Freeman, H. Hsieh, J. Y. Ja, J. E. Mahoney, L. T. Ngo, and S. L. Sze, "A user programmable reconfigurable logic array," *Proc. IEEE Custom Integrated Circuits Conf.*, pp. 233–235, 1986.

[27] A. Casotto, F. Romeo, and A. Sangiovanni-Vincentelli, "A parallel simulated annealing algorithm for the placement of macro cells," *IEEE Trans. Computer-Aided Design*, vol. CAD-6, no. 5, Sept. 1987.

[28] S. T. Chakradhar and A. Raghunathan, "Bottleneck removal algorithm for dynamic compaction and test cycles reduction," *Proc. European Design Automation Conf. (EURO-DAC)*, 1995.

[29] H. M. Chan, P. Mazumder, and K. Shahookar, "Macro-cell and module placement by genetic optimization with bit-map represented crossover operators," *Integration: An International VLSI Journal*, pp. 49–77, Dec. 1991.

[30] H. Chan and P. Mazumder, "Genetic algorithms and graph partitioning," *Proc. AAAI Conf.*, Sydney, Australia, Nov. 1994.

[31] H. Chan and P. Mazumder, "A systolic architecture for high-speed hypergraph partitioning using a genetic algorithm," Chapter in *Progress in Evolutionary Computation*, vol. 956, Heildelberg: Springer-Verlag, pp. 109–126, 1995.

[32] S. Chang "The generation of minimal trees with a Steiner topology," *J. ACM*, vol. 19, no. 4, pp. 699–711, Oct. 1972.

[33] K. Chen, J. Cong, Y. Ding, A. B. Kahng, and P. Trajmar, "DAG-Map: Graph-based FPGA technology mapping for delay optimization," *IEEE Design and Test of Computers*, pp. 7–20, 1992.

[34] N. P. Chen "New algorithms for Steiner tree on graphs," *Proc. Int. Symp. Circuits and Systems*, pp. 1217–1219, 1983.

[35] C. Cheng and E. S. Kuh, "Module placement based on resistive network optimization," *IEEE Trans. on Computer-Aided Design*, vol. 3, no. 7, pp. 218–225, 1984.

[36] K. T. Cheng and V. D. Agrawal, "An economical scan design for sequential logic test generation," *Proc. 19th Int. Symp. Fault-Tolerant Computing*, pp. 28–35, 1989.

[37] W. -T. Cheng, "The BACK algorithm for sequential test generation," *Proc. Int. Conf. Computer Design*, pp. 66–69, Oct. 1988.

[38] V. Chickermane and J. H. Patel, "An optimization based approach to the partial scan design problem," *Proc. Int. Test Conf.*, pp. 377–386, 1990.

[39] H. Cho, S.-W. Jeong, and F. Somenzi, "Synchronizing sequences and symbolic traversal techniques in test generation," *Journal of Electronic Testing: Theory and Application*, vol. 4, no. 1, pp. 19–31, 1993.

[40] S. Chopra, Edgar R. Gorres, and M. R. Rao, "Solving the Steiner tree problem on a graph using branch and cut," *Operations Research Society of America Journal of Computing*, vol. 4, no. 3, pp. 320–335, 1992.

[41] T. Chou and K. Roy, "Statistical estimation of sequential circuit activity," *Proc. Int. Conf. Computer-Aided Design*, pp. 34–37, 1995.

[42] T. Chou and K. Roy, "Accurate power estimation of CMOS sequential circuits," *IEEE Trans. VLSI Systems*, vol. 4, no. 3, pp. 369–380, Sept. 1996.

[43] J. P. Cohoon, S. U. Hegde, W. N. Martin, and D. S. Richards, "Punctuated equilibria: a parallel genetic algorithm," *Proc. Second Int. Conf. Genetic Algorithms*, J. J. Grefenstette, editor, pp. 148–154, Hillsdale, NJ: Lawrence Erlbaum Associates, 1987.

[44] J. P. Cohoon, S. U. Hegde, W. N. Martin, and D. S. Richards, "Distributed genetic algorithms for the floor plan design problem," *IEEE Trans. Computer Aided Design*, vol. 10, no. 4, pp. 483–492, April 1991.

[45] F. Corno, P. Prinetto, M. Rebaudengo, M. Sonza Reorda, and E. Veiluva, "A portable ATPG tool for parallel and distributed systems," *Proc. VLSI Test Symp.*, pp. 29–34, 1995.

[46] F. Corno, P. Prinetto, M. Rebaudengo, and M. Sonza Reorda, "A genetic algorithm for automatic test pattern generation for large synchronous sequential circuits," *IEEE Trans. Computer Aided Design*, vol. 15, no. 8, pp. 991–1000, August 1996.

[47] W. M. Dai, B. Eschermann, E. S. Kuh, and M. Pedram, "Hierarchical placement and floorplanning in BEAR," *IEEE Trans. Computer-Aided Design*, pp. 1335–1349, vol. 8, no. 12, 1989.

[48] C. R. Darwin, *On the Origin of Species by Means of Natural Selection*, London: John Murray, 1859.

[49] L. Davis "Applying Adaptive Algorithms to Epistatic Domains," *Proc. Int. Joint Conf. on Artificial Intelligence*, 1985.

[50] L. Davis, *Handbook of Genetic Algorithms*, New York: Van Nostrand Reinhold, 1991.

[51] L. Davis (ed.), *Genetic Algorithms and Simulated Annealing*, London: Pitman Press, 1987.

[52] T. E. Davis and J. C. Principe, "A simulated annealing like convergence theory for the simple genetic algorithm," *Proc. Fourth Int. Conf. Genetic Algorithms*, pp. 174–181, 1991.

[53] K. A. De Jong, "An analysis of the behavior of a class of genetic adaptive systems," Ph.D. Thesis, University of Michigan, 1975.

[54] K. A. De Jong, "Genetic algorithms are NOT function optimizers," in L. D. Whitley, ed., *Foundations of Genetic Algorithms 2*, San Mateo, CA: Morgan Kaufmann, pp. 5–17, 1993.

[55] K. A. De Jong and J. Sarma, "Generation gaps revisited," in L. D. Whitley, ed., *Foundations of Genetic Algorithms 2*, San Mateo, CA: Morgan Kaufmann, pp. 19–28, 1993.

[56] S. Devadas, K. Keutzer, and J. White, "Estimation of power dissipation in CMOS combinational circuits using Boolean function manipulation," *IEEE Trans. Computer-Aided Design*, pp. 373–383, March 1992.

[57] R. Dionne and M. Florian, "Exact and approximate algorithms for optimal network design," *Networks*, vol. 9, pp. 37–59, 1979.

[58] W. E. Donath "Complexity theory and design automation," *Proc. 17th Design Automation Conf.*, pp. 412–419, 1980.

[59] K. A. Dowsland, "Hill-climbing simulated annealing and the Steiner problem in graphs," *Eng. Opt.*, vol. 17, pp. 91–107, 1991.

[60] S. E. Dreyfuss and R. A. Wagner, "The Steiner problem in graphs," *Networks*, vol. 1, pp. 195–207, 1971.

[61] C. W. Duin and A. Volgenant, "Reduction tests for the Steiner problem in graphs," *Networks*, vol. 19, pp. 549–567, 1989.

[62] M. D. Durand, "Parallel simulated annealing accuracy vs. speed in placement," *IEEE Design and Test of Computers*, June 1989, pp. 8–34.

[63] S. Dutta and W. Deng, "A probability-based approach to VLSI circuit partitioning," *Proc. Design Automation Conf.*, pp. 100–105, 1996.

[64] E. B. Eichelberger and T. W. Williams, "A logic design structure for LSI testability," *J. Design Automat. Fault-Tolerant Comput.*, vol. 2, pp. 165–178, May 1978.

[65] H. Esbensen, "A genetic algorithm for macro cell placement," *Proc. European Design Automation Conf.*, pp. 52–57, 1992.

[66] H. Esbensen, "A macro-cell global router based on two genetic algorithms," *Proc. European Design Automation Conf.*, pp. 428-433, 1994.

[67] H. Esbensen and P. Mazumder, "SAGA: Unification of genetic algorithm with simulated annealing and its application to macro-cell placement," *Proc. IEEE Int. Conf. VLSI Design*, Calcutta, India, Jan. 1994.

[68] H. Esbensen and P. Mazumder, "A genetic algorithm for the Steiner problem in a graph," *Proc. European Design and Test Conf.*, Paris, pp. 402–406, Mar. 1994.

[69] L. J. Eshelman and J. D. Schaffer, "Real-coded genetic algorithms and interval-schemata," in L. D. Whitley, ed., *Foundations of Genetic Algorithms 2*, San Mateo, CA: Morgan Kaufmann, pp. 187–202, 1993.

[70] C. M. Fiduccia and R. M. Mattheyses, "A linear-time heuristic for improving network partitions," *Proc. Design Automation Conf.*, pp. 175–181, 1982.

[71] L. R. Foulds and V. J. Rayward-Smith, "Steiner problems in graphs: algorithms and applications," *Eng. Opt.*, vol. 7, pp. 7–16, 1983.

[72] R. Francis, J. Rose, and Z. Vranesic, "Chortle-crf: Fast technology mapping for lookup table-based FPGAs," *Proc. Design Automation Conf.*, pp. 227–233, 1991.

[73] R. J. Francis, J. Rose, and Z. Vranesic, "Technology mapping of lookup table-based FPGAs for performance," *Proc. IEEE Int. Conf. Computer-Aided Design*, pp. 568–571, 1991.

[74] M. R. Garey and D. S. Johnson, "The rectilinear Steiner tree problem is NP-complete," *SIAM Journal of Applied Mathematics*, vol. 32, no. 4, pp. 826–834, 1977.

[75] M. R. Garey and D. S. Johnson, *Computers and Intractability, A Guide to the Theory of NP-Completeness*, New York: W. H. Freeman and Company, 1979.

[76] A. Ghosh, S. Devadas, and A. R. Newton, "Test generation for highly sequential circuits," *Proc. Int. Conf. Computer-Aided Design*, pp. 362–365, Nov. 1989.

[77] A. Ghosh, S. Devadas, K. Keutzer, and J. White, "Estimation of average switching activity in combinational and sequential circuits," *Proc. Design Automation Conf.*, pp. 253–259, 1992.

[78] D. E. Goldberg and R. Lingle "Alleles, loci and the traveling salesman problem," *Proc. Int. Conf. Genetic Algorithms and their Applications*, 1985.

[79] D. E. Goldberg, *Genetic Algorithms in Search, Optimization, and Machine Learning*, Reading, MA: Addison-Wesley, 1989.

[80] D. E. Goldberg, "Real-coded genetic algorithms, virtual alphabets, and blocking," IlliGAL Report 90001, Illinois Genetic Algorithms Laboratory, Dept. of General Engineering, University of Illinois, Urbana, IL, 1990.

[81] D. E. Goldberg and K. Deb, "A comparative analysis of selection schemes used in genetic algorithms," in G. Rawlins, ed., *Foundations of Genetic Algorithms*, San Mateo, CA: Morgan Kaufmann, pp. 69–93, 1991.

[82] D. E. Goldberg, B. Kork, and K. Deb, "Messy genetic algorithms: Motivation, analysis, and first results," *Complex Systems*, vol. 3, pp. 493–530, 1989.

[83] D. E. Goldberg, K. Deb, and B. Korb, "Messy genetic algorithms revisited: Studies in mixed size and scale," *Complex Systems*, vol. 4, pp. 415–444, 1990.

[84] P. Goel, "RAPS test pattern generator," *IBM Technical Disclosure Bulletin*, vol. 21, no. 7, pp. 2787–2791, December 1978.

[85] P. Goel, "An implicit enumeration algorithm to generate tests for combinational logic circuits," *IEEE Trans. Computers*, vol. C-30, no. 3, pp. 215–222, March 1981.

[86] S. Goto, "An efficient algorithm for the two-dimensional placement problem in electrical circuits," *IEEE Trans. on Circuits and Systems*, vol. 25, no. 4, pp. 645–652, 1981.

[87] J. W. Greene and K. J. Supowit, "Simulated annealing without rejected moves," *Proc. IEEE Int. Conf. Computer-Aided Design*, pp. 658–663, 1984.

[88] J. J. Grefenstette, "Optimization of control parameters for genetic algorithms," *IEEE Trans. Systems, Man, and Cybernetics*, vol. SMC-16, no. 1, Jan/Feb. 1986.

[89] R. Gupta and M. A. Breuer, "Ordering storage elements in a single scan chain," *Proc. Int. Conf. Computer-Aided Design*, pp. 408–411, Nov. 1991.

[90] S. L. Hakami, "Steiner's problem in graphs and its implications," *Networks*, vol. 1, pp. 113–133, 1971.

[91] M. Hanan, "On Steiner's problem with rectilinear distance," *SIAM Journal of Applied Mathematics*, vol. 14, no. 2, pp. 255–265, 1966.

[92] S. Hellebrand and H. Wunderlich, "Tools and devices supporting the pseudo-exhaustive test," *Proc. European Conf. on Design Automation*, pp. 13–17, 1990.

[93] A. Herrigel and W. Fichtner, "An analytic optimization technique for placement of macro-cells", *Proc. Design Automation Conf.*, pp. 376–381, 1989.

[94] J. Hesser, R. Männer, and O. Stucky, "Optimization of Steiner trees using genetic algorithms," *Proc. 3rd Int. Conf. Genetic Algorithms*, pp. 231–236, 1989.

[95] J. H. Holland, *Adaptation in Natural and Artificial Systems*, Ann Arbor, MI: University of Michigan Press, 1975.

[96] E. Horowitz and S. Sahni, *Fundamentals of Computer Algorithms*, Rockville, MD: Computer Science Press, pp. 56–61, 1984.

[97] M. S. Hsiao, E. M. Rudnick, and J. H. Patel, "Alternating strategies for sequential circuit ATPG," *Proc. European Design and Test Conf.*, pp. 368–374, 1996.

[98] M. S. Hsiao, E. M. Rudnick, and J. H. Patel, "Automatic test generation using genetically-engineered distinguishing sequences," *Proc. VLSI Test Symp.*, pp. 216–223, 1996.

[99] M. S. Hsiao, E. M. Rudnick, and J. H. Patel, "Sequential circuit test generation using dynamic state traversal," *Proc. European Design and Test Conf.*, pp. 22–28, 1997.

[100] M. S. Hsiao, E. M. Rudnick, and J. H. Patel, "Fast algorithms for static compaction of sequential circuit test vectors," *Proc. VLSI Test Symp.*, 1997.

[101] M. S. Hsiao and S. T. Chakradhar, "State relaxation based subsequence removal for fast static compaction in sequential circuits," *Proc. Design, Automation and Test in Europe (DATE) Conf.*, pp. 577–582, 1998.

[102] M. S. Hsiao, E. M. Rudnick, and J. H. Patel, "Application of genetically engineered finite-state-machine sequences to sequential circuit ATPG," *IEEE Trans. Computer-Aided Design*, vol. 17, no. 3, pp. 239–254, March 1998.

[103] F. K. Hwang "On Steiner minimal trees with rectilinear distance," *SIAM J. Appl. Math,* vol. 30, pp. 104–114, 1976.

[104] F. K. Hwang "An $O(n\ log\ n)$ algorithm for suboptimal rectilinear Steiner trees," *IEEE Trans. Circuits and Systems,* vol. CAS-26, no. 1, pp. 75–77, 1979.

[105] L. Hyafia and L. Rivest, "Graph partitioning and constructing optimal decision trees are NP complete problems," Report No. 33, IRIA-Laboria, Rocquencourt France, 1973.

[106] R. Jayaraman and R. Rutenbar, "Floorplanning by annealing on a hypercube multiprocessor," *Proc. IEEE Int. Conf. Computer-Aided Design*, pp. 346–349, 1987.

[107] Y.-M. Jiang, K.-T. Cheng, and A. Krstic, "Estimation of maximum power and instantaneous current using a genetic algorithm," *Proc. Custom Integrated Circuits Conf.*, 1997.

[108] M. Jones and P. Bannerjee, "Performance of a parallel algorithm for standard cell placement on the Intel hypercube," *Proc. Design Automation Conf.*, 1987.

[109] B. A. Julstrom, "A genetic algorithm for the rectilinear Steiner problem," *Proc. 5th Int. Conf. Genetic Algorithms*, pp. 474–480, 1993.

[110] A. Kapsalis, V. J. Rayward-Smith, and G. D. Smith, "Solving the graphical Steiner tree problem using genetic algorithms," *Journal of the Operational Research Society*, vol. 44, no. 4, pp. 397–406, 1993.

[111] R. M. Karp, "Reducibility among combinatorial problems," in R. E. Miller, J. W. Thatcher (eds.), *Complexity of Computer Computations*, New York: Plenum Press, pp. 85–103, 1972.

[112] K. Karplus, "Xmap: A technology mapper for table-lookup field-programmable gate arrays," *Proc. Design Automation Conf.*, pp. 240–243, 1991.

[113] T. P. Kelsey, K. K. Saluja, and S. Y. Lee, "An efficient algorithm for sequential circuit test generation," *IEEE Trans. Computers*, vol. 42, no. 11, pp. 1361–1371, Nov. 1993.

[114] B. W. Kernighan and S. Lin, "An efficient heuristic procedure for partitioning graphs," *Bell Systems Technical Journal*, vol. 49, pp. 291–307, 1970.

[115] S. Kirkpatrick, C. D. Gelatt, Jr., and M. P. Vecchi, "Optimization by simulated annealing," *Science*, vol. 220, no. 4598, pp. 671–680, May 1983.

[116] R. M. Kling "Placement by simulated evolution," M.S. thesis, Coordinated Science Lab, College of Engr., University of Illinois at Urbana-Champaign, June 1987.

[117] R. M. Kling and P. Banerjee, "ESP: A new standard cell placement package using simulated evolution," *Proc. ACM/IEEE Design Automation Conf.*, pp. 60–66, 1987.

[118] U. R. Kodres, "Partitioning and card selection," in *Design Automation of Digital Systems*, M. A. Breuer, ed., pp. 173–212, 1972.

[119] V. Kommu, "Enhanced genetic algorithms in constrained search spaces with emphasis on parallel environments," Ph.D. Thesis, University of Iowa, Dec. 1993.

[120] G. A. Korn and T. M. Korn, *Mathematical Handbook for Scientists and Engineers*, New York: McGraw-Hill Book Company, Inc., 1961.

[121] L. Kou, G. Markowsky, and L. Berman, "A fast algorithm for Steiner trees," *Acta Informatica*, vol. 15, pp. 141–145, 1981.

[122] S. A. Kravitz and R. A. Rutenbar, "Placement by simulated annealing on a multiprocessor," *IEEE Trans. Computer Aided Design*, vol. CAD-6, no. 4, pp. 534–549, July 1987.

[123] H. Kriplani, "Worst case voltage drops in power and ground busses of CMOS VLSI circuits," Ph.D. Thesis, University of Illinois, Nov. 1993.

[124] H. Kriplani, F. Najm, P. Yang, and I. Hajj, "Resolving signal correlations for estimating maximum currents in CMOS combinational circuits," *Proc. Design Automation Conf.*, pp. 384–388, 1993.

[125] B. Krishnamurthy, "An improved min-cut algorithm for partitioning VLSI networks," *IEEE Trans. Computers*, vol. C-33, pp. 438–446, 1984.

[126] D. Krishnaswamy, M. S. Hsiao, V. Saxena, E. M. Rudnick, J. H. Patel, and P. Banerjee, "Parallel genetic algorithms for simulation-based sequential circuit test generation," *Proc. Int. Conf. VLSI Design*, pp. 475–481, Jan. 1997.

[127] B. Kröger, P. Schwenderling, and O. Vornberger, "Genetic packing of rectangles on transputers," *Transputing '91*, vol. 2, IOS Press, 1991.

[128] J. Kruskal, "On the shortest spanning subtree of a graph and the traveling salesman problem," *Proc. Amer. Math. Soc.*, vol. 7, no. 1, pp. 48–50, 1956.

[129] W. -J. Lai, C.- P. Kung, and C.- S. Lin, "Test time reduction in scan designed circuits," *Proc. European Conf. Design Automation*, pp. 489–493, 1993.

[130] Y. Lai, K. R. Pan, and M. Pedram, "FPGA synthesis using function decomposition," *Proc. Int. Conf. on Computer Design*, pp. 30–35, 1994.

[131] J. Lam and J.-M. Delsome, "Performance of a new annealing schedule," *Proc. ACM/IEEE Design Automation Conf.*, pp. 306–311, 1988.

[132] T. J. Lambert and K. K. Saluja, "Methods for dynamic test vector compaction in sequential test generation," *Proc. Int. Conf. VLSI Design*, pp. 166–169, 1996.

[133] E. L. Lawler, *Combinatorial Optimization: Networks and Matroids,* New York: Holt, Rinehart and Winston, 1976.

[134] D. H. Lee and S. M. Reddy, "A new test generation method for sequential circuits," *Proc. Int. Conf. Computer-Aided Design*, pp. 446–449, Nov. 1991.

[135] S. Y. Lee and K. K. Saluja, "An algorithm to reduce test application time in full scan designs," *Proc. Int. Conf. Computer-Aided Design*, pp. 17–20, Nov. 1992.

[136] S. Y. Lee and K. K. Saluja, "Sequential test generation with reduced test clocks for partial scan designs," *Proc. VLSI Test Symp.*, pp. 220–225, 1994.

[137] J. van Leeuwen, "Graph algorithms," in: J. van Leeuwen, ed., *Handbook of Theoretical Computer Science,* vol. A: *Algorithms and Complexity,* Amsterdam: Elsevier, 1990.

[138] F. T. Leighton "Complexity Issues in VLSI," Cambridge, MA: MIT Press, 1983.

[139] R. Lisanke, F. Brglez, A. J. Degeus, and D. Gregory, "Testability-driven random test-pattern generation," *IEEE Trans. Computer-Aided Design*, vol. 6, no. 6, pp. 1082–1087, Nov. 1987.

[140] A. Lucena and J. E. Beasley, "A branch and cut algorithm for the Steiner problem in graphs," working paper, The Management School, Imperial College, England, July 1992.

[141] H. -K. T. Ma, S. Devadas, A. R. Newton, and A. Sangiovanni-Vincentelli, "Test generation for sequential circuits," *IEEE Trans. Computer-Aided Design*, vol. 7, no. 10, pp. 1081–1093, Oct. 1988.

[142] S. Manich and J. Figueras, "Maximizing the weighted switching activity in combinational CMOS circuits under the variable delay model," *Proc. European Design & Test Conf.*, pp. 597–602, 1997.

[143] S. Manne, A. Pardo, R. I. Bahar, G. D. Hachtel, F. Somenzi, E. Macii, and M. Poncino, "Computing the maximum power cycles of a sequential circuit," *Proc. Design Automation Conf.*, pp. 23–28, 1995.

[144] T. E. Marchok, Aiman El-Maleh, W. Maly, and J. Rajski, "Complexity of sequential ATPG," *Proc. European Design and Test Conf.*, pp. 252–261, March 1995.

[145] R. Marlett, "An effective test generation system for sequential circuits," *Proc. Design Automation Conf.*, pp. 250–256, June 1986.

[146] B. Mathew and D. G. Saab, "Partial reset: An alternative DFT approach," *VLSI Design*, vol. 1, no. 4, pp. 299–311, 1994.

[147] E. J. McCluskey and S. B. Nesbat, "Design for autonomous test," *IEEE Trans. Computers*, vol. C-30, pp. 866–874, Nov. 1981.

[148] S. Mohan and P. Mazumder, "Wolverines: Standard cell placement on a network of workstations," *IEEE Trans. Computer-Aided Design*, vol. 12, no. 9, pp. 1312–1326, 1993.

[149] S. P. Morley and R. A. Marlett, "Selectable length partial scan: A method to reduce vector length," *Proc. Int. Test Conf.*, pp. 385–392, Oct. 1991.

[150] H. Mühlenbein, M. Gorges-Schleuter, O. Krämer, "Evolution algorithms in combinatorial optimization," *Parallel Computing*, vol. 7, pp. 65–85, 1988.

[151] H. Muhlenbein, M. Schomish, and J. Born, "The parallel genetic algorithm as a function optimizer," *Proc. Fourth Int. Conf. Genetic Algorithms*, pp. 271–278, 1991.

[152] R. Murgai, N. Shenoy, R. K. Brayton, and A. Sangiovanni-Vincentelli, "Improved logic synthesis algorithms for table look up architectures," *Proc. IEEE Int. Conf. Computer-Aided Design*, pp. 564–567, 1991.

[153] R. Murgai, N. Shenoy, R. K. Brayton, and A. Sangiovanni-Vincentelli, "Performance directed synthesis for table look up programmable gate arrays," *Proc. IEEE Int. Conf. Computer-Aided Design*, pp. 572–575, 1991.

[154] F. N. Najm, "A survey of power estimation techniques in VLSI circuits," *IEEE Trans. VLSI Systems*, vol. 2, no. 4, pp. 446–455, Dec. 1994.

[155] F. N. Najm, S. Goel, and I. N. Hajj, "Power estimation in sequential circuits," *Proc. Design Automation Conf.*, pp. 635–640, 1995.

[156] F. N. Najm and M. Y. Zhang, "Extreme delay sensitivity and the worst-case switching activity in VLSI circuits," *Proc. Design Automation Conf.*, pp. 623–627, 1995.

[157] W. Nam-Sung, "A heuristic method for FPGA technology mapping based on the edge visibility," *Proc. Design Automation Conf.*, pp. 248–251, 1991.

[158] S. Narayanan, R. Gupta, and M. Breuer, "Configuring multiple scan chains for minimum test time," *Proc. Int. Conf. Computer-Aided Design*, pp. 4–8, Nov. 1992.

[159] T. M. Niermann, "Techniques for sequential circuit automatic test generation," Technical Report CRHC-91-8, Coordinated Science Laboratory, University of Illinois at Urbana-Champaign, March 1991.

[160] T. M. Niermann and J. H. Patel, "HITEC: A test generation package for sequential circuits," *Proc. European Conf. Design Automation (EDAC)*, pp. 214–218, Feb. 1991.

[161] T. M. Niermann, W. -T. Cheng, and J. H. Patel, "PROOFS: A fast, memory-efficient sequential circuit fault simulator," *IEEE Trans. Computer-Aided Design*, pp. 198–207, Feb. 1992.

[162] T. M. Niermann, R. K. Roy, J. H. Patel, and J. A. Abraham, "Test compaction for sequential circuits," *IEEE Trans. Computer-Aided Design,*, vol. 11, no. 2, pp. 260–267, Feb. 1992.

[163] T. M. Niermann and J. H. Patel, "Method for automatically generating test vectors for digital integrated circuits," U.S. Patent No. 5,377,197, December, 1994.

[164] Y. Nishizaki, M. Igusa, and A. Sangiovanni-Vincentelli, "Mercury: A new approach to macro-cell global routing," *Proc. IFIP 10/WG 10.5 Int. Conf. VLSI*, Munich, 1989.

[165] I. M. Oliver, D. J. Smith, and J. R. C. Holland "A study of permutation crossover operators on the traveling salesman problem," *Proc. Int. Conf. Genetic Algorithms and their Applications*, pp. 224–230, 1985.

[166] H. Onodera, Y. Taniguchi, and K. Tamaru, "Branch-and-bound placement for building block layout," *Proc. 28th Design Automation Conf.*, pp. 433–439, 1991.

[167] K. R. Pan, "Multi-level logic synthesis based on function decomposition," Ph.D. Thesis, University of Southern California, May 1996.

[168] K. R. Pan, Y. Lai, and M. Pedram, "LUT-based FPGA synthesis for low power," *Int. Workshop on Power and Timing Modeling, Optimization and Simulation*, 1994.

[169] J. Park, C. Oh, and M. R. Mercer, "Improved sequential ATPG based on functional observation information and new justification methods," *Proc. European Design and Test Conf.*, 1995.

[170] S. Parkes, P. Banerjee, and J. Patel, "ProperHITEC: A portable, parallel, object-oriented approach to sequential test generation," *Proc. Design Automation Conf.*, pp. 717–721, 1994.

[171] S. Parkes, J. A. Chandy, and P. Banerjee, "A library-based approach to portable, parallel, object-oriented programming: Interface, implementation and application," *Proc. Supercomputing'94*, pp. 69–78, 1994.

[172] J. Plesnik, "A bound for the Steiner tree problem in graphs," *Math. Slovaca*, vol. 31, pp. 155–163, 1981.

[173] I. Pomeranz, L. N. Reddy, and S. M. Reddy, "COMPACTEST: A method to generate compact test sets for combinational circuits," *Proc. Int. Test Conf.*, pp. 194–203, Oct. 1991.

[174] I. Pomeranz and S. M. Reddy, "Application of homing sequences to synchronous sequential circuit testing," *IEEE Trans. Computers*, vol. 43, no. 5, pp. 569–580, 1994.

[175] I. Pomeranz and S. M. Reddy, "LOCSTEP: A logic simulation based test generation procedure," *Proc. Fault Tolerant Computing Symp.*, June 1995.

[176] I. Pomeranz and S. M. Reddy, "On generating compact test sequences for synchronous sequential circuits," *Proc. European Design Automation Conf. (EURO-DAC)*, pp. 105–110, 1995.

[177] I. Pomeranz and S. M. Reddy, "On static compaction of test sequences for synchronous sequential circuits," *Proc. Design Automation Conf.*, pp. 215–220, 1996.

[178] I. Pomeranz and S. M. Reddy, "Dynamic test compaction for synchronous sequential circuits using static compaction techniques," *Proc. Int. Symp. Fault-Tolerant Computing*, pp. 53–61, 1996.

[179] I. Pomeranz and S. M. Reddy, "On improving genetic optimization based test generation," *Proc. European Design and Test Conf.*, pp. 506–511, 1997.

[180] D. K. Pradhan and J. Saxena, "A design for testability scheme to reduce test application time in full scan," *Proc. VLSI Test Symp.*, pp. 55–60, April 1992.

[181] P. Prinetto, M. Rebaudengo, and M. Sonza Reorda, "An automatic test pattern generator for large sequential circuits based on genetic algorithms," *Proc. Int. Test Conf.*, pp. 240–249, Oct. 1994.

[182] A. Raghunathan and S. T. Chakradhar, "Acceleration techniques for dynamic vector compaction," *Proc. Int. Conf. Computer-Aided Design*, pp. 310–317, 1995.

[183] A. Raghunathan and S. T. Chakradhar, "Dynamic test sequence compaction for sequential circuits," *Proc. Int. Conf. VLSI Design*, pp. 170–173, 1996.

[184] V. J. Rayward-Smith and A. Clare, "On finding Steiner vertices," *Networks*, vol. 16, pp. 283–294, 1986.

[185] S. M. Reddy and R. Dandapani, "Scan design using standard flip-flops," *IEEE Design & Test*, pp. 52–54, Feb. 1987.

[186] M. W. Roberts and P. K. Lala, "An algorithm for the partitioning of logic circuits," *IEE Proc.* vol. 131, pt. E, no.4, pp. 113–118, July 1984.

[187] J. Rose, W. Snelgrove, and Z. Vranesic, "Parallel standard cell placement algorithms with quality equivalent to simulated annealing," *IEEE Trans. Computer Aided Design*, vol. CAD-7, no. 3, pp. 387–396, March 1988.

[188] E. M. Rudnick, J. G. Holm, D. G. Saab, and J. H. Patel, "Application of simple genetic algorithms to sequential circuit test generation," *Proc. European Design and Test Conf.*, pp. 40–45, Feb. 1994.

[189] E. M. Rudnick, J. H. Patel, G. S. Greenstein, and T. M. Niermann, "Sequential circuit test generation in a genetic algorithm framework," *Proc. Design Automation Conf.*, pp. 698–704, June 1994.

[190] E. M. Rudnick and J. H. Patel, "A genetic approach to test application time reduction for full scan and partial scan circuits," *Proc. Eighth Int. Conf. VLSI Design*, pp. 288–293, Jan. 1995.

[191] E. M. Rudnick and J. H. Patel, "Combining deterministic and genetic approaches for sequential circuit test generation," *Proc. Design Automation Conf.*, pp. 183–188, June 1995.

[192] E. M. Rudnick and J. H. Patel, "State justification using genetic algorithms in sequential circuit test generation," Coordinated Science Laboratory, University of Illinois, Urbana, IL, Tech. Report CRHC-96-01/UILU-ENG-96-2201, Jan. 1996.

[193] E. M. Rudnick and J. H. Patel, "Simulation-based techniques for dynamic test sequence compaction," *Proc. Int. Conf. Computer-Aided Design*, pp. 67–73, 1996.

[194] E. M. Rudnick, J. H. Patel, G. S. Greenstein, and T. M. Niermann, "A genetic algorithm framework for test generation," *IEEE Trans. Computer-Aided Design*, vol. 16, no. 9, pp. 1034–1044, Sept. 1997.

[195] D. G. Saab, Y. G. Saab, and J. A. Abraham, "CRIS: A test cultivation program for sequential VLSI circuits," *Proc. Int. Conf. Computer-Aided Design*, pp. 216–219, Nov. 1992.

[196] D. G. Saab, Y. G. Saab, and J. A. Abraham, "Iterative [simulation-based genetics + deterministic techniques] = complete ATPG," *Proc. Int. Conf. Computer-Aided Design*, pp. 40–43, Nov. 1994.

[197] D. G. Saab, Y. G. Saab, and J. A. Abraham, "Automatic test vector cultivation for sequential VLSI circuits using genetic algorithms," *IEEE Trans. Computer-Aided Design*, vol. 15, no. 10, pp. 1278–1285, Oct. 1996.

[198] Y. G. Saab and V. B. Rao, "Combinatorial optimization by stochastic evolution," *IEEE Trans. Computer-Aided Design*, vol. 10, no. 4, pp. 525–535, April 1991.

[199] S. Sahni and A. Bhatt "The complexity of design automation problems," *Proc. 17th Design Automation Conf.*, pp. 402–411, 1980.

[200] L. A. Sanchis, "Multi-way network partitioning," *IEEE Trans. Computers*, vol. C-38, pp. 62–81, Jan. 1989.

[201] T. Sasao, "FPGA design by generalized functional decomposition," *Logic Synthesis and Optimization*, Boston, MA: Kluwer Academic Publishers, pp. 233–258, 1993.

[202] P. Sawkar and D. Thomas, "Area and delay mapping for table-lookup based field programmable gate arrays," *Proc. Design Automation Conf.*, pp. 368–373, 1992.

[203] H. D. Schnurmann, E. Lindbloom, and R. G. Carpenter, "The weighted random test-pattern generator," *IEEE Trans. Computers*, vol. 24, no. 7, pp. 695–700, July 1975.

[204] M. H. Schulz and E. Auth, "Essential: An efficient self-learning test pattern generation algorithm for sequential circuits," *Proc. Int. Test Conf.*, pp. 28–37, Aug. 1989.

[205] D. G. Schweikert and B. W. Kernighan, "A proper model for the partitioning of electrical circuits," *Proc. ACM/IEEE Design Automation Workshop*, pp. 57–62, 1972.

[206] C. Sechen, *VLSI Placement and Global Routing Using Simulated Annealing*, Boston, MA: Kluwer Academic Publishers, 1988.

[207] S. Seshu and D. N. Freeman, "The diagnosis of asynchronous sequential switching systems," *IRE Trans. Electronic Computing*, vol. 11, pp. 459–465, Aug. 1962.

[208] K. Shahookar and P. Mazumder, "A genetic approach to standard cell placement with meta-genetic parameter optimization," *Proc. IEEE European Design Automation Conf.*, Glasgow, UK, pp. 370–378, March 1990.

[209] K. Shahookar and P. Mazumder, "Standard cell placement and the genetic algorithm," Chapter in *Advances in Computer-Aided Engineering Design, Vol. II*, I. N. Hajj (editor), Greenwich, CT: Jai Press, pp. 159–234, 1990.

[210] K. Shahookar and P. Mazumder, "A genetic approach to standard cell placement with meta-genetic parameter optimization," *IEEE Trans. Computer-Aided Design*, vol. 9, no. 5, pp. 500–511, May 1990.

[211] K. Shahookar and P. Mazumder, "VLSI cell placement techniques," *ACM Computing Surveys*, vol. 23, no. 2, pp. 143–220, June 1991.

[212] K. Shahookar, P. Mazumder, and S. M. Reddy, "Gate matrix placement by genetic algorithm combined with beam search," *Proc. IEEE Int. Conf. VLSI Design*, Bombay, India, Jan. 1993.

[213] K. Shahookar, W. Khamisani, P. Mazumder, and S.M. Reddy, "Genetic beam search for gate matrix placement," *IEE Proceedings-E: Computers and Digital Techniques*, vol. 141, no. 2, pp. 123–128, March 1994.

[214] K. Shahookar and P. Mazumder, "Genetic multiway partitioning," *Proc. IEEE Int. Conf. VLSI Design*, New Delhi, India, pp. 365–369, Jan. 1995.

[215] M. L. Shore, L. R. Foulds, and P. B. Gibbons, "An algorithm for the Steiner problem in graphs," *Networks*, vol. 12, pp. 323–333, 1982.

[216] C. Small, "Shrinking devices put the squeeze on system packaging," *EDN.*, pp. 41–46, February 1994.

[217] T. J. Snethen, "Simulator-oriented fault test generator," *Proc. Design Automation Conf.*, pp. 88–93, 1977.

[218] M. Srinivas and L. M. Patnaik, "A simulation-based test generation scheme using genetic algorithms," *Proc. Int. Conf. VLSI Design*, pp. 132–135, Jan. 1993.

[219] R. Srinivasan, S. K. Gupta, and M. A. Breuer, "An efficient partitioning strategy for pseudo-exhaustive testing," *Design Automation Conf.*, pp. 242–248, 1993.

[220] P. Suaris and G. Kedem, "Quadrisection: A new approach to standard cell layout," *Proc. IEEE Intl. Conf. on Computer-Aided Design*, pp. 474–477, 1987.

[221] H. Takahashi and A. Matsuyama, "An approximate solution for the Steiner problem in graphs," *Mathematica Japonica*, vol. 24, no. 6, pp. 573–577, 1980.

[222] R. Tanese, "Distributed genetic algorithms," *Proc. Third Int. Conf. Genetic Algorithms*, pp. 398–405, 1989.

[223] C. Teng, A. M. Hill, and S. Kang, "Estimation of maximum transition counts at internal nodes in CMOS VLSI circuits," *Proc. Int. Conf. Computer-Aided Design*, pp. 366–370, 1995.

[224] E. Trischler, "Incomplete scan path with an automatic test generation methodology," *Proc. Int. Test Conf.*, pp. 153–162, 1980.

[225] R. Tsay, E. S. Kuh, and C. Hsu, "Module placement for large chips based on sparse linear equations," *Intl. Journal of Circuit Theory Application*, vol. 16, pp. 411–423, 1988.

[226] C. Tsui, J. Monteiro, M. Pedram, A. Despain, and B. Lin, "Power estimation methods for sequential logic circuits," *IEEE Trans. VLSI Systems*, vol. 3, no. 3, pp. 404–416, September 1995.

[227] T. Uchino, F. Minami, T. Mitsuhashi, and N. Goto, "Switching activity analysis using Boolean approximation method," *Proc. Int. Conf. Computer-Aided Design*, pp. 20–25, 1995.

[228] J. G. Udell, Jr. and E. J. McCluskey, "Efficient circuit segmentation for pseudo-exhaustive test," *Proc. Int. Test Conf.*, pp. 148–151, 1987.

[229] M. Upton, K. Samii, and S. Sugiyama, "Simulated annealing placement for mixed macro cell and standard cell layouts," *Proc. Int. Workshop on Layout Synthesis*, vol. 1, 1990.

[230] B. Vinnakota and N. K. Jha, "Synthesis of sequential circuits for parallel scan," *Proc. European Conf. Design Automation*, pp. 366–370, 1992.

[231] C. Wang, K. Roy, and T. Chou, "Maximum power estimation for sequential circuits using a test generation based technique," *Proc. Custom Integrated Circuits Conf.*, 1996.

[232] B. W. Wah, A. Ieumwananonthachai, L. C. Chu, and A. Aizawa, "Rational scheduling of experiments and generalization in genetics-based learning," *IEEE Trans. Knowledge and Data Engineering*, 1995.

[233] Y.-C. Wei and C.-K. Cheng, "Ratio cut partitioning for hierarchical designs," *IEEE Trans. Computer-Aided Design*, vol. 10, pp. 911–921, July 1991.

[234] D. Whitley, "The GENITOR algorithm and selection pressure: Why rank-based allocation of reproductive trials is best," *Proc. Third Int. Conf. Genetic Algorithms*, pp. 116–121, 1989.

[235] M. J. Y. Williams and J. B. Angell, "Enhancing testability of large-scale integrated circuits via test points and additional logic," *IEEE Trans. Computers*, vol. C-22, pp. 46–60, Jan. 1973.

[236] P. Winter, "Steiner problem in networks: A survey," *Networks*, vol. 17, pp. 129–167, 1987.

[237] P. Winter and J. MacGregor Smith, "Path-distance heuristics for the Steiner problem in undirected networks," *Algorithmica*, vol. 7, pp. 309–327, 1992.

[238] N. Woo, "A heuristic method for FPGA technology mapping based on edge visibility," *Proc. Design Automation Conf.*, pp. 248–251, 1991.

[239] Wu E., "PEST: A tool for implementing pseudo-exhaustive self test," *AT&T Technical Journal*, pp. 87–100, Jan./Feb. 1991.

[240] H.-J. Wunderlich, "Multiple distributions for biased random test patterns," *IEEE Trans. Computer-Aided Design*, vol. 9, no. 6, pp. 584–593, June 1990.

[241] J. S. Yih and P. Mazumder, "A neural network design for circuit partitioning," *IEEE Trans. Computer-Aided Design*, vol. 9, no. 10, Oct. 1990.

INDEX

algorithm
    clustering 42
    constructive 20, 42
    graph 21, 25, 43
    iterative improvement 21, 42
    neural 46
    numerical optimization 25
    ratio-cut 45
    stochastic 21, 45
    unified 87
ALT-TEST 190–198
automatic placement 18–20, 22, 71
    distributed 259–260, 263
automatic routing 18, 26, 72, 135
automatic test generation 16, 19, 32, 159
    combinational circuits 160, 163–164
    parallel 280–292
    sequential circuits 162–163, 172
    simulation-based 162, 164
balance 38
BB 103–105
BEAR 104–105
bipartite graph 39
bipartitioning 22, 58, 67
cells 39
Chortle 144, 151–154
chromosome 3, 7
circuit segmentation 142–143, 146
classifier system 305
communication pattern 276

COMPACTEST 183–184
configurable logic blocks (CLBs) 141, 143
CRIS 163–164, 171, 173, 196–197
crossover 4, 7, 10, 50, 62, 73, 76, 95, 117, 148, 164, 174, 273
    probability 4, 12, 148, 257, 274–275
    PMX 26, 76–77, 82
    cycle 26, 77–78, 82
    heuristic 163
    order 26, 76–77, 82
    two-point 10
    uniform 10, 174
data decomposition 284–285, 290
defect 159–160
design for testability 159, 161, 178
    full scan 161, 178
    partial scan 181
design verification 18
detailed routing 26
DIGATE 199–200, 210
distance graph 110
distance network heuristic 115–119
distinguishing sequences 199, 201, 205
DNA 306
duplicate check 7, 52
dynamic state traversal 201, 203
electronic design automation 18
encoding 12, 48, 55, 61, 72, 74, 91, 115, 146, 164, 167, 174–175, 181, 188, 214–215, 234

Index

epoch length 260, 272–277
equivalent fault collapsing 160
evolution
    evolution strategy 306
    evolutionary algorithm 25, 306
    evolutionary programming 306
exploitation 12, 273–274, 307
exploration 12, 273–274, 307
external net 41
FASTEST 216
fault 160
    activation 201, 206
    coverage 19, 211
    equivalent fault collapsing 160
    multistage fault activation 203
    propagation 201
    sample 175, 182
    simulation 19, 160, 162–165, 194, 209, 213, 216
    single stuck-at fault model 160
    untestable 186, 209
Fiduccia-Mattheyses algorithm 21, 43–44, 57, 59, 62, 66
finite state machine sequences 199
fitness 2, 4
fitness function 4, 51, 56, 94, 117, 149, 163, 168–170, 189, 195, 206, 208–209
    approximate 165, 171, 175–176, 194
fitness scaling 13–14, 46, 48
force-directed placement 25
formal verification 18
FPGA 20, 30, 141, 143
full scan 161, 178
GASP 72, 73, 81, 84–85
GA-COMPACT 216–223
GA-HITEC 186–191, 197
GATEST 34, 171–178, 212
gate decomposition 144
GATTO 34, 163–164, 286
gene 3
generation 3
generation gap 4, 176

generations, number of 168
genetic algorithm 2–3, 46, 163–164
genetic operators 9
genotype 92, 95–98
genotype tree 98
global routing 26, 135
graph reduction 112, 114, 119
greedy search 42
Hamming distance 163, 237
HITEC 178, 184–186, 190–193, 195–197, 210, 212, 214, 217–223
hybrid genetic algorithm 38, 60, 67
inheritance 3
internal net 41
inversion 4, 14–16, 73, 97, 98, 117, 257
    probability 4, 7, 15, 82
ISCAS benchmarks 171–172, 240
iterated shortest path heuristic 119
Kernighan-Lin algorithm 21, 43, 66
load balancing 277, 279
local improvement 62
local optimization 38, 60, 62–66
logic synthesis 18, 31
lookahead mechanism 44
macro cell placement 22, 26
macro cell routing 14, 27
macro cells 22, 24, 85, 90
meta-genetic optimization 26, 78, 81
meta-level GA 78–83, 195
migration 259, 264, 272–274, 277, 284, 287–288, 291
    rate 264, 273
min-cut placement 20, 25, 38
minimum spanning tree 111, 131
multiway partitioning 22, 44, 55, 58
multistage fault activation 203
mutation 4, 7, 12, 53, 56, 74–75, 96, 117, 148, 168
    probability 4, 7, 12–13, 62, 74, 82, 168, 174–175, 257
natural selection 2
nets 39
neural network 46
offspring 4, 7, 12

one-point crossover 10
optimization 19
parallel genetic algorithm 3
parents 4
partial scan 181
partitioning 20, 38, 68
    bipartitioning 22, 58, 67
        multiway partitioning 22, 44, 55, 58
        ratio-cut partitioning 22, 45, 60
peak n-cycle power 230–231, 235, 243–247
peak single-cycle power 230, 235, 241–243, 246–247
peak sustainable power 230, 232, 237, 243–247
peak switching frequency per node 240
PGATEST 281, 284–293
pin 22
PLA 22
placement
    macro cell 22, 26
    min-cut 20, 25, 38
    force-directed 25
    standard cell 22, 26
PODEM algorithm 160–161
polar graph
    horizontal 29
    vertical 29
population 3
    size 3, 149, 168, 174, 175, 266, 269
power dissipation
    average 34, 229
    peak 35, 229
power estimation 18, 34
preprocessing 64
priority tree encoding 92, 96
probabilistic gain computation 44
PROOFS 171, 218
pseudo-exhaustive testing 150–151, 154–155
RAM 22, 218
ranking 14, 163–164

ratio-cut partitioning 22, 45, 60
rectilinear adjacency graph 29
rectilinear Steiner problem 109
reproduction 4
ROM 22
roulette wheel selection 9
routing
    detailed routing 26
    global routing 26, 135
    macro cell routing 14, 27
    routing area estimation 92
    routing channel 23, 28
    routing graph 27–28
SAGA 26, 86–90, 98, 101–104
schema 4–5, 7, 10–11, 15, 64
segmentation 142, 150
selection 4, 7, 9, 61, 174
    roulette wheel 9
    stochastic universal selection 9
    tournament 9
selection pressure 10, 13
semiperimeter 72
sequential depth 172
shortest path heuristic 110, 117
sigma scaling 49
simple GA 3, 5, 16, 33, 164, 194, 202, 234
simulated annealing 21, 25–26, 30, 45–46, 80–81, 84–86, 109
    parallel 254
simulation 18, 20
single stuck-at fault model 160
single time frame mode 206
Squeeze 216–223
standard cell 22, 38, 71
standard cell placement 22, 26
state justification 186, 188, 202, 204
steady-state GA 3, 7–8, 38, 46, 163–164
Steiner problem in a graph 30, 109–110
Steiner tree 109, 116–117
Steiner vertices 109
stochastic universal selection 9

STRATEGATE 34, 201–212
string 3
subpopulation-based GA 284, 288, 291
survival of the fittest 2–4
table lookups (TLUs) 142
technology mapping 30–31, 141–142,
        151–153
termination 150
test application time 178, 182, 212
test generation
    automatic test generation 16, 19,
        32, 159
    hybrid 184–190
test sequence compaction 212
    dynamic 212–213
    static 212
TimberWolf 84–85, 103
total replacement algorithm 3, 5
tournament selection 9, 174
unit delay 233, 249
validation 148–149
variable delay 233, 249
Verilog 18
VHDL 18
VLSI design 18, 22, 38
weighted DFS reordering 64
weighted graph 38
wire length 71–72, 86
Wolverines 254
zero delay 233, 249

ABOUT THE AUTHORS

Pinaki Mazumder received the BSEE degree from the Indian Institute of Science in 1976, the M.Sc. degree in computer science from the University of Alberta, Canada, in 1985, and the Ph.D. degree in electrical engineering from the University of Illinois at Urbana-Champaign in 1987. Presently, he is a Professor in the Department of Electrical Engineering and Computer Science at The University of Michigan, Ann Arbor. Prior to this, he was a Research Assistant with the Coordinated Science Laboratory, University of Illinois at Urbana-Champaign for two years and was with Bharat Electronics Ltd., India for over six years, where he developed several integrated circuits for consumer electronics. During the summers of 1985 and 1986, he was a Member of the Technical Staff in the Indian Hill branch of AT&T Bell Laboratories. During 1996–1997, he spent his sabbatical leave as a visiting faculty at Stanford University, the University of California at Berkeley, and Nippon Telephone and Telegraph, Japan.

His research interests include VLSI testing, physical design automation, and ultrafast circuit design. He has published over 100 papers on these topics in archival journals and proceedings of international conferences. Dr. Mazumder has lead his research group's efforts in VLSI testing and built-in self-repair techniques and has developed silicon compilers for RAMs, ROMs, and PLAs with built-in self-repairable capabilities. He has coauthored with Dr. Kanad Chakraborty a book: *Testing and Testable Design of High-Density Random-Access Memories*. Dr. Mazumder has also done quite extensive work in the area of VLSI physical design automation, and ultrafast circuits and CAD tools for nano- and quantum electronic devices.

Elizabeth M. Rudnick received the B.S. degree in chemical engineering in 1983 and M.S. and Ph.D. degrees in electrical engineering in 1990 and 1994 from the University of Illinois at Urbana-Champaign. She has worked at Motorola, Sunrise Test Systems, and Advanced Micro Devices in the areas of test generation, design verification, electronic design automation, and yield enhancement. She is currently an Assistant Professor in the Department of Electrical and Computer Engineering and a Research Assistant Professor in the Coordinated Science Laboratory at the University of Illinois, Urbana-Champaign. Her research interests include test generation, defect diagnosis, design verification, and design for testability.

Contributing Chapter Authors

Prithviraj Banerjee received the B.Tech. degree in electronics and electrical engineering from the Indian Institute of Technology, Kharagpur, India, in 1981, and the M.S. and Ph.D degrees in electrical engineering from the University of Illinois at Urbana-Champaign in 1982 and 1984, respectively. Dr. Banerjee is currently the Walter P. Murphy Chaired Professor of Electrical and Computer Engineering, Chairman of the Department of Electrical and Computer Engineering, and Director of the Center for Parallel and Distributed Computing at Northwestern University in Evanston, IL. Previously, he was the Director of the Computational Science and Engineering program and Professor of Electrical and Computer Engineering and the Coordinated Science Laboratory at the University of Illinois, Urbana-Champaign. Dr. Banerjee's research interests are in parallel algorithms for VLSI design automation and compilers for distributed memory parallel machines.

Henrik Esbensen received the B.Sc. in mathematics in 1989, the M.Sc. in computer science in 1991, and the Ph.D. in computer science in 1994, all from Aarhus University, Denmark. From 1994 to 1996, he was a Research Fellow in the EECS department at the University of California, Berkeley. Since 1996, he has been with Avant! Corporation, Fremont, CA. His interests include physical design, logic synthesis, graph algorithms, parallel algorithms, and complexity.

Michael S. Hsiao received the B.S. degree in computer engineering in 1992 and the M.S. and Ph.D. degrees in electrical engineering in 1993 and 1997 from the University of Illinois at Urbana-Champaign. During his undergraduate studies, he worked as a co-op engineer at Digital Equipment Corporation and National Semiconductor Corporation from 1988 to 1990. He is currently an Assistant Professor in the Department of Electrical and Computer Engineering at Rutgers University. His research interests include automatic test generation, fault simulation, design for testability, design verification, power estimation of VLSI circuits, and computer architecture.

Venkataramana Kommu received the B.Sc. Engg. degree from the Birla Institute of Technology, India in 1986 and the Ph.D. degree from the University of Iowa, Iowa City in 1993. Currently, he is with Synopsys Inc., CA, where he is working in the static timing verification group. Prior to Synopsys, he was working in Cadence Design Systems, Inc., in the area of logic synthesis. Prior to pursuing his doctoral degree, he worked in Hindustan Instrumentations Ltd. India in the area of digital design. His interests include logic synthesis, timing modeling and extraction, and design verification.

Dilip Krishnaswamy received the B. Tech. degree in electronics and communication engineering from the Indian Institute of Technology, Madras, India in 1991, the M.S. degree in computer science from Syracuse University, New York in 1993, and the Ph.D. degree in electrical engineering from the University of Illinois at Urbana-Champaign in 1997. He worked as a summer intern with the parallel applications group at the IBM T. J. Watson Research Center, Yorktown Heights, NY in 1994 and with the IC Verification Group at Cadence Design Systems in San Jose, CA in 1996. He is currently a member of the Logic Test Methodology group at Intel Corporation in Sacramento, CA. His research interests are in VLSI testing and fault simulation, parallel processing, CAD for VLSI systems, VLSI design, and image processing.

Janak H. Patel received the B.Sc. degree in physics from Gujarat University, India. He also received the B.Tech. degree from the Indian Institute of Technology, Madras, India, and M.S. and Ph.D. degrees from Stanford University, Stanford, CA, all in electrical engineering. He is currently a Professor of Electrical and Computer Engineering and Computer Science, and a Research Professor with the Coordinated Science Laboratory at the University of Illinois at Urbana-Champaign, where he is Co-Director of the Center for Reliable and High-Performance Computing. He is a co-founder of Sunrise Test Systems, an ATG and testability software company. His research interests are in the areas of automatic test generation, design for testability, fault simulation, and diagnosis. Dr. Patel is a Fellow of the IEEE and a recipient of the 1998 IEEE Emanuel R. Piore Award.

Irith Pomeranz received the B.Sc. degree in computer engineering and the D.Sc. degree from the Department of Electrical Engineering, Technion–Israel Institute of Technology in 1985 and 1989, respectively. She is currently a Professor in the Department of Electrical and Computer Engineering, University of Iowa, Iowa City. Her research interests include testing of VLSI circuits, design for testability, synthesis, and design verification.

Khushro Shahookar received the BSEE degree from the University of Engineering and Technology, Lahore, Pakistan in 1986 and the MSEE and Ph.D. degrees from the University of Michigan, Ann Arbor in 1989 and 1993. His main interests are VLSI fabrication technology and automated VLSI layout algorithms.

Mohan Sundarar received the BSEE degree from the Indian Institute of Technology, Madras, India in 1985 and the MSEE and Ph.D. degrees from the University of Michigan, Ann Arbor in 1991 and 1994. Currently, he is with Xilinx Corporation, San Jose, CA. Prior to doing his doctoral degree, he worked at Indian Telephone Industries, Bangalore, India from 1985 to 1989. His interests include FPGA design, distributed CAD tool design, and quantum circuit design.